Electrodynamics:
A Concise Introduction

Springer

New York
Berlin
Heidelberg
Barcelona
Budapest
Hong Kong
London
Milan
Paris
Santa Clara
Singapore
Tokyo

James Blake Westgard

Electrodynamics: A Concise Introduction

With 121 Figures

 Springer

James Blake Westgard
Department of Physics
Indiana State University
Terre Haute, IN 47809
USA

Cover illustration: The cover illustration shows three radiation patterns from an accelerating charge. At small velocities, as compared to the velocity of light, the observed distribution is proportional to the square of the component of the acceleration orthogonal to the line of sight of the observer. At higher velocities, the distribution is thrown more toward the direction of the velocity, and the diagram illustrates linear breaking—radiation and synchrotron radiation distributions. The diagram was created with the symbolic mathematics program, *Mathematica.*

Library of Congress Cataloging-in-Publication Data
Westgard, James B.
    Electrodynamics: a concise introduction/James B. Westgard.
      p.   cm.
    Includes bibliographical references and index.
    ISBN 0-387-94585-7 (hc: alk. paper)
    1. Electrodynamics. I. Title.
    QC631.W47 1995
    530.1'41—dc20   95-37687

Printed on acid-free paper.

Production managed by Francine McNeill; manufacturing supervised by Jeffrey Taub.
Typeset in TeX from the author's Microsoft Word files.
Printed and bound by R.R. Donnelley and Sons, Harrisonburg, VA.
Printed in the United States of America.

9 8 7 6 5 4 3 2 1

ISBN 0-387-94585-7 Springer-Verlag New York Berlin Heidelberg SPIN 10425511

Preface

This textbook is intended for advanced undergraduates or beginning graduates. It is based on the notes from courses I have taught at Indiana State University from 1967 to the present. The preparation needed is an introductory calculus-based course in physics and its prerequisite calculus courses. Courses in vector analysis and differential equations are useful but not required, since the text introduces these topics.

In writing this book, I tried to keep my own experience as a student in mind and to write the kind of book I liked to read. That goal determined the choice of topics, their order, and the method of presentation. The organization of the book is intended to encourage independent study. Accordingly, I have made every effort to keep the material self-contained, to develop the mathematics as it is needed, and to present new material by building incrementally on preceding material. In organizing the text, I have taken care to give explicit cross references, to show the intermediate steps in calculations, and to give many examples. Provided they are within the mathematical scope of this book, I have preferred elegant mathematical treatments over more ad hoc ones, not only for aesthetic reasons, but because they are often more profound and indicate connections to other branches of physics. I have emphasized physical understanding by presenting mechanical models.

This book is organized somewhat differently from the traditional textbook at this level. The first three chapters (about 100 pages) present the experimental foundation of the Maxwell field equations in three-dimensional notation, while detailed applications are reserved for later chapters. This organization presents an immediate overview of the subject, treats electricity and magnetism on an equal footing, and allows exploitation of many symmetries between the two—even at this level of discussion. This organization offers flexibility to instructors, who may choose to emphasize certain applications over others, with the assurance that students have the necessary foundation. These chapters also serve as a review for well-prepared students, who may proceed more quickly to later sections.

Chapter 4 introduces special relativity and the concept of a local Lorentz transformation, which is then used to formulate the Maxwell field equations in four dimensions. In this treatment, the logical connection between relativity and electrodynamics is very clear and immediate, and the local transformation concept lays a foundation for general relativity and quantum field theory. After expressing electrodynamics in the four-dimensional form, subsequent calculations are made in either three- or four-dimensional form, depending on the application. Being able to choose is a great advantage, since a calculation in four-dimensional notation can be much easier than the same calculation done in three-dimensional notation. The remaining chapters deal with steady-state fields, radiation in dielectric media, charged particle motion, radiation by moving charges, and classical electron theory.

I have introduced a few innovations into the content of this book. In Chapter 7 the Lorentz equations of motion of a particle in an electromagnetic field are reformulated as a relativistic fluid model. In this formulation a conserved vector field G^k is introduced, and the equations of motion become integral equations with infinite series solutions. These solutions generate corresponding diagrams that are analogous to the Feynman diagrams in quantum electrodynamics. I have also incorporated two recent advances in computer software by including the numerical solution of differential equations, and by using a symbolic mathematics program, *Mathematica*. Many of the illustrations in the text were generated by *Mathematica* calculations, and short program listings (notebooks) on special topics are provided in the appendixes. If they have *Mathematica* available, students may use these notebooks in the exercises.

I am deeply grateful to my own teachers, and to my students who have contributed to this work by their enthusiasm and comments. I therefore dedicate this book to them, as a way of passing the torch from one generation to the next.

Terre Haute, Indiana   James Blake Westgard

Contents

Preface   *v*

CHAPTER 1   **Introduction to Electrodynamics**   *1*
    1-1   A Brief History of Electromagnetism   *1*
    1-2   Vectors   *14*
    1-3   Vector Calculus   *18*
    1-4   Curvilinear Coordinate Systems   *26*
    1-5   Integration of Vectors   *37*
    1-6   Delta Functions   *46*
    1-7   Selected Bibliography   *48*
    1-8   Problems   *49*
    App. 1-1   Vector Operators   *50*
    App. 1-2   Curvilinear Coordinates   *59*

CHAPTER 2   **Experimental Foundation**   *71*
    2-1   Fields   *72*
    2-2   Coulomb's Law   *73*
    2-3   Ampère's Laws for the Magnetic Field   *78*
    2-4   Faraday's Induction Law   *84*
    2-5   Maxwell's Equations   *87*
    2-6   A Mechanical Model of the Electromagnetic Field   *90*
    2-7   The Michelson–Morley Experiment   *95*
    2-8   Systems of Units   *98*
    2-9   Selected Bibliography   *102*
    2-10   Problems   *102*
    App. 2-1   Field Plotter in 2-D   *105*

CHAPTER 3   **Dielectric and Magnetic Materials and Boundary Conditions**   *111*
    3-1   Dielectric Materials   *112*
    3-2   Currents   *117*
    3-3   Magnetic Materials   *121*
    3-4   Boundary Conditions   *126*
    3-5   Some Mathematical Aspects of Boundary Conditions   *130*

3-6   Selected Bibliography   *138*
3-7   Problems   *138*

CHAPTER 4   **Electromagnetic Equations**   *141*
4-1   Tensor Notation   *142*
4-2   Integral Theorems   *149*
4-3   Relativity: A New Kinematics   *152*
4-4   The Electromagnetic Field   *164*
4-5   Electromagnetic Potentials and Gauge Conditions   *173*
4-6   Lorentz Transformed Fields   *175*
4-7   The Lagrangian Method   *178*
4-8   Selected Bibliography   *181*
4-9   Problems   *182*
App. 4-1   Lorentz Transformation of the Field Tensor   *184*

CHAPTER 5   **Electromagnetic Fields in Steady States**   *187*
5-1   Steady-State Equations   *187*
5-2   Multipole Expansion   *189*
5-3   Laplace's Equation: Separation of Variables   *196*
5-4   The Cauchy–Riemann Equations and Conformal Mapping   *208*
5-5   Numerical Solutions by Finite-Element Analysis   *214*
5-6   Magnetic Fields   *217*
5-7   Selected Bibliography   *219*
5-8   Problems   *220*
App. 5-1   Relaxation Solution of Laplace's Equation   *222*
App. 5-2   Gram–Schmidt Orthogonalization   *224*
App. 5-3   Schwartz Transformations   *226*
App. 5-4   Fourier Series   *231*
App. 5-5   Laplace Equation   *240*

CHAPTER 6   **Radiation and Optics in Dielectric Media**   *247*
6-1   Wave Equation in Uniform Media   *247*
6-2   Spherical Waves   *252*
6-3   Radiation in Conductive and Dispersive Media   *261*
6-4   Refraction and Reflection at a Dielectric Boundary   *267*
6-5   Momentum and Energy   *274*
6-6   Huygens' Principle and Diffraction   *282*
6-7   Selected Bibliography   *292*
6-8   Problems   *293*
App. 6-1   Bessel and Legendre Functions   *294*
App. 6-2   Multipole Radiation Patterns   *300*
App. 6-3   Single Slit Diffraction by Helmholtz Integral   *305*
App. 6-4   The Electromagnetic Stress Tensor   *306*

CHAPTER 7   **Particle Motion in Electromagnetic Fields**   *311*

7-1   Uniform Fields   *311*
7-2   Numerical Solutions   *314*
7-3   An Example: Particle Optics   *320*
7-4   Velocity Field Model of Single Particle Kinematics   *324*
7-5   Cross Sections   *337*
7-6   Selected Bibliography   *341*
7-7   Problems   *342*
App. 7-1   Finite-Element Solution of a Differential Equation Using a Spreadsheet   *342*
App. 7-2   Taylor Series Solutions of Differential Equations   *345*

CHAPTER 8   **Radiation by Moving Charges**   *349*

8-1   Multipole Expansion   *350*
8-2   A Physical Model   *358*
8-3   Frequency Analysis of Radiation   *360*
8-4   Calculating with Delta Functions   *368*
8-5   Radiation by Charged Particles   *371*
8-6   Selected Bibliography   *388*
8-7   Problems   *389*
App. 8-1   Radiation by a Fast Charged Particle   *390*
App. 8-2   Tensor Potentials   *396*

CHAPTER 9   **Beyond the Classical Theory**   *401*

9-1   Radiation Reaction   *402*
9-2   Classical Models of the Electron   *405*
9-3   Quantization   *414*
9-4   Unification with Weak Interactions   *424*
9-5   Selected Bibliography   *429*

Index   *431*

CHAPTER 1

Introduction to Electrodynamics

> The history of science is science itself; the history
> of the individual, the individual.
>
> Johann Wolfgang von Goethe
> *Mineralogy and Geology*

> Mathematics, rightly viewed, possesses not only
> truth, but supreme beauty—a beauty cold and
> austere, like that of sculpture.
>
> Bertrand Russell
> *The Study of Mathematics* (1902)

1-1   A BRIEF HISTORY OF ELECTROMAGNETISM

The history of electromagnetism is intertwined with the revolutions in astronomy and mechanics, and with the history of science and intellectual history in general. It is convenient to distinguish three periods in the history which we might call *early*, *classical*, and *modern*, each initiated by certain clusters of critical, distinguishing discoveries. Curiously, these clusters occurred around the centennial years 1600, 1800, and 1900.

The Early Period: 1600–1800.   Of the tumultuous social and intellectual changes of the late Renaissance, none had a more profound effect on Western thought than the scientific revolution of the seventeenth century, which overturned traditional beliefs about the nature of the physical universe and promoted development of the scientific method. In explorations of unknown parts of the world, in religious beliefs, in political structures, in the arts, and in the physical sciences, new ideas were replacing traditional authority. Scholarship before the seventeenth century, including the study of the natural sciences, was characterized by axiomatic, deductive logic, derived from the ancient philosophers, especially Plato and Aristotle, and modeled on the methods of Euclid. This authoritarian approach, consistent with established religious and political practices, prevailed through the Middle Ages and into the Renaissance. At this time, geographical explorations and new inventions such as those by Galileo Galilei stimulated precise observation and measurement of nature. Many of the physical and astronomical concepts derived from Aristotle and Ptolemy were shown to be wrong, demonstrating the limitations of the purely axiomatic method. A new start was needed, free from the ancient dogmas. The early period is thus characterized by the rise of the scientific method, founded largely on the vision of Francis Bacon and Rene Descartes, exemplified by Galileo, and advanced by Isaac Newton. In this new mode of thinking, induction joins deduction as an equal, mathematics becomes the preferred language, and experiment decides between competing theories.

Several important events mark the beginning of the early period of electromagnetism. William Gilbert's *de Magnete*, published in 1600, was the first systematic description of experiments done with magnets and dielectrics. Francis Bacon's *The Advancement of Learning*, published in 1605, described and defined the new methodology and promoted its application in acquiring useful knowledge. In 1609, Galileo constructed a telescope with sufficient power to open new vistas in astronomy to human observation.

The advances in electromagnetism are contemporaneous with the great period of development of the science of mechanics. It begins with Galileo's development of kinematics and ends with the analytical mechanics of Lagrange. In mathematics, it spans the analytical geometry of Descartes and the analyses of Euler and Lagrange. Important electromagnetic advances in this period include: the distinction between conductors and insulators; the discovery that lightning is an electrical discharge; the discoveries of positive and negative electric charge; static electric and magnetic laws of force; and the biological effects of electricity. Instruments were devised which allowed controlled observation and precise measurement, among them the torsion balance, the Leyden jar or condenser, and various mechanical devices for generating static charges.

During the early period, boundaries between science and other disciplines were indistinct, as were divisions between branches of the sciences. Many of the most important scientific investigations were conducted by talented amateurs—"Renaissance men." Newton, best known as a mathematician and theoretical physicist, also performed important experimental investigations in several areas, especially optics. The

Royal Society of London, founded in 1660 to advance the growing interest in science, brought together leaders in science from England and the Continent. The success of science in interpreting Nature was a major contributor in creating the philosophical framework of the time. Exhilarated by the promises of the new science, the French *philosophes* proclaimed the eighteenth century to be the Age of Enlightenment.

Benjamin Franklin is an exemplar of the amateur scientist of the time. Born in Boston in 1706 and apprenticed to his half-brother as a printer, he settled in Philadelphia where he established his own printing and publishing business. His financial success allowed him to retire in 1748 and to devote himself to other pursuits, including philanthropy, public service, and science. He invented bifocal glasses, the Franklin stove, and the lightning rod. He performed many electrical experiments and enunciated the one-fluid theory of electricity, which included conservation of electric charge. He observed that electric forces vanish inside a hollow conductor. From this observation, his friend Joseph Priestly inferred in 1767 that the electrical force, like gravity, is proportional to the inverse square of the distance.

The electrostatic force law was determined directly by Charles Augustin Coulomb in 1785 by means of the torsion balance, which he had invented. (The force law was also determined by Henry Cavendish slightly earlier but was not published by him.) John Michell, using a torsion balance (invented independently), determined the force between two magnetic poles. Thus at the end of the preclassical period, mathematically precise electrostatic and magnetostatic laws were determined, putting them at last on a level with Newton's gravitational law. Now mathematical development could begin.

Investigations into the nature of light were also under way, independent of electromagnetic considerations since there was no hint of any connection between them. In 1675, Roemer discovered that light has a finite velocity and made a rough estimate of its value. Other optical discoveries were: Snell's law of refraction; the polarization of light; and double refraction. Several competing theories of light were proposed. A particle model promoted by the followers of Newton temporarily prevailed over a wave model associated with Christiaan Huygens, perhaps because of the lack of experimental evidence for a wave theory, and because it was not yet understood how waves could account for beams of light.

For scientists of this time the familiar forces were contact forces; in contrast, the forces that magnets and electrical charges exerted were mysterious—some writers even declared them occult. In order to conceive these forces as contact forces, Descartes introduced the concept of the *aether*,[1] a kind of invisible fluid filling all space and transmitting

---

[1] We use this spelling to distinguish it from the anesthetic, spelled "ether."

the electric, magnetic, and gravitational forces. The early aether theories were not mathematical, and hence did not attempt to produce mathematical results, such as the inverse square law. Newton himself refrained from making hypotheses (whether about aether or other nonobservable things), restricting himself to mathematical inquiry. Although it was eventually made superfluous by relativity, the aether theory was useful in preparing scientists to accept electric and magnetic fields, which could be made mathematically precise and conformable to relativity.

The Classical Period: 1800–1900. Like the early period, the classical period in the history of electromagnetism occurred during a time of intense intellectual and political change. The development of democracy, stemming from the American Revolution, the industrial revolution, and the rise of classical capitalism were accompanied by achievements in thermodynamics, optics, and electromagnetism.[2] This period also saw changes in the role of scientists in society; science became more institutional, and scientists professionalized. The classical period in the development of electromagnetism begins with two events, Thomas Young's diffraction experiments in 1801, demonstrating the wave properties of light, and Alessandro Volta's invention in 1800 of the voltaic pile (the electric battery).

Extremely precocious as a child, Young is said to have read the Bible through twice by the age of four and learned Greek by the age of five. His knowledge of Greek enabled him to decipher one of the two Egyptian scripts on the Rosetta stone. Thus he and Jean Francois Champollion, who deciphered the other script, are considered cofounders of the science of Egyptology. Young's double slit experiment was the first conclusive evidence that light is a wave. His other scientific achievements include development of theories of color vision and of elasticity. "Young's modulus" is named for him.

Volta's invention of the electric battery, building on Luigi Galvani's discovery of the *galvanic effect*, perfectly illustrates the role of serendipity in science. Galvani was professor of anatomy at the University of Bologna. In 1780, he found that in the course of dissection, a frog's leg muscle contracted when the nerve was touched by a scalpel. As he was familiar with similar effects of stimulation by static electrical discharges, he proposed that this effect was due to electrical causes. A crucial feature of the experiment was that the frog was held to an iron plate by brass hooks, i.e., two different metals were involved. After Galvani's report of these experiments, others immediately began to investigate the effect. Among the researchers was Alessandro Volta,

---

[2] This century encompassed the Victorian age and saw enormous intellectual achievements in economics, biology, geology, linguistics, chemistry, and mathematics.

professor of natural philosophy at the University of Pavia, who showed that Galvani's electrical effect was not fundamentally a biological effect at all. The frog could be replaced by any conducting solution which provided contact between the two different metals. By trying different metals in the experiment, he established the electrochemical potentials for the metals. But the most significant contribution was his invention of the electric battery. By making a stack of alternating disks of copper, zinc, and moistened pasteboard, he could produce a large and steady electric current.[3] This accomplishment was of first-rate importance; since experiments could now be done using large constant currents, it made possible all the subsequent advances in magnetism. Benefits appeared immediately: Carlisle and Nicholson discovered that a steady current decomposes water into hydrogen and oxygen. Thus electrochemistry was established: electroplating and the electrolytic purification of chemical elements result from this discovery.

Some relationship between electricity and magnetism had long been suspected, but serious obstacles impeded attempts to combine them. For example, they did not seem to act on the same materials: magnets acted on other magnets and iron but nothing else, while electric forces acted on any object. A connection between electricity and magnetism was discovered by Hans Christian Oersted in 1819. Passing a current through a wire and using a small compass as a magnet, Oersted discovered what eventually would be known as the magnetic field. By moving the compass to different positions, he was able to map the magnetic field around the wire. He had at last determined the relationship between electricity and magnetism: an electric current creates a magnetic field. Oersted published his findings in July, 1820. His discoveries were largely qualitative, but other researchers soon put them into mathematical form. By October of that year, Jean-Baptiste Biot and Félix Savart reported the first of several quantitative treatments, and Andre-Marie Ampère had begun investigations from which he was able to state the Oersted effect in exact mathematical form. Electromagnetism had reached a level where the powerful mathematical methods developed for mechanics could be applied, and indeed to attract the attention of the great mathematicians of the time such as Karl Friedrich Gauss. Still, a problem with the unification into electromagnetism remained— it was asymmetrical. Since electrical effects could induce magnetism, shouldn't magnetic effects also induce electricity? The search for the inverse effect was on; it was discovered by Faraday in 1831.

In 1812, the year Napoleon began his disastrous invasion of Russia, a twenty-year-old bookbinder's apprentice, Michael Faraday, attended four lectures by Humphry Davy of the Royal Institution. It is not altogether obvious which was the more significant event! Faraday had been

---

[3] Presumably either the pasteboard or the water contained some electrolyte.

reading scientific works in his master's shop and had performed some of the experiments he read about.[4] He took careful notes on Davy's lectures, bound them (who better than a bookbinder), sent them to Davy, and asked for a job as his assistant. When a vacancy occurred a short time later, Davy hired him. Davy and the new apprentice scientist left for a tour of Continental scientific establishments within the year. After their tour they began research at the Royal Institution, where Faraday quickly rose from assistant to collaborator. When Davy became president of the Royal Society in 1825, Faraday took his place as director of the laboratory at the Royal Institution. Faraday's path to success is nearly impossible to imagine today. Furthermore, it is unthinkable today that in the middle of a major war, the greatest scientist of one nation would be allowed to visit the laboratories of its enemy. Scientific exchanges such as those between George III and Napoleon were not repeated in 1914.

Michael Faraday was unquestionably one of the greatest experimental scientists of all time; it is also fair to call him one of the greatest inventors. He is also representative of a new kind of scientist, neither academic like Newton, nor a self-supported amateur like Franklin. He was one of the first institutional scientists, and the Royal Institution was a forerunner of the great national and industrial laboratories of today. The importance of his contributions is equaled only by the range of topics he investigated. The law of electromagnetic induction, the law of electrolysis, an electric motor, an electric generator, the transformer, self-inductance and mutual inductance, liquification of chlorine and other gases, critical temperature in phase transitions, magneto-optic effects, electric polarization, dielectric constant, and paramagnetism, the physics and chemistry of plasmas—all are his discoveries or inventions. Although not mathematically trained, he nevertheless contributed a theoretical construct of paramount importance, that of electric and magnetic fields and the concept of lines of flux.

Oersted and Faraday showed electricity and magnetism to be related, but this did not unify the two phenomena into a single theory. That was to be the great contribution to electromagnetism by James Clerk Maxwell. Maxwell was born in 1831 in Edinburgh. He was educated at the Edinburgh Academy and at Edinburgh and Cambridge universities. His first published paper, written at age 14, was read to the Royal Society of Edinburgh; we may say he was a fairly precocious young man. In 1854 he took honors in the Mathematics Tripos examination, thus showing early promise of mathematical abilities which would complement and extend Faraday's uncanny physical insight. He was elected

---

[4] Science and the world owe much to the easy-going nature of Faraday's master, Mr. Riebau.

a Fellow of Trinity College at Cambridge in 1854, Professor of Natural Philosophy Marischal College at Aberdeen in 1856, and was Chair of Physics, Kings College, London, 1860–1865. These were the crucial years in Maxwell's development of electromagnetism; as early as 1854 he wrote to William Thomson (Lord Kelvin) stating his intention to work on this problem. His results were published in a series of papers between 1855 and 1865. His 1861 paper, "On Physical Lines of Force," presented a mechanical model of the aether, involving rotating spheres and idler wheels. This model was a direct descendent of Faraday's lines-of-force model. However, in his culminating 1864 paper, "A Dynamical Theory of the Electromagnetic Field," the mechanical model had vanished, leaving only the famous Maxwell field equations in essentially present-day form.

But what was Maxwell's specific contribution, bearing in mind that several of the equations are direct consequences of the discoveries of Coulomb, Ampère, and Faraday? Just as Newton had used the work of his predecessors, Maxwell had also "stood on the shoulders of giants." His contributions are several, and crucial. In the first place, he gave mathematical legitimacy to Faraday's field concept. This concept, so familiar today, was a major breakthrough at that time. The alternative mathematical formalism required expressing the equations in terms of forces between currents and charges acting at a distance with no intermediary. Such models inadequately explain the time delay between the source of radiation and its detector, and the related question of energy conservation. Maxwell's second crucial contribution is the vacuum polarization current, which is the electric counterpart of Faraday's magnetic induction. Without this term the equations do not have wave solutions or satisfy charge conservation. Maxwell's third major contribution is devising a consistent mechanical model of the electric and magnetic fields and the related equations. Undoubtedly, this mechanical or geometrical model helped promote acceptance of the equations, considering the strong interest in technology (and its influence on physics) at this time. The mid-nineteenth century produced a virtual flood of mechanical innovations, and a mechanical picture of electromagnetism must have been congenial to those who would have found it difficult to accept otherwise. These equations were far more than a summary, since they would stimulate much further research, both in corroboration and in application. They form the physical foundation of modern electrical technology.

It is difficult to overemphasize the importance of Maxwell's achievement. The small and beautifully symmetric set of equations summarized all of the known features of electricity and magnetism discovered up to his time. The equations showed that not only are electric and magnetic fields related, but are actually aspects of a single field, the electromagnetic field. Moreover, the unification was a double one, since the laws of optics also result from the electromagnetic equations. The nearest equivalent to Maxwell's equations in importance was Newton's laws of mechanics, published two centuries earlier.

Important as they are, the electromagnetic equations form only a part of Maxwell's work. He was one of the founders of statistical mechanics and kinetic theory. He won the Adams prize at Cambridge in 1856 for his work on the structure of the rings of Saturn. In 1871, he became Professor of Experimental Physics at Cambridge, where under his direction the plans for the Cavendish Laboratory were drawn. This laboratory was to have enormous influence on physics in the early twentieth century. Maxwell died in 1879. He was only forty-eight.

Conclusive experimental verification of the Maxwell equations was provided in 1886 by Heinrich Hertz, who had discovered a means of generating and detecting electromagnetic radiation. The Hertz detector consisted of a spark gap in a loop of wire with attached conducting flat plates which acted as both capacitor and antenna. The transmitter was an identical loop, connected to an induction coil to produce the high voltage spark. Since the two circuits were identical, they were tuned to the same frequency. Detection consisted of observing a spark in the detector loop coincident with the spark in the transmitter. Using this apparatus, Hertz was able to verify some of the main predictions of the Maxwell equations, particularly the existence of the electromagnetic radiation, transverse polarization, reflection, refraction, and an estimate of the velocity.

Maxwell's and Faraday's mechanical models of the electromagnetic field stimulated further model building. One model proposed an analogy with sound and other waves, which were disturbances traveling in some appropriate medium. By analogy, the electric and magnetic field strengths in the proposed model should correspond to the pressure or displacement of the medium in sound waves. This medium, which must fill all space since it carried the light from the stars, was referred to as the *luminiferous aether*. And the aether must have peculiar properties. It must be very stiff in order to account for the great velocity of light, but still allow the planets to move through it with negligible resistance, since they exhibit no significant deceleration.

Such was the state of electromagnetism in 1887 when Albert Michelson and Edward Morley first performed their celebrated experiment on the velocity of light. Michelson had recently invented a device which could determine the velocity of light to unprecedented accuracy. Michelson and Morley proposed to use it to determine the velocity of the Earth through the aether, which would incidentally support its existence. By Galilean kinematics the velocity of an object as seen by an observer is the difference between the object's velocity and the observer's velocity (relative to some fixed coordinate system). That being so, the measured velocity of a light wave should depend on whether the observer and the wave move in the same or in opposite directions. However, a particular feature of the Maxwell equations is that they predict a velocity for the electromagnetic waves which depends only on the electric and magnetic coupling constants and is thus a constant itself, independent of the velocity of the observer. This feature contradicts Galilean kinematics. The Michelson and Morley experiment was designed to distinguish between the two predictions. The interferometer was set up with one arm parallel to the Earth's motion (in the East–West direction) and the other at right angles to it (in the North–South direction), thus utilizing the rotational velocity of the Earth. The velocity of the East–West arm of the interferometer thereby reverses direction every 12 hours, and measurement of the shift in the interferometer pattern for 24 hours should indicate the direction of the aether drift and determine its magnitude. They found no evidence of any aether drift, either in 1887 or in any subsequent measurement made with improved instrumentation. The null result brought into question the existence of the aether, at least as an actual substance filling all space. As Michelson said of it, the luminiferous aether is to some extent a hypothetical substance, and if it consists of matter at all, it must be very rare and elastic. *It entirely escapes all our senses of perception.*[5]

In 1892, George FitzGerald attempted to salvage the aether theory by suggesting that Michelson and Morley's null result was a consequence of the longitudinal arm contracting by just the right amount while the transverse arm remained fixed. There was, of course, no provision for this contraction in Galilean kinematics. The contraction of the inferometer arm was the "FitzGerald contraction." The change in length was generalized by Hendrick Antoon Lorentz and Jules Henri Poincaré, who showed that the contraction could be justified by the field equations when a temporal transformation was included as well. The Lorentz transformations became an integral part in the theory of relativity. As far as aether itself was concerned, it was no longer taken seriously, since relativity theory rendered it unobservable and therefore superfluous.

The discovery of the electron in 1897 by Joseph John Thomson was perhaps the last major discovery in classical electromagnetism, but it was to have a profound impact on quantum physics. Its discovery ended the controversy between the one fluid- and the two-fluid theory of electric charge. Since all matter was shown to contain electrons, and positive charges as well (later found to reside in the nucleus), electrification and most electric currents could be explained as transfer of the relatively mobile electrons. Electrical discharges in rarified gases had been an active area of scientific pursuit since Faraday's time. Initially

---

[5] Livingston, p. 99.

the interest was simply to investigate the current-carrying mechanism in gases. Geissler's invention of the mercury vacuum pump made it possible to investigate conduction in highly rarified gases. It was found that *cathode rays*, i.e., the discharge from the negative electrode, caused the glass walls to fluoresce. Solid objects in the path of the beam cast shadows, thus implying that the cathode rays moved in straight lines. Furthermore, the cathode rays were deflected by electric and magnetic fields and must therefore be negatively charged. It was found that there was no such emanation from the anode (although a hole punched through the cathode allowed positive charges called *canal rays* to pass through the hole). Using a rapidly rotating mirror to determine the velocity of the cathode rays, Thomson found the velocity considerably less than the speed of light, and dependent on the potential across the tube; clearly the cathode rays were not a form of electromagnetic radiation. A series of investigations with electric and magnetic fields transverse to the cathode ray velocity showed that they had the same ratio of charge to mass whatever the gas in the tube. Corresponding investigations of the canal rays by Wien showed strong dependence on the gas used and also a much smaller charge to mass ratio. These investigations implied that extremely light, negatively charged particles are a component of all the gases (and by extension of all matter). These new particles were called *electrons*, from the Greek word for amber. Their charge was determined approximately by measuring the motion of charged water droplets in a Wilson cloud chamber. In 1909, Robert Millikan was able to do a much more precise measurement of the charge using oil droplets of uniform size.

The Modern Period: 1900–Present.   The cluster of discoveries initiating the third period of electromagnetism is large and diverse. It includes J.J. Thomson's discovery of the electron in 1897, Max Planck's discovery of quantization in 1900, and Albert Einstein's theory of relativity in 1905. We might also include Wilhelm Röntgen's discovery of x-rays in 1895 and Antoine Henri Becquerel's discovery of radioactivity in 1896. These developments would change the nature of science virtually beyond the recognition of earlier scientists. Science until 1900 was essentially deterministic. The universe was regarded as a great machine like a clock, and if one were sufficiently clever and industrious, its future state could be predicted from knowledge of its former state with any desired accuracy. The year 1900 marks the advent of uncertainty in physical science.

Electromagnetic unification was a great event in science, for its beautiful simplifications, for its added insights into understanding nature, and for the technology it would create. But still greater unifications were to come: the unification of matter, energy, and momentum, and the unification of space with time. The theory of special relativity had been waiting to happen since the 1880s. The symmetries inherent in Maxwell's equations and in the Lorentz force on a moving particle, and the implications of the Michelson–Morley experiment, were clear signposts pointing the way. FitzGerald, Lorentz, and Poincaré were headed in that direction, and if nothing else, the inevitable invention of high-energy accelerators would have forced its attention on science. At some energy the discrepancies with Newtonian mechanics would simply become too large to ignore. One can easily imagine that special and general relativity might have been created as gradual readjustments to classical physics by many individuals. Its actual history is more dramatic.

When reading biographies of the great scientists, one often finds evidence of great strength of character, quite apart from their intellectual capacity. A sincere passion for truth, perseverance, skepticism of authority, intellectual courage—Albert Einstein had these in great measure. He was born in Ulm in 1879, the son of a businessman. The family moved to Munich in 1880, where Einstein received his early schooling. He was a brilliant student in the topics that interested him, even though the regimentation and the rote teaching methods of the time were not to his liking. His independence of mind led to some friction with his teachers in his gymnasium years. To a lesser extent this resistance to the constraints of formal schooling continued through his university years; for this reason he was partly self-taught in the very topics that were to make him famous. When he was fifteen his father moved the family to Milan, leaving Albert behind to finish his education. Before the year was out he left school and rejoined his family in Milan. After resuming his schooling in Switzerland, he received his diploma at age seventeen. In 1900 he graduated from the Federal Institute of Technology in Zurich, and became a Swiss citizen in 1901. His antipathy to regimentation continued throughout his life.

Einstein accepted employment in the Swiss patent office in 1902, and in this setting he began his series of revolutionary publications. Two of them, both published in 1905, are of particular importance here: the paper quantizing the photoelectric effect, and the first paper on special relativity. He began a short peripatetic academic career, accepting in turn, appointments at the University of Zurich in 1909, at the German University in Prague in 1911, and at the Polytechnic in Zurich in 1912. By 1912 his abilities had become widely recognized, and he accepted the directorship of the research branch of the Kaiser Wilhelm Institute in Berlin. In this position he was free to devote himself to his research. It was here that the general theory was completed. He remained in Berlin until 1933, when he fled the National Socialist regime. He then joined the Institute for Advanced Study at Princeton.

If one must strip relativity down to its bare essential, it is geometry. That is particularly obvious in the general theory, where the entire

Riemannian formalism is essential, but it is no less true of the special theory. In the geometry of relativistic kinematics, space and time do not have separate existence as they have in Newtonian kinematics, but are unified into four coequal dimensions, where four-dimensional Minkowski geometry replaces three-dimensional Euclidean geometry. In this context, the contraction proposed by FitzGerald to explain the Michelson–Morley results becomes a kind of rotation in the x–t plane, i.e., the Lorentz transformation, implying a corresponding dilation of time intervals. This dilation is now easily observed in the increased lifetimes of moving radioactive particles. Since space has become four dimensional, momentum must have four components as well, where the fourth component is energy. Thus momentum and energy are unified into a single quantity. One consequence of the unification of momentum and energy is the famous equivalence between mass and energy, $E = mc^2$; thus any object having mass has energy just by virtue of that mass. This equivalence is the source of the enormous energy released by certain nuclear reactions.

But the special theory, wonderful as it was in unifying disparate phenomena, was insufficient. Einstein early saw that the equivalence between mass and energy not only implied that massive objects have energy even at rest, but that the converse must also hold: Any object containing energy, including light, must have its equivalent mass as well. Thus a beam of light must be deflected by a gravitational field, like a beam of particles with mass. Gravity had entered the picture.

Speculations on the role of mass in physics led to the equivalence principle and ultimately to the general theory, which was published in essentially final form in 1916. Experimental verification of general relativity had to wait until the eclipse of 1919, in which light from a star was observed to be deflected by the sun's gravitational field. General relativity showed that gravity was "only" a manifestation of geometry. But what a geometry! It is intrinsically curved, but in four dimensions. The curvature is local and differs from one place to another, and in one point in time to the next. Space is no longer merely empty space, a plenum in which particles could move and collide but which itself had no interaction with the particles. Instead it has properties of its own, allowing it to transmit the interaction between particles. When space is understood in this sense, geometry can be said to be the successor to aether.

After 1919 Einstein applied relativity to cosmology, and to attempts at the unification of gravity with electromagnetism. The former has been extremely successful: the language of modern cosmology is essentially relativity. The latter was not successful: important experimental facts about elementary particles and fields were unknown to Einstein and his contemporaries. Indeed, that unification has not been possible to this day, but instead electromagnetism has been unified with weak interactions. The final word in the unification of forces has not yet been spoken.

For all its spectacular aspects, relativity is fundamentally classical physics. Relativity deals with classical quantities like velocity and energy, and its mathematics, while advanced, would have been recognizable to earlier mathematicians. Relativity was largely the product of one man with contributions from others. Quantum physics is quite different in having many authors, having noncommuting variables, and particularly in having fundamental uncertainty built into it.

Quantization was introduced in physics by Max Planck in 1900, in the course of removing an inconsistency in the theory of electromagnetic cavity radiation. The price paid for this "improvement" was a major revision of the kinematic principles developed by Galileo and all of his successors. For example, the energy of a quantized system is an integral multiple of a minimum quantity of energy called the *quantum*. A system could contain any integer number of quanta, but never a noninteger number. Furthermore, the quantum depended on the frequency of the radiation, not on the square of the amplitude that Maxwell's theory required. Quantization gave excellent agreement with experiment, but the new conceptual difficulties it requires made many skeptics, including Planck himself. Then in 1905 Einstein showed that energy quantization also accounted for the peculiar properties of the photoelectric effect. Nils Bohr developed a quantum theory of the atom, correctly predicting atomic spectral and x-ray frequencies. It was becoming clear that quantization was a fundamental principle of physics, and not an aspect of classical physics which might eventually be discovered.

For all its successes, the early quantum mechanics had no generally applicable equations of motion for electrons or other particles. This situation changed in 1925, when Erwin Schrödinger and Werner Heisenberg, working independently, produced two very different systems of quantum equations of motion. Subsequently, it was shown that the Schrödinger *wave mechanics* and Heisenberg *operator mechanics* were, in fact, different aspects of a common theory. Also in 1925, experiments by Samuel Goudsmit and George Uhlenbeck showed that electrons have a half-integer intrinsic angular momentum, called *spin*, and an associated intrinsic magnetic moment. The existence of these intrinsic properties indicated that electrons cannot be conceived as point charges, as they are in classical treatments. The Schrödinger and Heisenberg equations were not relativistic, and the magnetic moments had to be added ad hoc to the interaction potentials. Then, in 1928 Paul Dirac obtained a relativistic equation of motion, with the electron spin as an integral feature. But the Dirac equation also predicted positively charged particle with the same mass as ordinary electrons (*positrons*). Four years later a positron was observed by Carl Anderson in a cosmic ray interaction.

When radioactivity was discovered in 1896, it was found to include a negatively charged component called beta rays. These were later found to be high-energy electrons. The particles observed in beta decays conserved neither energy nor angular momentum! In order to rescue these conservation principles, Enrico Fermi proposed a new massless, chargeless particle called the neutrino that carries off the missing energy and angular momentum. Since the strength of this interaction is much weaker than electromagnetic interactions, they are called *weak interactions*. During the period from 1940 to 1980 many new particles were discovered in cosmic ray interactions and in the high-energy accelerator facilities built during that time. Theorists were able to group the new-found particles into families and find a mathematical expression for the weak forces. All weak interaction coupling strengths (expressed, e.g., by the reaction rates) were found to be essentially the same, and it was found that parity was not conserved in weak interactions, in contrast to electromagnetic interactions.

The history of electromagnetism as a separate topic ends with its unification with weak interactions. In the late 1960s Sheldon Glashow, Abdus Salam, and Stephen Weinberg found a way to combine the weak and electromagnetic interactions into a unified form. In this theory, weak forces between interacting particles with mass are carried by intermediate particles, much as the electromagnetic forces are carried by the photon. Weak intermediate bosons were found in accelerator experiments in 1983 at precisely the predicted masses.

1-2   VECTORS

Many quantities in physics are characterized by having a specific direction as well as magnitude; such quantities are called vectors. Familiar examples are force, velocity, and acceleration. Electromagnetism uses many vector quantities and may fairly be referred to as a vector theory. Although it is possible to express its equations entirely in terms of components, they become unnecessarily cumbersome in that form and are hard to work with. We assume the reader is familiar with the algebra of vectors from elementary mechanics, but we give a short review of some essential features needed in Chapter 2. We will discuss vector calculus in the next section, and more advanced features in Chapter 4.

Define a vector to be an ordered set of N quantities, called components, having the linear properties:

1. *Two vectors are added by adding their corresponding components*:

$$U^j = V^j + W^j, \qquad j = 1 \ldots N. \tag{1-2.1}$$

2. *Multiplying a vector by a constant multiplies each component by the constant*:

$$c[V^j] = [cV^j], \qquad j = 1 \ldots N. \tag{1-2.2}$$

A consequence of these properties is that any linear combination of vectors is itself a vector. The number, N, is the dimension of the vector. Subtraction is similarly defined by subtracting the corresponding components. Then a null vector (with zero components) results from subtracting a vector from itself. In the elementary definition, vector components may be imagined to be numbers; however, it is occasionally useful to consider vectors with components such as matrices, differential operators, or vectors themselves. We only require that they have the linear properties of equations (1-2.1) and (1-2.2).

In the preceding equations we let the general component V^j denote the vector itself; thus the presence of the index (j) signals that the quantity is a vector. There are two other common notations for vectors. It is frequently convenient to display the components as a column matrix:

$$V^j = \begin{bmatrix} V^1 \\ \vdots \\ V^N \end{bmatrix}, \tag{1-2.3}$$

where the index is a superscript, or the transpose row matrix:

$$V_j = [V_1, \ldots, V_N], \tag{1-2.4}$$

where the index is a subscript.[6] The other notation for vectors is the use of boldface symbols. In this notation equation (1-2.1) becomes:

$$\mathbf{U} = \mathbf{V} + \mathbf{W}. \tag{1-2.1'}$$

Boldface notation is commonly reserved for three-dimensional vectors.

The *inner product* or *dot product* of two vectors is defined by the sum of the products of corresponding components:

$$S \equiv \sum_{j=1}^{N} V_j W^j, \tag{1-2.5}$$

---

[6] In Section 1-4 we will also use subscripts and superscripts to distinguish between covariant and contravariant vectors.

where S is a *scalar*, not a vector. The *magnitude* of the vector is expressed by the inner product of the vector with itself:

$$|\mathbf{V}| = \left[\sum_{j=1}^{N} V_j V^j \right]^{1/2}.$$
(1-2.6)

In three dimensions the dot product of two vectors becomes:

$$\mathbf{V} \cdot \mathbf{W} \equiv V_1 W_1 + V_2 W_2 + V_3 W_3.$$
(1-2.7)

If the angle between two vectors is θ, then the dot product can be written:

$$\mathbf{V} \cdot \mathbf{W} = |\mathbf{V}||\mathbf{W}| \cos \theta.$$
(1-2.8)

The dot product is commutative:

$$\mathbf{V} \cdot \mathbf{W} = \mathbf{W} \cdot \mathbf{V}.$$
(1-2.9)

It is frequently useful to express vectors as a linear combination of basis vectors. For example, in three dimensions, we may write:

$$\mathbf{V} = V_1 \mathbf{e}_1 + V_2 \mathbf{e}_2 + V_3 \mathbf{e}_3,$$
(1-2.10)

where the indices on the V label the components of the vector, $\mathbf{V}$, but the indices on the unit vectors label the vector, not the components. The most useful basis vectors are mutually orthogonal unit vectors, i.e., satisfying the condition:

$$\mathbf{e}_i \cdot \mathbf{e}_j = \delta_{ij},$$
(1-2.11)

where the *Kronecker delta symbol* is:

$$\delta_j^i \equiv \begin{cases} 1, & j = i, \\ 0, & j \neq i. \end{cases}$$
(1-2.12)

Thus the dot product of two orthogonal vectors vanishes. Orthogonal basis vectors having unit magnitude are called *orthonormal*. The vectors:

$$\mathbf{e}_1 = [1, 0, 0],$$
$$\mathbf{e}_2 = [0, 1, 0],$$
$$\mathbf{e}_3 = [0, 0, 1],$$
(1-2.13)

form a particularly simple system of basis vectors. They are not unique, however; the set:

$$\mathbf{a}_1 = [1, -1, 0],$$
$$\mathbf{a}_2 = [1, 1, 0],$$
$$\mathbf{a}_3 = [0, 0, 1],$$
(1-2.14)

is also an orthogonal basis. For an orthonormal basis, evaluating the linear coefficients is particularly easy, i.e., multiplying equation (1-2.10) by e_i gives:

$$V_i = \mathbf{V} \cdot \mathbf{e}_i. \tag{1-2.15}$$

In three dimensions we may define a second kind of product, the *cross product*:

$$\mathbf{U} = \mathbf{V} \times \mathbf{W} \equiv \begin{bmatrix} V_2 W_3 - V_3 W_2 \\ V_3 W_1 - V_1 W_3 \\ V_1 W_2 - V_2 W_1 \end{bmatrix}. \tag{1-2.16}$$

Taking components, the indices appear in cyclical order:

$$U_i = V_j W_k - V_k W_j, \qquad i, j, k = 1, 2, 3. \tag{1-2.17}$$

If the angle between the two vectors is θ, then the magnitude of the cross product is:

$$|\mathbf{V} \times \mathbf{W}| = |V||W| \sin \theta. \tag{1-2.18}$$

The direction of the cross product is normal to the plane determined by the two vectors. The cross product of two parallel vectors vanishes. *The cross product is not commutative*:

$$\mathbf{V} \times \mathbf{W} = -\mathbf{W} \times \mathbf{V}. \tag{1-2.19}$$

EXAMPLE

The area of a parallelogram whose sides are given by $\mathbf{U}$ and $\mathbf{V}$ is given by the cross product:

$$\mathbf{A} = \mathbf{U} \times \mathbf{V}. \tag{1-2.20}$$

Thus the area is represented by a vector whose magnitide equals the area of the parallelogram and whose direction is perpendicular to it. This representation of the area is sometimes useful for writing the surface element in surface integrals. Some care must be taken in choosing the order of $\mathbf{U}$ and $\mathbf{V}$ in the product, unless only the magnitude is wanted.

Since there are two kinds of products, two kinds of triple products are possible. The scalar triple product is:

$$S = \mathbf{U} \cdot (\mathbf{V} \times \mathbf{W}). \tag{1-2.21}$$

It can be neatly expressed as the determinant composed of the components of the vectors:

$$\mathbf{U} \cdot (\mathbf{V} \times \mathbf{W}) = \begin{vmatrix} U_1 & U_2 & U_3 \\ V_1 & V_2 & V_3 \\ W_1 & W_2 & W_3 \end{vmatrix}. \tag{1-2.22}$$

Since the dot and cross operators may be exchanged, the triple cross product is often written without parentheses:

$$\mathbf{U} \cdot (\mathbf{V} \times \mathbf{W}) = (\mathbf{V} \times \mathbf{W}) \cdot \mathbf{U} \equiv \mathbf{U} \cdot \mathbf{V} \times \mathbf{W}. \tag{1-2.23}$$

The order of the three vectors may be permuted cyclically without changing the value, whereas reverse cyclical order reverses the sign of the product:

$$\mathbf{U} \cdot \mathbf{V} \times \mathbf{W} = \mathbf{W} \cdot \mathbf{U} \times \mathbf{V} = \mathbf{V} \cdot \mathbf{W} \times \mathbf{U} = -\mathbf{U} \cdot \mathbf{W} \times \mathbf{V}. \tag{1-2.24}$$

EXAMPLE

The volume of a parallelopiped whose sides are formed by the vectors. This is useful to express the volume element in integration:

$$dV = d\mathbf{a} \cdot d\mathbf{b} \times d\mathbf{c}. \tag{1-2.25}$$

A second kind of triple vector product is a vector:

$$\mathbf{T} = \mathbf{U} \times (\mathbf{V} \times \mathbf{W}). \tag{1-2.26}$$

The product, $\mathbf{T}$, lies in the plane defined by the two vectors inside the parentheses ($\mathbf{V}$ and $\mathbf{W}$). This can be seen by the expansion:

$$\mathbf{U} \times (\mathbf{V} \times \mathbf{W}) = (\mathbf{U} \cdot \mathbf{W})\mathbf{V} - (\mathbf{U} \cdot \mathbf{V})\mathbf{W}. \tag{1-2.27}$$

This product is neither commutative nor associative:

$$\mathbf{U} \times (\mathbf{V} \times \mathbf{W}) \neq (\mathbf{U} \times \mathbf{V}) \times \mathbf{W}. \tag{1-2.28}$$

1-3   VECTOR CALCULUS

The mechanical equations of motion are written succinctly as vector equations where the displacement, velocity, acceleration, and many other dynamical variables are expressed by vectors, parametrized by time. Let the parametric equations of the trajectory of a particle be given by:

$$\mathbf{r} = \mathbf{r}(t), \tag{1-3.1}$$

where the radius vector, $\mathbf{r}$, represents position of the particle and time is the parameter of motion; then the velocity is given by the derivative of $\mathbf{r}$ with respect to time:

$$
\mathbf{v}(t) = \underset{\Delta t \to 0}{\text{Limit}} \frac{\Delta \mathbf{r}(t)}{\Delta t}
$$

$$
= \left[\frac{dx}{dt}, \frac{dy}{dt}, \frac{dz}{dt} \right]. \tag{1-3.2}
$$

That is, the derivative of the vector is expressed by the derivatives of its components. Similarly, the acceleration is:

$$
\mathbf{a}(t) = \frac{d\mathbf{v}(t)}{dt}. \tag{1-3.3}
$$

Since force is also a vector, Newton's law may be written in vector notation:

$$
\mathbf{F}(\mathbf{r}, \mathbf{v}, t) = m \frac{d^2\mathbf{r}(t)}{dt^2}. \tag{1-3.4}
$$

Figure 1-1 shows the position and velocity vectors, and their relation to the trajectory. Note that the velocity is tangent to the trajectory.

The force in equation (1-3.4) is a function of $\mathbf{r}$ and t; such quantities are referred to as fields. Examples of physical fields are well known; they include: charge density $\rho(r)$ which is a scalar, and electric $\mathbf{E}(r)$ and magnetic fields $\mathbf{B}(r)$ which are vectors. The differential properties of these fields, as e.g., their partial derivatives, often have physical significance.

Let us consider a scalar field, $\Phi(x, y, z)$, and its variation along some arbitrary path in space. For this path the differential pathlength is:

$$
ds^2 = dx^2 + dy^2 + dz^2. \tag{1-3.5}
$$

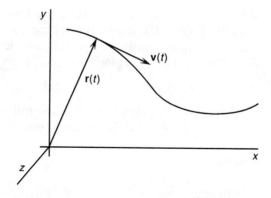

FIGURE 1-1. THE PARTICLE TRAJECTORY. The radius vector represents the position of the particle. Position, velocity, and acceleration are parametric functions of time.

The *directional derivative* of Φ along the path is given by its total derivative:

$$\frac{d\Phi}{ds} = \sum_{j=1}^{3} \frac{\partial\Phi}{\partial x^j} \frac{dx^j}{ds} = \mathbf{n} \cdot \nabla\Phi, \qquad (1\text{-}3.6)$$

expressed as the dot product of two vectors. The unit vector:

$$\mathbf{n} = \frac{d\mathbf{r}}{ds} = \left[\frac{dx}{ds}, \frac{dy}{ds}, \frac{dz}{ds} \right] \qquad (1\text{-}3.7)$$

is tangent to the path, and thus represents its local direction. The second vector:

$$\mathbf{grad}\ \Phi = \left[\frac{\partial\Phi}{\partial x}, \frac{\partial\Phi}{\partial y}, \frac{\partial\Phi}{\partial z} \right]$$

$$= \nabla\Phi \qquad (1\text{-}3.8)$$

is the *gradient* of the function $\Phi(xyz)$. Its magnitude is the maximum directional derivative of Φ on all of the paths that pass through the point (x, y, z). It points in the direction of the maximum slope at that point. We define the differential vector operator "del":

$$\nabla = \left[\frac{\partial}{\partial x}, \frac{\partial}{\partial y}, \frac{\partial}{\partial z} \right]. \qquad (1\text{-}3.9a)$$

We will use several notations for this operator:

$$\nabla \equiv \frac{\partial}{\partial x^j} \equiv \partial_j. \qquad (1\text{-}3.9b)$$

EXAMPLE

If the height of a hill is given by the function, $\Phi = \Phi(x, y)$, then the gradient points in the direction of maximum steepness seen by someone standing at point (x, y); its magnitude is the maximum of all the directional derivatives passing through that point:

$$\text{max slope} = |\nabla\Phi|. \qquad (1\text{-}3.10)$$

To a hiker walking along a path on the hill, the slope of the path at point (x, y) is the directional derivative in equation (1-3.6).

EXAMPLE

The function $\phi = \text{Exp}[-(x^2 + 0.8y^2)]xy$, and the parabolic path, $y = x^2$, are shown in Figure 1-2(a). Then the gradient and the directional derivative are shown in Figures 1-2(b) and 1-2(c). The calculations and

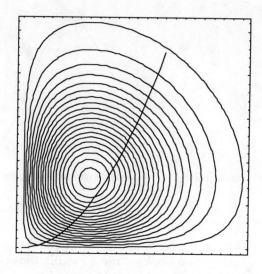

Figure 1-2a.

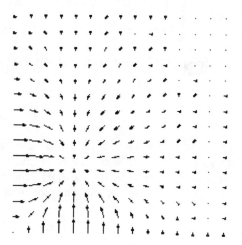

Figure 1-2b.

Figure 1-2a, b.   The function $\phi(x, y)$ discussed in the text is plotted here as a contour plot, and its gradient is shown as a field of arrows indicating direction and magnitide. Each contour represents a constant value of the function. At each point (x, y) the direction of the gradient is orthogonal to the contour lines. The arrows point uphill.

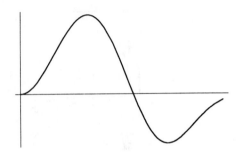

Figure 1-2c.   The directional derivative along the parabolic path shown in Figure 1-2(a).

the plots were made with the *Mathematica* computer application. A listing (notebook) of these calculations is given in Appendix 1-1.

The gradient of this function is:

$$\nabla\phi = \begin{bmatrix} \text{Exp}[-(x^2 + 0.8x^2)]y - 2\,\text{Exp}[-(x^2 + 0.8x^2)]x^2 y \\ \text{Exp}[-(x^2 + 0.8x^2)]x - 1.6\,\text{Exp}[-(x^2 + 0.8x^2)]xy^2 \end{bmatrix} \qquad (1\text{-}3.11)$$

and the unit tangent vector along the parabolic path $y = x^2$ is:

$$\mathbf{n} = \begin{bmatrix} \dfrac{1}{\sqrt{[1 + 4x^2]}}, & \dfrac{2x}{\sqrt{[1 + 4x^2]}} \end{bmatrix}. \qquad (1\text{-}3.12)$$

The gradient of a scalar field characterizes the differential or local properties of the field, Specifically, it gives the rates of change in different directions at each point. Similarly, the differential properties of a vector field are expressed by the array of its derivatives:

$$\partial_j V_k = \begin{bmatrix} \partial_x V_x & \partial_x V_y & \partial_x V_z \\ \partial_y V_x & \partial_y V_y & \partial_y V_z \\ \partial_z V_x & \partial_z V_y & \partial_z V_z \end{bmatrix}. \qquad (1\text{-}3.13)$$

This array can be expressed as the sum of a symmetric term and an antisymmetric term. That is:

$$\partial_j V_k = \frac{1}{2}(\partial_j V_k + \partial_k V_j) + \frac{1}{2}(\partial_j V_k - \partial_k V_j). \qquad (1\text{-}3.14)$$

The two terms are linearly independent of each other, and are thereby mutually exclusive. The second term (antisymmetric) can be expressed as the cross product with the operator, called the *curl*:

$$\text{curl } \mathbf{V} = \nabla \times \mathbf{V} = \begin{bmatrix} \partial_y V_z - \partial_z V_y \\ \partial_z V_x - \partial_x V_z \\ \partial_x V_y - \partial_y V_x \end{bmatrix}. \tag{1-3.15}$$

It is expressed as a vector, since it has three distinct components. The first (symmetric) term in equation (1-3.14) contains, in its diagonal, the *divergence*:

$$\text{div } \mathbf{V} = \nabla \cdot \mathbf{V} = \partial_x V_x + \partial_y V_y + \partial_z V_z. \tag{1-3.16}$$

The independence of the divergence and curl can also be seen from two vector identities. Suppose the vector $\mathbf{V}$ is obtained from the curl of some vector $\mathbf{U}$, then div $\mathbf{V}$ vanishes:

$$\nabla \cdot \mathbf{V} = \nabla \cdot (\nabla \times \mathbf{U}) \equiv 0, \tag{1-3.17}$$

but curl $\mathbf{V}$ does not, in general. On the other hand, if $\mathbf{V}$ is the gradient of some scalar field Φ, its curl vanishes:

$$\nabla \times \mathbf{V} = \nabla \times (\nabla \Phi) \equiv 0, \tag{1-3.18}$$

but its divergence does not, in general.

A vector's divergence represents its differential outward (divergent) tendency, while the curl represents its differential rotational tendency. Fields with these properties are illustrated in Figure 1-3 where the first vector $\mathbf{V} = [x, y]/r$ is a purely divergent field, while the second vector $\mathbf{W} = [\sin(y), -\sin(x)]$ is purely rotational. A three-dimensional plot of the divergence is shown in Figure 1-4.

The divergence has a simple, intuitive interpretation in one dimension: Consider an incompressible fluid flowing slowly in a pipe of constant diameter. Then, if the pipe has leaks along its length, the velocity of flow is a function of the distance down the pipe, i.e., the velocity is a (one-dimensional) field.

$$\text{div } \mathbf{V} \approx \frac{\Delta V_x}{\Delta x} \rightarrow \frac{dV_x}{dx}. \tag{1-3.19}$$

The rate of leakage from the pipe is the divergence in a literal sense.

EXAMPLE

Consider the divergence of the two vectors:

$$\mathbf{V} = \frac{\mathbf{r}}{r} = \left[\frac{x}{r}, \frac{y}{r}, \frac{z}{r} \right], \tag{1-3.20}$$

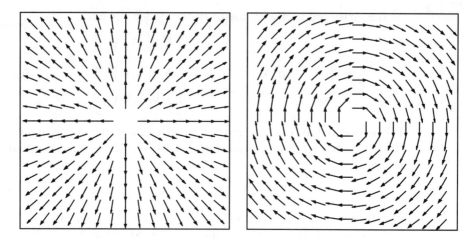

Figure 1-3. Purely divergent and purely rotational vector fields.   The fields are given by the equations: $\mathbf{V} = [x, y]/r$ and $\mathbf{W} = [\sin(y), -\sin(x)]$ and plotted over the ranges $-2, 2$ in both x and y.

which is purely radial, and:

$$\mathbf{W} = [y, -x, 0], \tag{1-3.21}$$

which is purely rotational. The divergence and curl of these two vectors are:

$$\nabla \cdot \mathbf{V} = \frac{1}{r}, \tag{1-3.22}$$

$$\nabla \times \mathbf{V} = 0, \tag{1-3.23}$$

$$\nabla \cdot \mathbf{W} = 0, \tag{1-3.24}$$

$$\nabla \times \mathbf{W} = [0, 0, -2]. \tag{1-3.25}$$

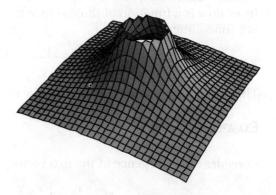

Figure 1-4.   A plot of the divergence of the $\mathbf{V} = [x/r, y/r]$. That is, div $\mathbf{V} = 1/r$.

TABLE 1-1.   DIFFERENTIAL VECTOR IDENTITIES

W and V are arbitrary vectors, ϕ is an arbitrary scalar, and r is the radius vector.

$$\nabla \cdot (\Phi V) \equiv \Phi \nabla \cdot V + \nabla \Phi \cdot V \tag{1-3.27}$$

$$\nabla \times (\Phi V) \equiv \Phi \nabla \times V + \nabla \Phi \times V \tag{1-3.28}$$

$$\nabla \cdot (V \times W) \equiv \nabla \times V \cdot W - V \cdot \nabla \times W \tag{1-3.29}$$

$$\nabla \times (V \times W) \equiv V(\nabla \cdot W) - (V \cdot \nabla)W + (W \cdot \nabla)V - W(\nabla \cdot V) \tag{1-3.30}$$

$$\nabla \cdot (\nabla \times V) \equiv 0 \tag{1-3.31}$$

$$\nabla \times (\nabla \times V) \equiv \nabla(\nabla \cdot V) - \nabla^2 V \tag{1-3.32}$$

$$\nabla \times (\nabla \Phi) \equiv 0 \tag{1-3.33}$$

$$\nabla(V \cdot W) \equiv V \times (\nabla \times W) + W \times (\nabla \times V) + (V \cdot \nabla)W + (W \cdot \nabla)V \tag{1-3.34}$$

$$\nabla \cdot r = 3 \tag{1-3.35}$$

$$\nabla \times r \equiv 0 \tag{1-3.36}$$

$$(V \cdot \nabla)r \equiv V \tag{1-3.37}$$

$$\nabla^2 \frac{1}{|r|} = \nabla \cdot \left(\frac{-r}{r^3}\right) = -4\pi\delta^3(r) \tag{1-3.38}$$

It is apparent from the preceding discussion that the gradient operator has a dual nature: as a vector it has dot and cross products, but as a differential operator it acts on a scalar or vector function. The two aspects of the operator do not have to act on the same object. For example, in the operator:

$$(V \cdot \nabla)W = \left[V_x \frac{\partial}{\partial x} + V_y \frac{\partial}{\partial y} + V_z \frac{\partial}{\partial z} \right] W \tag{1-3.26}$$

the differentiation acts only on the vector **W**, but the dot operator acts only on the vector, **V**. Or they may act on the same object as in the divergence and curl operators. The divergence of the gradient is called the *Laplacian*.

There are many useful identities between the various combinations of products and operators discussed above. The most useful are given in Table 1-1. Proofs are left to the problems; in general, they can be accomplished directly by taking components. Proof of equation (1-3.38) is given in Section 1-6.

1-4 Curvilinear Coordinate Systems

In the preceding discussion, all the vectors and vector operators have been expressed in Cartesian coordinates. However, in many cases it is desirable to use some curvilinear coordinate system such as cylindrical or spherical. In this section, we consider how to express vectors and vector operators in curvilinear coordinates and how to convert from one system to another. The primitive definition of a vector was an ordered set of quantities. However, in considering their transformations to other coordinate systems, it is necessary to give an extra condition in the definition, depending on how the vector transforms under the coordinate transformation.

Suppose the coordinate system is transformed by the equations:

$$\xi^j = \xi^j(x^i) \tag{1-4.1a}$$

having an inverse transformation:

$$x^i = x^i(\xi^j). \tag{1-4.1b}$$

These equations form a set of parametric equations for the transformed coordinates. For example, pick constant values for ξ^2 and ξ^3; then ξ^1 is the parameter in equations (1-4.1b). Similar parametric equations hold for each of the other ξ^k. These curves are the transformed coordinates. This is shown in Figure 1-5.

We find basis vectors for the new coordinates by finding the tangent vectors along the coordinate curves. Recall that the velocity vector is tangent to the trajectory where t is the parameter; similarly, the transformed basis vector:

$$\mathbf{b}^k = \frac{\partial \mathbf{r}}{\partial \xi^k} \tag{1-4.2}$$

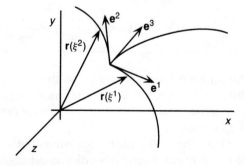

FIGURE 1-5. CONTRAVARIANT COORDINATES AND BASIS VECTORS.   Each curvilinear coordinate is a function of one of the parameters ξ^1, ξ^2, or ξ^3, which are the transformed coordinates. The basis vectors are obtained from the corresponding derivatives.

is tangent to the k-coordinate. We will call its magnitude, the contravariant *scale factor*, and give it the symbol:

$$h^k = |\mathbf{b}^k| \qquad (1\text{-}4.3)$$

and define the contravariant unit basis vectors:

$$\mathbf{e} = \frac{\mathbf{b}^k}{h^k}. \qquad (1\text{-}4.4)$$

EXAMPLE

When transforming to cylindrical coordinates, equations (1-4.1b) become:

$$x = r \cos \theta,$$
$$y = r \sin \theta, \qquad (1\text{-}4.5)$$
$$z = z.$$

The transformed basis vectors are:

$$\mathbf{b}^r = \frac{\partial \mathbf{r}}{\partial r} = [\cos \theta, \sin \theta, 0],$$

$$\mathbf{b}^\theta = \frac{\partial \mathbf{r}}{\partial \theta} = r[-\sin \theta, \cos \theta, 0], \qquad (1\text{-}4.6)$$

$$\mathbf{b}^z = \frac{\partial \mathbf{r}}{\partial z} = [0, 0, 1],$$

where $\mathbf{b}^r$ points radially, $\mathbf{b}^\theta$ is tangential to a circle of radius r, and $\mathbf{b}^z$ is unchanged. They are mutually orthogonal, as advertised, but are not unit vectors since $|\mathbf{b}^\theta| = r$. We see that these basis vectors depend on their location, specifically on the angle; in this, they differ greatly from Cartesian basis vectors which are constant vectors.

Vectors must transform in some corresponding way to the new coordinate system, i.e., the components of the vector take some new functional form in the new system. Consider the transformation of the displacement vector dx^j. Taking the differentials in equation (1-4.1) gives:

$$d\xi^j = \sum_{i=1} \frac{\partial \xi^j}{\partial x^i} \, dx^i. \qquad (1\text{-}4.7)$$

Thus components of the new displacement vector, $d\xi^j$, are linear combinations of the original displacement vector dx^j. The linear coefficients:

$$J_i^j = \frac{\partial \xi^j}{\partial x^i} \qquad (1\text{-}4.8)$$

form a 3 × 3 array (in three-dimensional space), called the *Jacobian matrix*. The determinant formed from this array is called the *Jacobian*. An inverse transformation exists only where the Jacobian differs from zero; we will assume the inverse exists except at isolated points. The Jacobian has several notations. For the transformation from Cartesian to spherical coordinates it may be written:

$$\mathbf{J} = \frac{\partial(r, \theta, \phi)}{\partial(x, y, x)}. \tag{1-4.9}$$

We let equation (1-4.7) be the model transformation for other vectors:

$$\bar{V}^j = \sum_{i=1} \frac{\partial \xi^j}{\partial x^i} V^i = \sum_{i=1} J^j_i V^i, \tag{1-4.10a}$$

where $\bar{V}$ represents the same vector in the new coordinates. Any vector that transforms by equation (1-4.10a) is called a *contravariant vector*. Other examples of contravariant vectors are velocity and acceleration.

Some discussion of notation is in order. A contravariant vector is denoted by a superscript index, and will be expressed as a column matrix as in equation (1-2.3). We used the bar over the vector to indicate the transformed vector; elsewhere it may be more convenient to use another letter. It is customary to use the *Einstein summation convention* in which the double index (i in this case) signifies summation over its range. In the Einstein notation, the transformation equation becomes:

$$\bar{V}^j = \frac{\partial \xi^j}{\partial x^i} V^i. \tag{1-4.10b}$$

EXAMPLE

For the transformation equations from cylindrical to Cartesian coordinates, the Jacobian matrix is:

$$J^j_i = \begin{bmatrix} \cos\theta & -r\sin\theta & 0 \\ \sin\theta & r\cos\theta & 0 \\ 0 & 0 & 1 \end{bmatrix} \tag{1-4.11}$$

and the infinitesmal displacement vector in cylindrical coordinates is:

$$dx = \cos\theta \, dr - r\sin\theta \, d\theta,$$
$$dy = \sin\theta \, dr + r\cos\theta \, d\theta, \tag{1-4.12}$$
$$dz = dz.$$

Inverting these equations to get the differentials in cylindrical coordinates:

$$dr = \cos\theta\, dx + \sin\theta\, dy,$$

$$d\theta = -\frac{\sin\theta}{r}\, dx + \frac{\cos\theta}{r}\, dy, \qquad (1\text{-}4.13a)$$

$$dz = dz.$$

When desired, $\cos\theta$, $\sin\theta$, and r can be expressed as functions of x, y, and z by inverting the transformation equations (1-4.5) and substituting.

$$dr = \frac{x}{\sqrt{x^2 + y^2}}\, dx + \frac{y}{\sqrt{x^2 + y^2}}\, dy,$$

$$d\theta = -\frac{y}{x^2 + y^2}\, dx + \frac{x}{x^2 + y^2}\, dy, \qquad (1\text{-}4.13b)$$

$$dz = dz.$$

Not all vectors transform by the contravariant rule. The gradient of Φ is:

$$V_i = \frac{\partial\Phi}{\partial x^i} \qquad (1\text{-}4.14)$$

and is denoted by a subscript. The derivatives are with respect to the contravariant displacement vector, x^i. The transformation of the gradient differs from the contravariant transformation. It is:

$$\frac{\partial}{\partial\xi^j} = \frac{\partial x^i}{\partial\xi^j}\frac{\partial}{\partial x^i}. \qquad (1\text{-}4.15)$$

With the gradient as the model, a *covariant vector* is defined by having the transformation:

$$\bar{V}_j = \frac{\partial x^i}{\partial\xi^j} V_i. \qquad (1\text{-}4.16)$$

The covariant transformation is similar to the contravariant transformation except that the covariant Jacobian matrix:

$$J_j^i = \frac{\partial x^i}{\partial\xi^j} \qquad (1\text{-}4.17)$$

is the inverse of the contravariant Jacobian matrix:

$$J_k^i J_j^k = \frac{\partial x^i}{\partial\xi^k}\frac{\partial\xi^k}{\partial x^j} = \frac{\partial x^i}{\partial x^j} = \delta_j^i. \qquad (1\text{-}4.18)$$

For covariant vectors it is convenient to define the coordinate curves in the following way. In equations (1-4.1b) set just one of the parameters ξ_k to be constant, e.g., $\xi_1 = c$, and let the other two, ξ_2 and ξ_3, take on appropriate ranges of values. For consistency, we take covariant parameters. Then equations (1-4.1b) become parametric equations for a surface:

$$x^1 = x^1(c, \xi_2, \xi_3),$$

$$x^2 = x^2(c, \xi_2, \xi_3), \tag{1-4.19}$$

$$x^3 = x^3(c, \xi_2, \xi_3),$$

with ξ_2 and ξ_3 as parameters. We define a *family of surfaces*, by letting c take a range of values. We will call this family of surfaces the ξ_1-family. Two other families of surfaces can be chosen by choosing a fixed value for ξ_2 or ξ_3. The three families of surfaces intersect, but not necessarily orthogonally. See Figure 1-6.

By using these surfaces we can define a set of covariant basis vectors and covariant coordinates. Find the gradient at each point on the ξ_k-surface, and define the covariant basis vectors:

$$\mathbf{b}_k = \nabla \xi_k. \tag{1-4.20}$$

By definition, the gradient has the direction of the maximum change in the function. Therefore, the direction of the basis vector, defined by the ξ_k-surface, is perpendicular to the surface. The other two covariant basis vectors are defined analogously.

Now we can define the corresponding covariant coordinate. From the definition of the directional derivative we write the differential:

$$d\xi_k = \nabla \xi_k \cdot \frac{d\mathbf{r}}{ds} \, ds = \mathbf{b}_k \cdot \mathbf{n} \, ds, \tag{1-4.21}$$

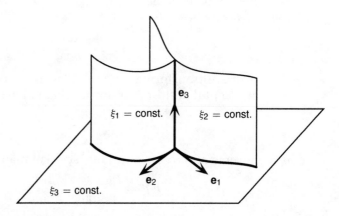

Figure 1-6. Covariant coordinates and basis vectors.   The intersection of two surfaces defines a coordinate in the new system. The basis vectors are tangent to the coordinates defined by the intersection. This basis is orthogonal.

where s is the pathlength parameter and the vector $\mathbf{n}$ has unit length. Now choose $\mathbf{n}$ parallel to $\mathbf{b}_k$, i.e., perpendicular to the ξ_k-surface and integrate this differential from a starting point on the ξ_k-surface. The parametric curve defined by this integral is the covariant coordinate, $\xi_k(s)$. By picking different starting points on the initial ξ_k-surface we can find a family of ξ_k-curves, all orthogonal the ξ_k-surface. The other two covariant coordinates are defined analogously.

Covariant unit basis vectors are obtained by dividing by the magnitude. That is:

$$\mathbf{e}_k = \frac{\mathbf{b}_k}{h_k},$$ (1-4.22)

where the covariant scale factor is:

$$h_k = |\mathbf{b}_k|.$$ (1-4.23)

The contravariant and covariant basis vectors are orthogonal:

$$\mathbf{b}^i \cdot \mathbf{b}_k = \frac{\partial \mathbf{r}}{\partial \xi^i} \cdot \nabla \xi^k = \sum_{n=1}^{3} \frac{\partial x^n}{\partial \xi^i} \frac{\partial \xi^k}{\partial x^n} = \frac{\partial \xi^k}{\partial \xi^i}.$$

$$= \delta_i^k$$ (1-4.24)

One consequence of this is that the covariant and contravariant scale factors are reciprocals:

$$h^j = h_j^{-1}.$$ (1-4.25)

Thus the unit basis vectors, *taken at the same point*, satisfy orthornormality relations like the Cartesian basis vectors:

$$\mathbf{e}^i \cdot \mathbf{e}^j = \mathbf{e}_i \cdot \mathbf{e}_j = \delta^{ij}$$ (1-4.26)

and the cross products:

$$\mathbf{e}^i \times \mathbf{e}^j = \mathbf{e}^k, \qquad i, j, k \text{ cyclical.}$$ (1-4.27)

These relations are very convenient when expressing vectors in terms of the unit vectors.

The surfaces defined by equations (1-4.19) need not be mutually orthogonal; in that case, the contravariant and covariant coordinates curves are different. We can see this as follows. The intersection of two surfaces satisfies the defining equations for both surfaces. Thus the curves defined by the intersections are single parameter curves, and are therefore contravariant coordinates. If the surfaces are not orthogonal, then the normal to the surface, $\mathbf{e}_k$, lies in a different direction from the tangent of the boundary curve, $\mathbf{e}^k$. The covariant and contravariant coordinate curves, whose tangents these basis vectors are, are therefore distinct.

Covariant and contravariant bases are reciprocal in the sense that the transformation equation (1-4.1a) defines contravariant coordinates, while its inverse, equation (1-4.1b), defines covariant coordinates. Equations (1-4.2) and (1-4.20) define contravariant and covariant basis vectors, of which the orthogonality in equation (1-4.24) is the consequence. Furthermore, as equation (1-4.18) shows, a covariant Jacobian matrix is the inverse of the corresponding contravariant matrix. For orthogonal surfaces in Euclidean space, the normals to the surfaces lie parallel to their intersections. *We will use orthogonal coordinate systems.* Then the two sets of unit vectors are identical:

$$\mathbf{e}_k = \mathbf{e}^k. \tag{1-4.28}$$

EXAMPLE

In cylindrical coordinates the three surfaces are: (1) cylinders of radius r concentric with the z-axis; (2) planes which pass through the z-axis at angles theta (relative to the x-axis); and (3) planes parallel to the x–y plane a height z above the z-axis. Thus the parameters r, θ, and z are the transformed coordinates. The covariant basis vectors for cylindrical coordinates are:

$$\mathbf{b}_r = \nabla r = \nabla \sqrt{x^2 + y^2} = \left[\frac{x}{\sqrt{x^2 + y^2}}, \frac{y}{\sqrt{x^2 + y^2}}, 0 \right] = [\cos\theta, \sin\theta, 0],$$

$$\mathbf{b}_\theta = \nabla\theta = \nabla \tan^{-1} \frac{y}{x} = \left[-\frac{y}{r^2}, \frac{x}{r^2}, 0 \right] = \frac{1}{r}[-\sin\theta, \cos\theta, 0], \tag{1-4.29}$$

$$\mathbf{b}_z = \nabla z = [0, 0, 1].$$

We can pick the scale factors by inspection; comparison with equations (1-4.6) shows these scale factors are reciprocal to the contravariant scale factors, as expected.

Not all physical quantities are vectors. Quantities such as the invariant mass of a particle are important physical quantities. These quantities must satisfy the requirement that they be independent of the coordinate system in which they are measured. In this case, the transformation equation is particularly simple; it is invariant or *scalar*:

$$\Phi(\xi^i) = \Phi(x^i), \tag{1-4.30}$$

where ξ^i and x^i both refer to the same point in space. The *inner product* of the contravariant vector V^j with the covariant vector U_j is defined:

$$\Phi = \sum V^j U_j = V^j U_j. \tag{1-4.31}$$

The inner product is an invariant form of the dot product, i.e., a scalar. This is easy to show. Using the covariant and contravariant transformation rules gives:

$$\bar{V}^j \bar{U}_j = \frac{\partial \xi^j}{\partial x^k} \frac{\partial x^i}{\partial \xi^j} V^k U_i = \frac{\partial x^i}{\partial x^k} V^k U_i = \delta_k^i V^k U_i$$

$$= V^k U_k,$$

(1-4.32)

which is obviously invariant. By contrast, the dot product formed by two contravariant vectors or by two covariant vectors is not invariant and therefore not scalar. An exception occurs when covariant and contravariant vectors are identical, as they are in orthogonal coordinates in Euclidean space.

Any vector in the space can be written as a linear combination of its basis vectors. But we have seen that there may be two different sets of basis vectors. The consequence of this is that one set of basis vectors can be expressed as a linear combination of the other. Every covariant vector has its associated contravariant vector. The transformation from contravariant to covariant coordinates may be written:

$$V_i = g_{ij} V^j = \frac{\partial x_i}{\partial x^j} V^j,$$

(1-4.33)

where the transformation coefficients, or *lowering operator*, g_{ij} are called the *metric tensor*. Remember, we sum over the index j. In this case, the transformed coordinates are the covariant coordinate set, $\xi_i = x_i$. There is, of course, an inverse raising transformation:

$$V^i = g^{ij} V_j = \frac{\partial x^i}{\partial x_j} V_j.$$

(1-4.34)

We see that the metric tensor can be expressed as a kind of Jacobian matrix. The raising operation is the inverse of the lowering operation, as can be seen by applying each operation to a vector:

$$V^i = g^{ij} V_j = g^{ij} g_{jk} V^k = \frac{\partial x^i}{\partial x_j} \frac{\partial x_j}{\partial x_k} V^k$$

$$= \delta_k^j V^k.$$

(1-4.35)

That is:

$$g^{ij} g_{jk} = \delta_k^i,$$

(1-4.36)

which is the identity operation, transforming a contravariant vector into itself. In this way, we find a kind of invariant dot product between two contravariant vectors, by first converting one to a covariant vector:

$$\mathbf{V} \cdot \mathbf{V} = V^i V_i = g^{ik} V_k V_i.$$

(1-4.37)

An important example of this is the pathlength interval along some curve:

$$ds^2 = d\mathbf{r} \cdot d\mathbf{r} = dx^j dx_j = g^{ik} dx_k dx_j.$$

(1-4.38)

In Cartesian coordinates the metric tensor may be expressed by the Kronecker delta:

$$g^{ij} = \delta^{ij} \tag{1-4.39}$$

giving the familiar expression:

$$ds^2 = dx^2 + dy^2 + dz^2. \tag{1-4.40}$$

For cylindrical coordinates the path interval is:

$$ds^2 = dr^2 + r^2\,d\theta^2 + dz^2 \tag{1-4.41}$$

with the metric tensor:

$$g^{ij} = \begin{bmatrix} 1 & 0 & 0 \\ 0 & r^2 & 0 \\ 0 & 0 & 1 \end{bmatrix}. \tag{1-4.42}$$

The metric tensor is diagonal when the coordinates are orthogonal. When the contravariant and covariant vectors are expressed in terms of their basis vectors, the path interval can be expressed in terms of the scale factors:

$$ds^2 = (d\xi^i \mathbf{b}^i) \cdot (d\xi^j \mathbf{b}^j) = \mathbf{b}^i \cdot \mathbf{b}^j\,d\xi^i\,d\xi^j. \tag{1-4.43}$$

For orthogonal coordinate systems the components of the metric tensor are:

$$g^{ij} = h^{i2}\delta^{ij}. \tag{1-4.44}$$

In curvilinear coordinates the gradient, divergence, and curl operators depend on the scale factors. We have two ways of expressing the differential of a function, from its functional dependence:

$$d\Phi = \frac{\partial \Phi}{\partial \xi^k}\,d\xi^k = \frac{\partial \Phi}{\partial \xi^k}\,\nabla \xi^k \cdot d\mathbf{r} = \frac{\partial \Phi}{\partial \xi^k}\,h_k \mathbf{e}_k \cdot d\mathbf{r} \tag{1-4.45}$$

and in terms of its gradient:

$$d\Phi = \nabla \Phi \cdot d\mathbf{r}. \tag{1-4.46}$$

Comparing these expressions enables us to evaluate the gradient in curvilinear coordinates. Since the components of the vector $d\mathbf{r}$ are arbitrary, we equate corresponding coefficients. Then, applying equation (1-4.25) gives the gradient:

$$\nabla \Phi = \left[\frac{1}{h^1}\frac{\partial \Phi}{\partial \xi^1}, \frac{1}{h^2}\frac{\partial \Phi}{\partial \xi^2}, \frac{1}{h^3}\frac{\partial \Phi}{\partial \xi^3} \right]. \tag{1-4.47}$$

EXAMPLE

In cylindrical coordinates the gradient is:

$$\nabla \Phi = \frac{\partial \Phi}{\partial r} \mathbf{e}_r + \frac{1}{r} \frac{\partial \Phi}{\partial \theta} \mathbf{e}_\theta + \frac{\partial \Phi}{\partial z} \mathbf{e}_z. \tag{1-4.48}$$

The divergence operator acts on the operand as both vector and as differential operator. However, as we saw above, the basis vectors are themselves functions of the coordinates and have derivatives. Consequently, the differential operator acts on the basis vectors as well as on the components, and must be considered in calculating the divergence in curvilinear coordinates. For orthogonal bases, the basis vectors satisfy:

$$\mathbf{e}_i = \mathbf{e}_j \times \mathbf{e}_k, \tag{1-4.49}$$

where the indices are in cyclical order. Using equation (1-4-47) the first term in the divergence is:

$$\nabla \cdot \mathbf{V}|_1 = \nabla \cdot \left(V^1 \mathbf{e}_1 \right) = \nabla \cdot \left(h^2 h^3 V^1 \frac{\mathbf{e}_1}{h^2 h^3} \right)$$

$$= \frac{\mathbf{e}_1}{h^2 h^3} \cdot \nabla (h^2 h^3 V^1) + (h^2 h^3 V^1) \nabla \cdot \left(\frac{\mathbf{e}_2 \times \mathbf{e}_3}{h^2 h^3} \right) \tag{1-4.50}$$

$$= \frac{1}{h^1 h^2 h^3} \cdot \frac{\partial (h^2 h^3 V^1)}{\partial \xi^1} + (h^2 h^3 V^1) \nabla \cdot (h_2 \mathbf{e}_2 \times h_3 \mathbf{e}_3).$$

The second term vanishes since:

$$\nabla \cdot (h_2 \mathbf{e}_2 \times h_3 \mathbf{e}_3) = \nabla \cdot (\mathbf{b}_2 \times \mathbf{b}_3) = \nabla \cdot (\nabla \xi^2 \times \nabla \xi^3)$$

$$= \nabla \times \nabla \xi^2 \cdot \nabla \xi^3 - \nabla \times \nabla \xi^3 \cdot \nabla \xi^2$$

$$= 0 \tag{1-4.51}$$

using equation (1-3.29). Then adding the other two terms, the divergence in curvilinear coordinates becomes:

$$\nabla \cdot \mathbf{V} = \frac{1}{h^1 h^2 h^3} \left[\frac{\partial (h^2 h^3 V^1)}{\partial \xi^1} + \frac{\partial (h^3 h^1 V^2)}{\partial \xi^2} + \frac{\partial (h^1 h^2 V^3)}{\partial \xi^3} \right]. \tag{1-4.52}$$

From the gradient and divergence, we get the Laplacian in curvilinear coordinates:

$$\nabla^2 = \frac{1}{h^1 h^2 h^3} \left[\frac{\partial}{\partial \xi^1} \left(\frac{h^2 h^3}{h^1} \frac{\partial}{\partial \xi^1} \right) + \frac{\partial}{\partial \xi^2} \left(\frac{h^3 h^1}{h^2} \frac{\partial}{\partial \xi^2} \right) \right.$$

$$\left. + \frac{\partial}{\partial \xi^3} \left(\frac{h^1 h^2}{h^3} \frac{\partial}{\partial \xi^3} \right) \right]. \tag{1-4.53}$$

In Cartesian coordinates, where the scale factors are unity, the gradient, divergence, and Laplacian reduce to the familiar expressions.

There is an analogous calculation for the curl. For conciseness, let us initially take the first component of the vector:

$$\nabla \times \mathbf{V}|_1 = \nabla \times (V^1 \mathbf{e}_1) = \nabla \times \left(h^1 V^1 \frac{\mathbf{e}_1}{h_1} \right)$$

$$= \nabla(h^1 V^1) \times \frac{\mathbf{e}_1}{h^1} + (h^1 V^1)\nabla \times \left(\frac{\mathbf{e}_1}{h^1} \right)$$

$$= \nabla(h^1 V^1) \times \frac{\mathbf{e}_1}{h_1} + (h^1 V^1)\nabla \times \mathbf{b}_1$$

$$= \frac{1}{h^1} \nabla(h^1 V^1) \times \mathbf{e}_1 \tag{1-4.54}$$

using equations (1-4.20), (1-3.28), and (1-3.33), and recalling that for orthogonal systems, the contravariant and covariant unit vectors are identical. Expanding the gradient and multiplying gives:

$$\nabla \times \mathbf{V}|_1 = \frac{1}{h_1} \nabla(h^1 V^1) \times \mathbf{e}_1$$

$$= \frac{1}{h^1} \left(\frac{1}{h^1} \frac{\partial(h^1 V^1)}{\partial \xi^1} \mathbf{e}_1 + \frac{1}{h^2} \frac{\partial(h^1 V^1)}{\partial \xi^2} \mathbf{e}_2 + \frac{1}{h^3} \frac{\partial(h^1 V^1)}{\partial \xi^3} \mathbf{e}_3 \right) \times \mathbf{e}_1$$

$$= \frac{1}{h^3 h^1} \frac{\partial(h^1 V^1)}{\partial \xi^3} \mathbf{e}_2 - \frac{1}{h^1 h^2} \frac{\partial(h^1 V^1)}{\partial \xi^2} \mathbf{e}_3. \tag{1-4.55}$$

To this we must add the contributions from the other two terms, which we obtain by symmetry:

$$\nabla \times \mathbf{V} = \frac{1}{h^3 h^1} \frac{\partial(h^1 V^1)}{\partial \xi^3} \mathbf{e}_2 - \frac{1}{h^1 h^2} \frac{\partial(h^1 V^1)}{\partial \xi^2} \mathbf{e}_3$$

$$+ \frac{1}{h^1 h^2} \frac{\partial(h^2 V^2)}{\partial \xi^1} \mathbf{e}_3 - \frac{1}{h^3 h^2} \frac{\partial(h^2 V^2)}{\partial \xi^3} \mathbf{e}_1$$

$$+ \frac{1}{h^2 h^3} \frac{\partial(h^3 V^3)}{\partial \xi^2} \mathbf{e}_1 - \frac{1}{h^1 h^3} \frac{\partial(h^3 V^3)}{\partial \xi^1} \mathbf{e}_2. \tag{1-4.56}$$

Taking the third component as representative, the curl becomes:

$$\nabla \times \mathbf{V} = \frac{1}{h^1 h^2} \left[\frac{\partial(h^2 V^2)}{\partial \xi^1} - \frac{\partial(h^1 V^1)}{\partial \xi^2} \right] \mathbf{e}_3 + \cdots. \tag{1-4.57}$$

Writing the curl in the form of a determinant is a convenient way of remembering its form:

$$\nabla \times \mathbf{V} = \frac{1}{h^1 h^2 h^3} \begin{vmatrix} h^1 \mathbf{e}_1 & h^2 \mathbf{e}_2 & h^3 \mathbf{e}_3 \\ \dfrac{\partial}{\partial \xi^1} & \dfrac{\partial}{\partial \xi^2} & \dfrac{\partial}{\partial \xi^3} \\ h^1 V^1 & h^2 V^2 & h^3 V^3 \end{vmatrix}, \tag{1-4.58}$$

where the operators in the second row act on the expressions in the third row. This expression takes the familiar form in Cartesian coordinates. Table 1-2 summarizes the vector operators in curvilinear coordinates.

1-5 INTEGRATION OF VECTORS

In three dimensions, we may define path integrals, surface integrals, and volume integrals. In this section we will consider these integrals in Cartesian and curvilinear coordinates. In general, integrating a function between two points in two or more dimensions is more complicated than it is in one, because it is necessary to specify the path as well as the end points.

The Path Integral. The differential displacement along a given path is:

$$d\mathbf{r} = \left[\frac{dx^1}{ds}, \frac{dx^2}{ds}, \frac{dx^3}{ds} \right] ds = \mathbf{n}\, ds, \tag{1-5.1}$$

where, $ds = |d\mathbf{r}|$, is the pathlength parameter and $\mathbf{n}$ is a unit vector tangent to the path. We define the *path integral* of a vector function $\mathbf{V}(x, y, z)$ by the expression:

$$P = \int_{r_0}^{\mathbf{r}} \mathbf{V} \cdot \mathbf{n}\, ds = \int_{r_0}^{\mathbf{r}} \mathbf{V} \cdot d\mathbf{r}. \tag{1-5.2}$$

In evaluating these integrals along the path it is, in general, necessary to express the differentials, dx, dy, and dz in terms of the path of integration. In Cartesian coordinates the path integral is:

$$\int_{\mathbf{r}_0}^{\mathbf{r}} \mathbf{V} \cdot d\mathbf{r} = \int_{x_0}^{x} V_x\, dx + \int_{y_0}^{y} V_y\, dy + \int_{z_0}^{z} V_z\, dz. \tag{1-5.3}$$

TABLE 1-2.   VECTOR OPERATORS IN CURVILINEAR COORDINATES

The second line illustrates the operators in spherical coordinates.

Scale factors	$[h^1, h^2, h^3]$ $[1, r, r \sin \theta]$

$\nabla \Phi$

$$\left[\frac{1}{h^1} \frac{\partial \Phi}{\partial \xi^1}, \frac{1}{h^2} \frac{\partial \Phi}{\partial \xi^2}, \frac{1}{h^3} \frac{\partial \Phi}{\partial \xi^3} \right]$$

$$\left[\frac{\partial \Phi}{\partial r}, \frac{1}{r} \frac{\partial \Phi}{\partial \theta}, \frac{1}{r \sin \theta} \frac{\partial \Phi}{\partial \phi} \right]$$

$\nabla \cdot \mathbf{V}$

$$\frac{1}{h^1 h^2 h^3} \left[\frac{\partial (h^2 h^3 V^1)}{\partial \xi^1} + \frac{\partial (h^3 h^1 V^2)}{\partial \xi^2} + \frac{\partial (h^1 h^2 V^3)}{\partial \xi^3} \right]$$

$$\frac{1}{r^2 \sin \theta} \left[\frac{\partial (r^2 \sin \theta V^r)}{\partial r} + \frac{\partial (r \sin \theta V^\theta)}{\partial \theta} + \frac{\partial (r V^\phi)}{\partial \phi} \right]$$

$\nabla^2 \Phi$

$$\frac{1}{h^1 h^2 h^3} \left[\frac{\partial}{\partial \xi^1} \left(\frac{h^2 h^3}{h^1} \frac{\partial}{\partial \xi^1} \right) + \frac{\partial}{\partial \xi^2} \left(\frac{h^3 h^1}{h^2} \frac{\partial}{\partial \xi^2} \right) + \frac{\partial}{\partial \xi^3} \left(\frac{h^1 h^2}{h^3} \frac{\partial}{\partial \xi^3} \right) \right]$$

$$\frac{1}{r^2 \sin \theta} \left[\sin \theta \frac{\partial}{\partial r} \left(r^2 \frac{\partial \Phi}{\partial r} \right) + \frac{\partial}{\partial \theta} \left(\frac{1}{\sin \theta} \frac{\partial \Phi}{\partial \theta} \right) + \frac{1}{\sin \theta} \frac{\partial^2 \Phi}{\partial \phi^2} \right]$$

$\nabla \times \mathbf{V}$

$$\frac{1}{h^1 h^2 h^3} \begin{vmatrix} h^1 \mathbf{e}_1 & h^2 \mathbf{e}_2 & h^3 \mathbf{e}_3 \\ \dfrac{\partial}{\partial \xi^1} & \dfrac{\partial}{\partial \xi^2} & \dfrac{\partial}{\partial \xi^3} \\ h^1 V^1 & h^2 V^2 & h^3 V^3 \end{vmatrix}$$

$$\frac{1}{r^2 \sin \theta} \begin{vmatrix} \mathbf{e}_r & r\mathbf{e}_\theta & r \sin \theta \mathbf{e}_\phi \\ \dfrac{\partial}{\partial r} & \dfrac{\partial}{\partial \theta} & \dfrac{\partial}{\partial \phi} \\ V^r & r V^\theta & r \sin \theta V^\phi \end{vmatrix}$$

In curvilinear coordinates, the path of integration may expressed by parametric equations:

$$x^j = x^j \left(\xi^k \right). \tag{1-5.4}$$

In terms of the new coordinates, the differential displacement is:

$$d\mathbf{r} = \frac{\partial \mathbf{r}}{\partial \xi^k} d\xi^k$$

$$= \mathbf{e}^k h^k \, d\xi^k. \tag{1-5.5}$$

Then we can integrate over the new variable ξ^k, with appropriate limits of integration.

EXAMPLE

Integrate the vector $\mathbf{V} = [y, -x]$, from the $(0, 0)$ to $(2, 4)$ along the rectangular path shown in Figure 1-7. On the two legs of the path, the differentials are $d\mathbf{r} = [dx, 0]$ and $d\mathbf{r} = [0, dy]$. Along this path, equation (1-5.3) becomes:

$$\int_{r_0}^{r} \mathbf{V} \cdot \mathbf{n} \, ds = \left[\int_0^2 y \, dx + \int_0^0 -x \, dy \right]_{\substack{dy=0 \\ y=0}} + \left[\int_2^2 y \, dx + \int_0^4 -x \, dy \right]_{\substack{dx=0 \\ x=2}}$$

$$= -8 \tag{1-5.6}$$

EXAMPLE

Integrate the vector in the previous example along the parabolic path, $y = x^2$, between the same end points $(0, 0)$ and $(2, 4)$. For this smooth curve we can express the second differential in terms of the first, i.e., $dy = 2x \, dx$. Then the integral becomes:

$$\int_{(0,0)}^{(2,4)} \mathbf{V} \cdot \mathbf{dr} = \left[\int y \, dx + \int -x \, dy \right]_{\substack{y=x^2 \\ dy=2x\,dx}} = \int_0^2 x^2 - 2x^2 \, dx$$

$$= -\left[\frac{1}{3} x^3 \right]_0^2 = -\frac{8}{3} . \tag{1-5.7}$$

Comparing these two examples shows that the integral depends on the path of integration as well as the end points.

The Surface Integral. We define the *surface integral*:

$$I = \iint \mathbf{V} \cdot d\mathbf{S}, \tag{1-5.8}$$

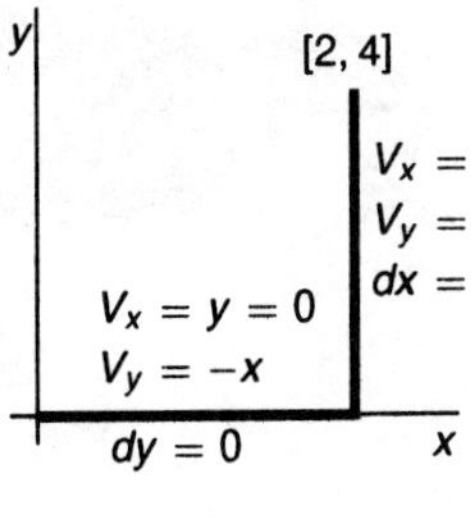

FIGURE 1-7. Path of integration.

where $d\mathbf{S}$ is the element of the surface area. It is a vector, directed perpendicular to the surface. In simple cases, such as a sphere, we know the direction of $d\mathbf{S}$ at each point on the surface; however, for more complicated surfaces it is very convenient to express it in terms of the curvilinear coordinates discussed above. A surface is specified by equations with two parameters:

$$\mathbf{r} = \mathbf{r}\left(\xi^1, \xi^2\right). \tag{1-5.9}$$

As we saw in Section 1-4, choosing values for ξ^1 and ξ^2 defines a point on the surface. Then, holding one of the parameters constant defines a curve on the surface. On this surface an infinitesmal element of area is the area of the quadrilateral bounded by two pairs of parametric curves. See Figure 1-8.

Since the parametric curves are not necessarily orthogonal, the surface element is given by the cross product:

$$d\mathbf{S} = d\mathbf{r}_1 \times d\mathbf{r}_2 = \frac{\partial \mathbf{r}}{\partial \xi_1} \times \frac{\partial \mathbf{r}}{\partial \xi_2} d\xi_1 \, d\xi_2 = \mathbf{e}^1 \times \mathbf{e}^2 h^1 h^2 \, d\xi_1 \, d\xi_2$$

$$= \mathbf{e}^3 h^1 h^2 \, d\xi_1 \, d\xi_2. \tag{1-5.10}$$

Because of the properties of the cross product, the direction of the surface element vector is orthogonal to the surface. Since the order of $d\xi^1$ and $d\xi^2$ is arbitrary, the cross product creates an ambiguity in the direction of the surface element. For closed surfaces, we choose the outward direction. The area of the surface is found by integrating the

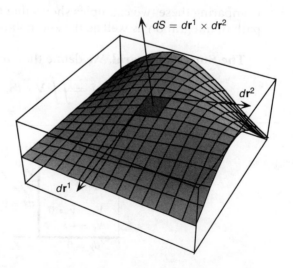

FIGURE 1-8.   The surface element in curvilinear coordinates.

unit normal vector:

$$A = \iint \mathbf{n} \cdot d\mathbf{S}. \tag{1-5.11}$$

EXAMPLE

In cylindrical coordinates the surface element on the cylinder is:

$$dS_r = \mathbf{e}^r \cdot d\mathbf{S} = r \, dz \, d\theta. \tag{1-5.12}$$

EXAMPLE

In spherical coordinates the surface element on the sphere is:

$$dS = \mathbf{e}^r \cdot d\mathbf{S} = r \sin \theta \, d\theta \, d\phi. \tag{1-5.13}$$

The Volume Integral. We can find the volume element in curvilinear coordinates by generalizing the discussion for surface elements to three dimensions. That is, consider an infinitesmal parallelopiped having sides given by $d\mathbf{r}^1$, $d\mathbf{r}^2$, and $d\mathbf{r}^3$. By equation (1-2.25) the volume, $dV = d^3r$, of this element is:

$$d^3r = d\mathbf{r}^1 \times d\mathbf{r}^2 \cdot d\mathbf{r}^3 = \begin{bmatrix} \dfrac{\partial x^1}{\partial \xi^1} d\xi^1 & \dfrac{\partial x^1}{\partial \xi^2} d\xi^2 & \dfrac{\partial x^1}{\partial \xi^3} d\xi^3 \\[2mm] \dfrac{\partial x^2}{\partial \xi^1} d\xi^1 & \dfrac{\partial x^2}{\partial \xi^2} d\xi^2 & \dfrac{\partial x^2}{\partial \xi^3} d\xi^3 \\[2mm] \dfrac{\partial x^3}{\partial \xi^1} d\xi^1 & \dfrac{\partial x^3}{\partial \xi^2} d\xi^2 & \dfrac{\partial x^3}{\partial \xi^3} d\xi^3 \end{bmatrix}$$

$$= J \, d\xi^1 \, d\xi^2 \, d\xi^3, \tag{1-5.14}$$

where J is the Jacobian for the transformation. In orthogonal coordinates the volume element reduces to:

$$d^3r = h^1 h^2 h^3 \, d\xi^1 \, d\xi^2 \, d\xi^3. \tag{1-5.15}$$

Then the volume integral of a function can be written:

$$\iiint F \, d^3r = \iiint F(\xi) J(\xi) \, d\xi^1 \, d\xi^2 \, d\xi^3. \tag{1-5.16}$$

EXAMPLE

The volume of a sphere is:

$$V = \iiint d^3r = \iiint_{(0,-\pi/2,0)}^{(R,\pi/2,2\pi)} r^2 \sin \theta \, dr \, d\theta \, d\phi$$

$$= 4\pi \frac{R^3}{3}. \tag{1-5.17}$$

Stokes's Theorem.   *The closed path integral of a vector field can be expressed as a surface integral:*

$$\oint \mathbf{V} \cdot d\mathbf{r} = \iint \nabla \times \mathbf{V} \cdot d\mathbf{S}, \qquad (1\text{-}5.18)$$

where the closed path in the line integral is the periphery, taken in the clockwise direction, of the arbitrary surface in the surface integral. The element of surface d**S** *is directed outward. This equation is known as* Stokes's theorem. *Figure 1-9 illustrates the path and surface differentials* d**r** *and* d**S**.

Proof of Stokes's theorem when the surface is a plane is straightforward. The path integral around the boundary curve can be written as the sum of closed path integrals over all the rectangular paths in the interior of the region shown in Figures 1-9 and 1-10:

$$\oint \mathbf{V} \cdot \mathbf{n}\, ds = \sum_{n=1}^{N} \left[\oint \mathbf{V} \cdot d\mathbf{r} \right]_n. \qquad (1\text{-}5.19)$$

In this sum, all of the internal integrals vanish because each side of every internal boundary is integrated twice, once in each direction. Only those parts of the integrals that include part of the boundary curve survive.

Now consider the sum in equation (1-5.19) where $\mathbf{V} = [V_x, V_y]$ and let the path of integration be an arbitrary loop in the x–y plane. For each of the rectangles there are four terms in each path integral taken

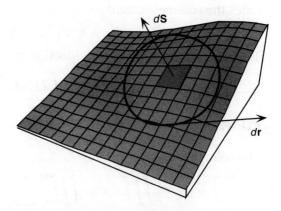

FIGURE 1-9. THE SURFACE AND BOUNDARY CURVE IN STOKES'S THEOREM.   The region of the surface integral is bounded by the closed loop of the path integral.

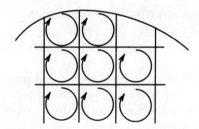

FIGURE 1-10. A SECTION OF A CLOSED REGION. The path integral on the boundary equals the sum of the path integrals of the internal sections, since all of the internal paths are integrated twice, once in each direction.

in the clockwise direction:

$$\sum_{n=1}^{N} \left[\oint \mathbf{V} \cdot d\mathbf{r} \right]_n = \sum_{n=1}^{N} \left[\int_{y}^{y+\Delta y} V_y(x, y)\, dy + \int_{x}^{x+\Delta x} V_x(x, y+\Delta y)\, dx \right.$$

$$+ \int_{y+\Delta y}^{y} V_y(x+\Delta x, y+\Delta y)\, dy +$$

$$\left. \int_{x+\Delta x}^{x} V_x(x+\Delta x, y+\Delta y)\, dx \right]$$

$$= \sum_{n=1}^{N} \left[\int \frac{\partial V_y}{\partial x} \Delta x\, dy - \int \frac{\partial V_x}{\partial y} \Delta y\, dx \right], \qquad (1\text{-}5.20)$$

where the minus sign comes from integrating backward from $x + \Delta x$ to x and from $y + \Delta y$ to y. In the limit, where N becomes infinite and Δx and Δy become infinitesmal, this expression becomes the double integral over the surface. The integrand is identical to the z component of curl $\mathbf{V}$:

$$\sum_{n=1}^{N} \left[\oint \mathbf{V} \cdot d\mathbf{r} \right]_n = \iint [\nabla \times \mathbf{V}]_z\, dx\, dy = \iint \nabla \times \mathbf{V} \cdot d\mathbf{S} \qquad (1\text{-}5.21)$$

and therefore equation (1-5.18) which is Stokes's theorem for a surface in the x–y plane.

To generalize to a curved surface, write the path integral (1-5.19) explicitly as a dot product. Then each of the three terms in the dot product has a form similar to that of equation (1-5.21), but lying in the x–y, y–z, and x–z planes. The limit of the sums is the surface integral over the curved surface. The result is Stokes's theorem.

Intuitively, we can imagine that the path integral is the inverse operation of the directional derivative. That is, integrating a vector field along some path gives a scalar:

$$\Phi(\mathbf{r}) = \int_{\mathbf{r}_0}^{\mathbf{r}} \mathbf{n} \cdot \nabla\Phi \, ds = \int_{\mathbf{r}_0}^{\mathbf{r}} \mathbf{V} \cdot \mathbf{n} \, ds \qquad (1\text{-}5.22)$$

which is a function of the end point of the integral, i.e., a scalar field. However, path integration differs from ordinary integration in that there is an infinity of paths between the initial and final points. When the path integral depends on the path, the result is not necessarily single valued.

What is the condition on the vector field that its path integral be a single-valued function of the end points? Suppose that the path integral of the vector field around *every* closed path (in some region) vanishes. Now pick two points on the path of integration, $\mathbf{r}_0$ and $\mathbf{r}$. Figure 1.11 shows a typical closed path between the points $\mathbf{r}_0$ and $\mathbf{r}$. These two points on the path split it into two parts, and if the integral around the closed path vanishes, then the integral from $\mathbf{r}_0$ to $\mathbf{r}$ on one of the parts must be the negative of the integral from $\mathbf{r}$ to $\mathbf{r}_0$ on the other half. That is, if the path integral around a *closed path* vanishes:

$$\oint \mathbf{V} \cdot \mathbf{n} \, ds = \left[\int_{\mathbf{r}_0}^{\mathbf{r}} \mathbf{V} \cdot d\mathbf{r} \right]_1 + \left[\int_{\mathbf{r}}^{\mathbf{r}_0} \mathbf{V} \cdot d\mathbf{r} \right]_2$$

$$= \left[\int_{\mathbf{r}_0}^{\mathbf{r}} \mathbf{V} \cdot d\mathbf{r} \right]_1 - \left[\int_{\mathbf{r}_0}^{\mathbf{r}} \mathbf{V} \cdot d\mathbf{r} \right]_2$$

$$= 0 \qquad (1\text{-}5.23)$$

then the integration is independent of the path. But if the path is arbitrary, we may choose the end point to be any point in space. Then the integral:

$$\Phi(\mathbf{r}) = \int_{\mathbf{r}_0}^{\mathbf{r}} \mathbf{V} \cdot d\mathbf{r} \qquad (1\text{-}5.24)$$

defines a scalar field which is a function only of the end point.

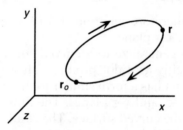

FIGURE 1-11. PATH OF INTEGRATION.

The following theorem is a direct consequence of the preceding calculations:

The necessary and sufficient condition that a vector field be integrable to a (single-valued) scalar function is that its curl vanishes.

Proof is as follows: If curl **V** = 0, then its closed path integral vanishes, and a scalar function Φ is defined as above. Conversely, if a scalar field Φ exists, its gradient is the vector field, **V** = grad Φ. But curl **V** = curl grad Φ = 0.

The Divergence Theorem. A second integral theorem relates a closed surface integral to the integral over the volume bounded by the surface. *This is the divergence theorem*:

$$\oiint \mathbf{V} \cdot d\mathbf{S} = \iiint \nabla \cdot \mathbf{V} \, d^3r. \tag{1-5.25}$$

Its proof is similar to the proof of Stokes's theorem. The argument is in summary:

1. Divide the volume into many small cube-shaped volumes filling the interior of the volume. For convenience, we will use Cartesian coordinates.
2. The sum of the surface integrals over the small volumes equals the surface integral over the boundary surface, since the internal surfaces are integrated twice, with opposite orientations of the area vector:

$$\oiint \mathbf{V} \cdot d\mathbf{S} = \sum_{n=1}^{N} \oiint \mathbf{V} \cdot d\mathbf{S}$$

$$= \sum_{n=1}^{N} \left[\oiint V_x \, dy \, dz + \oiint V_y \, dz \, dx \right.$$

$$\left. + \oiint V_z \, dx \, dy \right] \tag{1-5.26}$$

3. Express the components of **V** as Taylor expansions to first order. The closed integrals become the difference between integrations on upper and lower surfaces. For example, the first term is:

$$\sum_{n=1}^{N} \oiint V_x \, dy \, dz = \sum_{n=1}^{N} \left[\iint \left(V_x + \frac{\partial V_x}{\partial x} \Delta x \right) dy \, dz - \iint V_x \, dy \, dz \right]$$

TABLE 1-3.   **Integral Vector Identities**

Stokes's theorem	$\oint \mathbf{V} \cdot d\mathbf{r} = \iint \nabla \times \mathbf{V} \cdot d\mathbf{S}$
Divergence theorem	$\oiint \mathbf{V} \cdot d\mathbf{S} = \iiint \nabla \cdot \mathbf{V}\, d^3 r$
Green's theorem	$\oiint \Psi \nabla \Phi \cdot d\mathbf{S} = \iiint (\Psi \nabla^2 \Phi + \nabla \Psi \cdot \nabla \Phi)\, d^3 r$
Symmetrical form of Green's theorem	$\oiint (\Psi \nabla \Phi - \Phi \nabla \Psi) \cdot d\mathbf{S} = \iiint (\Psi \nabla^2 \Phi - \Phi \nabla^2 \Psi)\, d^3 r$

$$= \sum_{n=1}^{N} \iint \left(\frac{\partial V_x}{\partial x} \Delta x \right) dy\, dz. \tag{1-5.27}$$

4. In the limit when the N becomes infinite, the sum of terms becomes the volume integral:

$$\oiint \mathbf{V} \cdot d\mathbf{S} = \iiint \left(\frac{\partial V_x}{\partial x} + \frac{\partial V_y}{\partial y} + \frac{\partial V_z}{\partial z} \right) dx\, dy\, dz$$

$$= \iiint \nabla \cdot \mathbf{V}\, d^3 r. \tag{1-5.28}$$

This is the divergence theorem.

The most useful integral identities are listed in Table 1-3. The *Green's theorem* identity is obtained by applying $\mathbf{V} = \text{grad } \Phi$ and identity (1-3.27) to the divergence theorem. Green's theorem can be put in symmetric form, which is the basis for very elegant solutions of the electromagnetic field equations which will be discussed in Chapters 5 and 6.

1-6   Delta Functions

The Dirac delta function is useful in the solution of differential equations. It is defined by the integral property:

$$\int_a^b \delta(x)\, dx = \begin{cases} 0, & \text{if interval } a - b \text{ does not include } x = 0, \\ 1, & \text{if interval } a - b \text{ does include } x = 0. \end{cases} \tag{1-6.1}$$

Intuitively, the delta function should have the "values":

$$\delta(x) = \begin{cases} 0, & x \neq 0, \\ \infty, & x = 0. \end{cases} \qquad (1\text{-}6.2)$$

That is, the delta function is the integral of the unit step function shown in Figure 1-12; it is not an analytic function.[7] The most useful property of the δ-function is the integral:

$$\int_a^b f(x)\,\delta(x - x_0)\,dx = f(x_0), \qquad (1\text{-}6.3)$$

where $f(x)$ is an arbitrary function.

An angle measured in radians is defined by:

$$d\theta\,(\text{radians}) = \frac{ds}{r}, \qquad (1\text{-}6.4)$$

where r is the radius of the circle and ds is the differential circumference. For a complete circle the angle is:

$$\theta_{\text{whole circle}} = \frac{2\pi r}{r} = 2\pi. \qquad (1\text{-}6.5)$$

By analogy on a sphere we define the *solid angle* in terms of differential area. Suppose dS is the element of surface of a sphere of radius, r. Then we define the solid angle associated with dS as:

$$d\Omega = \frac{dS}{r^2}, \qquad (1\text{-}6.6)$$

where for the whole sphere:

$$\Omega_{\text{whole sphere}} = \frac{4\pi r^2}{r^2} = 4\pi\ (\text{steradians}). \qquad (1\text{-}6.7)$$

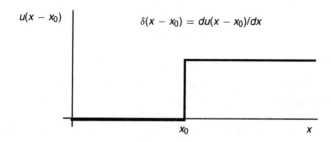

FIGURE 1-12.   The delta function is the derivative of the unit step function.

---

[7] A proper treatment of the Dirac delta function requires the theory of distributions. See, e.g., E. Butkov, *Mathematical Physics*.

Now let us extend the calculation to some surface which is not a sphere with its center at the origin. Then the solid angle is the projection in the direction of the radius vector:

$$d\Omega = \frac{\mathbf{r} \cdot d\mathbf{S}}{|\mathbf{r}|^3} = -\nabla \frac{1}{|\mathbf{r}|} \cdot d\mathbf{S}. \tag{1-6.8}$$

We can find an expression for the three-dimensional delta function:

$$\delta^3(r) = \delta(x)\,\delta(y)\,\delta(z) = -\frac{1}{4\pi}\,\nabla^2 \frac{1}{|\mathbf{r}|}. \tag{1-6.9}$$

First, note that the Laplacian vanishes except at the origin:

$$\nabla^2 \frac{1}{|\mathbf{r}|} = 0, \qquad r \neq 0. \tag{1-6.10}$$

Now apply the divergence theorem:

$$\iiint \nabla^2 \frac{1}{|\mathbf{r}|}\, d^3r = \oiint \frac{-\mathbf{r}}{|\mathbf{r}|^3} \cdot d\mathbf{S} = -\int d\Omega = -4\pi. \tag{1-6.11}$$

Therefore equations (1-6.10) and (1-6.11) express the essential properties of a delta function, equation (1-6.1). This is the proof of equation (1-3.38). This three-dimensional delta function has an important application in the solution of Poisson's equation.

Another representation of the delta function is the integral:

$$\int_{\infty}^{\infty} e^{i(k-k')x}\, dx = 2\pi\delta(k - k'). \tag{1-6.12}$$

Proof of this is given in Butkov using Fourier transforms; its reasonableness can be seen by taking the integral over a finite range:

$$\int_{-k\pi}^{k\pi} e^{i(n-m)x}\, dx = \begin{cases} 2\pi k, & n = m, \\ 0, & n \neq m, \end{cases} \quad k = \text{integer}, \tag{1-6.13}$$

in which $2\pi k$ becomes infinite as the range of integration becomes infinite.

1-7   SELECTED BIBLIOGRAPHY

S. Brush, Editor, *History of Physics*. Selected Reprints, American Association of Physics Teachers, 1988.

E. Butkov, *Mathematical Physics*, Addison-Wesley, 1968.

J. Heilbron, *Electricity in the 17th and 18th Centuries*, University of California Press, 1979.

B. Hoffmann, *Albert Einstein Creator and Rebel*, New American Library, 1972.

D. Livingston, *The Master of Light, A Biography of Albert A. Michelson*, University of Chicago Press, 1979.

J. Mathews and R. Walker, *Mathematical Methods of Physics*, 2nd ed., Benjamin, 1970.

M. Spiegel, *Vector Analysis*, Schaum, 1959.

J. Thomas, *Michael Faraday and the Royal Institution*, Adam Hilger, 1991.

E. Whittaker, *A History of the Theories of Aether and Electricity*, Dover, 1989.

S. Wolfram, *Mathematica, A System for Doing Mathematics by Computer*, 2nd ed., Addison-Wesley, 1991.

1-8 PROBLEMS

1. Show that if a vector has zero components in one coordinate system, they will be zero in all coordinate systems. (Hint: Transformations of vectors are linear operations.)

2. Verify identities (1-3.27), (1-3.29), (1-3.31), (1-3.33), (1-3.35), and (1-3.37) by direct calculation using vector components.

3. Verify identities (1-3.28), (1-3.30), (1-3.32), (1-3.34), and (1-3.36) by direct calculation using vector components.

4. Verify identities (1-3.31) and (1-3.33) explicitly in spherical coordinates.

5. A hyperboloidal coordinate system is defined by the equations:

$$u = z^2 - x^2 - y^2, \qquad -\infty < u < \infty,$$

$$v = 2z\sqrt{x^2 + y^2}, \qquad -\infty < v < \infty,$$

$$\phi = \arctan \frac{y}{x}, \qquad 0 \le \phi \le 2\pi,$$

in the variables x, y, and z and the parameters u, v, and ϕ.

A. Sketch the set of three curvilinear coordinates corresponding to parameters u, v, and ϕ.

B. Calculate the covariant basis vectors.

C. Show that these coordinates are orthogonal.

D. Calculate the scale factors and the Jacobian.

E. Calculate the gradient, divergence, curl, and Laplacian operators.

F. Calculate the metric tensor.

6. A paraboloidal coordinate system is defined by the equations:

$$x = 2uv \cos \phi,$$

$$y = 2uv \sin \phi,$$

$$z = (u^2 - v^2),$$

in the variables x, y, and z and the parameters u, v, and ϕ.

    A. Sketch the set of three curvilinear coordinates corresponding to parameters u, v, and ϕ.

    B. Calculate the contravariant basis vectors.

    C. Calculate the scale factors and the Jacobian.

    D. Calculate the infinitesimal surface and volume elements in terms of du, dv, and $d\phi$.

    E. Calculate the volume inside the paraboloid defined by $v = 1$ from $u = 0$ to $u = 1$.

    F. Calculate the area of the paraboloid defined by these conditions.

7. Verify Stokes's theorem for the vector $\mathbf{V} = [yz, -zx, xy]$ by direct calculation of the surface integral on the hemisphere of radius, $R = 1$ centered on the origin, and the path integral on the circular closed path in the x–y plane.

8. In discussing equation 1-6.10, we claimed that the Laplacian of $1/r$ vanishes except at the origin. Verify this by direct calculation in Cartesian and spherical coordinates.

APPENDIX 1-1   VECTOR OPERATORS

This notebook illustrates some of the properties of the vector operators: gradient, directional derivative, divergence, and curl. We use only two dimensions, in order to conveniently plot the scalar and vector fields. This notebook uses only Cartesian coordinates. Appendix 1-2 gives the operators in curvilinear coordinates. The user can make modifications in the vector and scalar functions to investigate further. Having a copy of Stephen Wolfram's book *Mathematica, a System for Doing Mathematics by Computer* is advisable.

Initialize

The Remove["Global'@"] command removes all user-defined symbols. It is not necessary for initial calculations, but removes garbage after trying out modifications. To remove the assignment for an individual symbol called "sym", for example, use the command:

```
sym = .
```

The symbol will still be recognized after this command is performed, but has no assigned value. The command, Remove[sym], removes sym entirely.

```
Remove ["Global'@"]
```

Load the *Mathematica* package for plotting vector fields.

```
Needs ["Graphics'PlotField'"]
```

Define Scalar Function

Choose a scalar function to use in the calculations. This one was chosen to have no singularities, but no particular symmetries. It is meant to be representative of other such functions.

```
phi = Exp[-(x^2+0.8*y^2)]*x*y
     2        2
  -x  - 0.8 y
 E               x y
```

Function Plots

Standard three-dimensional plot of a function $z = f(x, y)$

```
Plot3D[phi,{x,0,2.5},{y,0,2.8},PlotPoints->30]
```

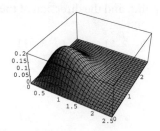

Contour plots show lines of equal function value. This way of plotting is familiar in topographical maps where the lines represent constant height. Contours are also familiar as equipotential lines.

```
ContourPlot[phi,{x,0,2.5},{y,0,2.8},
  Contours->20, ContourShading->False,PlotPoints->50]
```

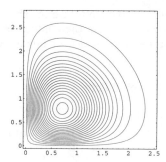

Gradient

Find the gradient of the scalar function, phi, defined above.

First, take the derivatives of phi, but do not show the calculations. With the semicolon on a line the calculation is performed but not displayed.

```
gradphixx = D[phi,x];
gradphiyy = D[phi,y];
```

Second, form the two-dimensional vector by making a list from the derivatives calculated above.

```
gradf = {gradphixx,gradphiyy}
```

```
     2        2              2      2
   -x  - 0.8 y           -x  - 0.8 y    2
 {E                 y - 2 E              x  y,
```

```
     2       2               2       2
   -x  - 0.8 y            -x  - 0.8 y     2
  E             x - 1.6 E                x y }
```

Plot the Gradient Field

Plot the gradient as an array of arrows. The length of the arrows represents the magnitude of the vector at that point, and the direction of the arrow represents the direction of the field at that point.

```
PlotGradientField[phi, {x, 0, 2.5}, {y, 0, 2.5}]
```

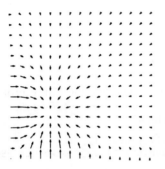

Compare the gradient field plot with the contour plot. The arrows in the gradient plot are orthogonal to the contours in the contour plot. In the context of electrostatics this means that the lines of electric flux are orthogonal to the equipotential surfaces.

Directional Derivatives on a Parametric Path

Evaluation of Gradient Along the Line $y = 1$

Evaluate the gradient along a specific value of y by substituting a specific value. Then plot it.

```
fx1 = gradphixx/.y->1
Plot[fx1, {x,0,3}]
```

```
           2                2
  -0.8 - x           -0.8 - x    2
 E            - 2 E              x
```

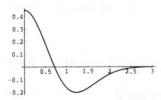

Define the Path and the Unit Tangent Vector

Define parametric equations for the path and form them into a vector. The method used here is to make some substitution rules for x and y in terms of the parameter s, and the substituting the rules into x and y. We will use this substitution rule again.

```
pathRules = {x->s, y->s^2}
```

$$\{x \to s, y \to s^2\}$$

```
rx = x/.pathRules;
ry = y/.pathRules;
path = {rx,ry}
```

$$\{s, s^2\}$$

Find the unit tangent to the path by taking the derivatives of x and y and dividing by their magnitude.

```
v = {D[rx,s],D[ry,s]};
vmagnitude = Sqrt[v.v]
tangent = v/vmagnitude
```

$$Sqrt[1 + 4 s^2]$$

$$\left\{ \frac{1}{Sqrt[1 + 4 s^2]}, \frac{2 s}{Sqrt[1 + 4 s^2]} \right\}$$

Show that the tangent has unit length. The command "Simplify[arg]" algebraically simplifies its argument.

```
tangentSquared = tangent.tangent;
Simplify[tangentSquared]
```

$$1$$

Plotting the Path

This is a parametric plot, i.e., it plots the equations $x = x(s)$ and $y = y(s)$ with s as the parameter.

```
ParametricPlot[path, {s,0,2}, Ticks->None]
```

Define Directional Derivative Along the Path

The quantities: gradf (the gradient of phi) and pathRules were calculated above. The directional derivative is the dot product of the gradient and the unit tangent vector at that point. We evaluate it for points on the path by applying the substitution rules.

```
directionalDerivative = tangent.gradf/.pathRules
(* Simplify the expression, show it, and plot it *)
simp = Simplify[directionalDerivative]
Plot[simp,{s,0,1.5}, Ticks->None]
```

$$\frac{E^{-s^2 - 0.8 s^4} s^2 - 2 E^{-s^2 - 0.8 s^4} s^4}{\text{Sqrt}[1 + 4 s^2]} +$$

$$\frac{2 s (E^{-s^2 - 0.8 s^4} s - 1.6 E^{-s^2 - 0.8 s^4} s^5)}{\text{Sqrt}[1 + 4 s^2]}$$

$$\frac{3.2 E^{-s^2 - 0.8 s^4} (0.9375 s^2 - 0.625 s^4 - 1. s^6)}{\text{Sqrt}[1 + 4 s^2]}$$

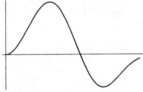

Divergence

Define and Plot the Vector Field

We define a representative vector field. This example was chosen to make its divergence easy to visualize. The exponential factor makes the field vanish at large distances from the origin. The factor, x, causes a noticeable change in the horizontal direction.

```
vectorField = Exp[-x^2-y^2]*{x, 0}
PlotVectorField[vectorField, {x, -2, 2}, {y, -2, 2}]
```

$$\{E^{-x^2 - y^2} x, 0\}$$

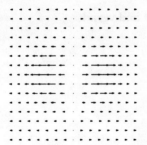

Calculate and Plot the Divergence of VectorField

The vector field, vectorField, is expressed in Mathematica as a list of two items. The components are addressed individually in calculations by

```
vectorfield[x] = vectorfield[[1]]
```

and

```
vectorfield[y] = vectorfield[[2]]
```

Then we find the divergence by taking the appropriate derivatives and summing them.

```
divField = D[vectorField[[1]],x] + D[vectorField[[2]],y]
divplot = ContourPlot[divField,{x,-2,2},{y,-2,2},
        Contours->20,PlotPoints->50]
```

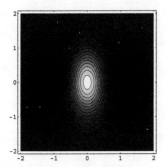

Comparing vectorField with divField shows that the divergence has its maximum where the vectorfield has its greatest changes. In this case, going along the horizontal axis shows three such locations, two negative changes and one positive change in the center.

Curl

Compare two very similar vector fields of unit magnitude. The first is purely divergent, and the second purely rotational, as is seen immediately by taking their divergence and curl.

Define the Vector Fields vcurl and vgrad

```
r = Sqrt[x^2+y^2]
vcurl = {y,-x}/r
vgrad = {x, y}/r
```

$$\text{Sqrt}[x^2 + y^2]$$

$$\left\{\frac{y}{\text{Sqrt}[x^2 + y^2]}, \ -\left(\frac{x}{\text{Sqrt}[x^2 + y^2]}\right)\right\}$$

$$\left\{\frac{x}{\text{Sqrt}[x^2 + y^2]}, \ \frac{y}{\text{Sqrt}[x^2 + y^2]}\right\}$$

Plot the Vector Fields

```
PlotVectorField[vcurl, {x, -2, 2}, {y, -2, 2}]
PlotVectorField[vgrad, {x, -2, 2}, {y, -2, 2}]
```

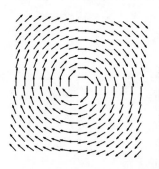

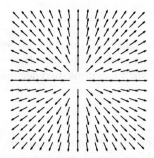

Calculate the Divergence and Curl of the Fields vcurl and vgrad

Since the vector fields are two dimensional, there is only one component of the curl.

```
curlvcurl = D[vcurl[[2]],x] -D[vcurl[[1]],y];
            Simplify[curlvcurl]
```

```
        1
-(-------------)
        2    2
  Sqrt[x  + y ]
```

```
divvcurl = D[vcurl[[1]],x]+D[vcurl[[2]],y];
           Simplify[divvcurl]
```

```
0
```

```
curlvgrad = D[vgrad[[2]],x] -D[vgrad[[1]],y];
            Simplify[curlvgrad]
```

```
0
```

```
divvgrad = D[vgrad[[1]],x]+D[vgrad[[2]],y];
           Simplify[divvgrad]
```

```
       1
-------------
       2    2
Sqrt[x  + y ]
```

Path Integral

Define the Path and the Vector Field

Define parametric equations for the path and form them into a vector. The method used here is to make some substitution rules for x and y in terms of the parameter s, and substituting the rules into x and y. We will use this substitution rule below.

Path

```
intPathRules = {x->Sin[2*s], y->Cos[s]}
rx = x/.intPathRules;
ry = y/.intPathRules;
intPath = {rx,ry}
```

```
{x -> Sin[2 s], y -> Cos[s]}
{Sin[2 s], Cos[s]}
```

Vector Field

```
vectorField = {x,y}
```

```
{x, y}
```

Plot the Path of Integration and the Vector Field

```
ParametricPlot[intPath,{s,0,Pi/2},AspectRatio->1]
```

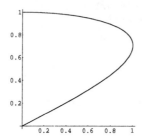

```
PlotVectorField[vectorField,{x,0,1},{y,0,1}]
```

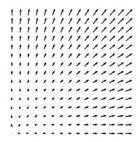

Find the Unit Tangent Vector for the Path

This calculation is identical to that done for the unit tangent of the directional derivative.

```
b = {D[rx,s],D[ry,s]};
bmagnitude = Sqrt[b.b]
unitb = b/bmagnitude
```

$$\text{Sqrt}[4\ \text{Cos}[2\ s]^2 + \text{Sin}[s]^2]$$

$$\left\{ \frac{2\ \text{Cos}[2\ s]}{\text{Sqrt}[4\ \text{Cos}[2\ s]^2 + \text{Sin}[s]^2]},\ -\left(\frac{\text{Sin}[s]}{\text{Sqrt}[4\ \text{Cos}[2\ s]^2 + \text{Sin}[s]^2]} \right) \right\}$$

Define the Integrand

```
intgr = vectorField.unitb/.intPathRules;
intgrand = Simplify[intgr]
```

$$-\left(\frac{\text{Cos}[s]\ \text{Sin}[s]}{\text{Sqrt}[4\ \text{Cos}[2\ s]^2 + \text{Sin}[s]^2]} \right) + \frac{\text{Sqrt}[2]\ \text{Sin}[4\ s]}{\text{Sqrt}[5 - \text{Cos}[2\ s] + 4\ \text{Cos}[4\ s]]}$$

Integrate and Plot

In general, the integral may not be possible to integrate exactly, but it can be evaluated numerically. We will express the integral as a table of numerical integrals evaluated up to the running end point sFin.

```
pathIntegral = Table[NIntegrate[intgrand,{s,0,sFin}],
                      {sFin,0,Pi/2,.1}]
```

{0, 0.00744511, 0.0291114, 0.0629035, 0.104954, 0.148694, 0.18249,

   0.185442, 0.134672, 0.0415866, -0.0621971, -0.159832, -0.244286,

   -0.311744, -0.359686, -0.386436}

```
ListPlot[pathIntegral,PlotJoined->True]
```

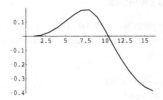

APPENDIX 1-2 CURVILINEAR COORDINATES

We want to transform from Cartesian coordinates (x, y, z) to curvilinear coordinates (u, v, w) by the equations: $u = u(x, y, z)$, $v = v(x, y, z))$, $w = w(x, y, z)$. For the transformation to be useful, the inverse transformation: $x = x(u, v, w)$, $y = y(u, v, w)$, $z = z(u, v, w)$ must also be defined. The basis vectors and scale factors are functions of the transformed coordinates. The example chosen here is cylindrical hyperbolic, where the third coordinate, z, extends out from the page. This coordinate system can be converted to a hyperboloidal system by rotating the hyperbolas around either the horizontal or the vertical axis.

As a check of this method, try the identity transformation:

$$x = u,$$

$$y = v,$$

$$z = w,$$

or the transformation into rotated Cartesian coordinates.

$$x = u + v,$$

$$y = u - v,$$

$$z = w,$$

In general, great care is required in choosing the parameter limits in curvilinear coordinates.

The first three sections are intended to illustrate the principles while not burdening the user with advanced aspects of *Mathematica*. The last section indicates how to perform more "heavy duty" applications by using the VectorAnalysis package.

Initialize

```
(* Remove all previous user-defined symbols *)
Remove ["Global'@"]
Needs ["Calculus'VectorAnalysis'"]
```

Transformation Equations and Inverses

The transformations are first given as substitution rules. Then, these can be converted into functions to be plotted and transformation equations that can be used for further manipulation.

```
(* User defined transformations *)
(* given as substitution rules*)
rulexx = xx -> 2*u*v;
ruleyy = yy -> u^2-v^2;
rulezz = zz -> w;
ruler = {rulexx,ruleyy,rulezz}
```

$$\{xx \rightarrow 2\ u\ v,\ yy \rightarrow u^2 - v^2,\ zz \rightarrow w\}$$

Plots of Transformation Functions

```
(* Functions for plotting *)
functs = {xx,yy,zz}/.ruler
```

$$\{2\ u\ v,\ u^2 - v^2,\ w\}$$

```
(* Transformed coordinates ranges *)
(* These determine the ranges of the plots *)
uumin = -2;
uumax = 2;
vvmin = -2;
vvmax = 2;
(* Cartesian coordinate ranges *)
xxmin = -4;
xxmax = 4;
yymin = -4;
yymax = 4;
(* Plot the equations *)
uuplot = ContourPlot [functs [[1]], {u,uumin,uumax},
        {v,vvmin,vvmax},ContourShading->False]
vvplot = ContourPlot [functs [[2]], {u,uumin,uumax},
        {v,vvmin,vvmax},ContourShading->False]
Show [uuplot,vvplot]
```

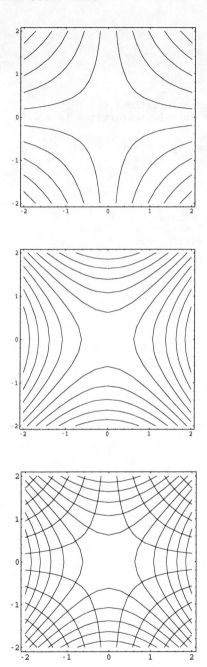

Inverse Transformations

```
(* Put the transformations in equation form *)
transformeqns = {x==xx,y==yy,z==zz}/.ruler;
TableForm[transformeqns]
```

```
x == 2 u v

     2    2
y == u  - v

z == w

(* Find inverse transformations *)
inverseRuleList = Solve[transformeqns,{u,v,w}];
                                 2    2            2    2
                 Sqrt[-y - Sqrt[x  + y ]] (y - Sqrt[x  + y ])
{{w -> z, u -> ----------------------------------------------,
                                  Sqrt[2] x


                         2    2
         Sqrt[-y - Sqrt[x  + y ]]
  v -> ---------------------------},
                Sqrt[2]


                       2    2              2    2
             (-y - Sqrt[x  + y ]) Sqrt[-y + Sqrt[x  + y ]]
{w -> z, u -> ----------------------------------------------,
                               Sqrt[2] x


                        2    2
          Sqrt[-y + Sqrt[x  + y ]]
   v -> -(---------------------------)},
                 Sqrt[2]


                        2    2            2    2
             Sqrt[-y - Sqrt[x  + y ]] (-y + Sqrt[x  + y ])
{w -> z, u -> ----------------------------------------------,
                             Sqrt[2] x


                        2    2
          Sqrt[-y + Sqrt[x  + y ]]
    v -> -(---------------------------)},
                 Sqrt[2]


                        2    2            2    2
             Sqrt[-y + Sqrt[x  + y ]] (y + Sqrt[x  + y ])
{w -> z, u -> ----------------------------------------------,
                             Sqrt[2] x


                        2    2
          Sqrt[-y + Sqrt[x  + y ]]
   v -> -(---------------------------}},
                 Sqrt[2]
```

It is typical for the transform equations not to have unique inverses. Therefore, it is necessary for the user to choose one set of the inverse substitution rules to use when necessary. For example:

```
inverseRuleChoice = inverseRuleList[[2]];
    TableForm[inverseRuleChoice]
```

```
w -> z

              2    2                     2    2
     (-y - Sqrt[x  + y ]) Sqrt[-y + Sqrt[x  + y ]]
u -> ------------------------------------------------
                      Sqrt[2] x


                    2    2
       Sqrt[-y + Sqrt[x  + y ]]
v -> -(-----------------------)
             Sqrt[2]
```

Basis Vectors, Jacobian, and Scale Factors

We solve the transformation equations for the Cartesian coordinates. This converts the transformation equations from implicit form to explicit form where necessary. Then we find the unnormalized basis vectors.

```
(* Define some temporary symbols *)
xxTemp = xx/.ruler;
yyTemp = yy/.ruler;
zzTemp = zz/.ruler;
(* Find the tangent vectors by differentiation *)
b = {{D[xxTemp,u],D[yyTemp,u],D[zzTemp,u]},
        {D[xxTemp,v],D[yyTemp,v],D[zzTemp,v]},
        {D[xxTemp,w],D[yyTemp,w],D[zzTemp,w]}};
(* Display the tangent vectors as a column *)
TableForm[b]

2 v    2 u    0
2 u    -2 v   0
0      0      1
```

Jacobian

This is the matrix formed from the derivatives of the Cartesian coordinates with respect to the transformed coordinates, as expressed above in the basis vectors. It is displayed as a matrix.

```
(* Find the Jacobian matrix *)
JMatrix = {b[[1]],b[[2]],b[[3]]};
        MatrixForm[JMatrix]

2 v    2 u    0
2 u    -2 v   0
0      0      1

(* Find the Jacobian *)
J = Det[JMatrix]

      2       2
-4 u   - 4 v
```

Scale Factors

```
h = {Sqrt[b[[1]].b[[1]]],
        Sqrt[b[[2]].b[[2]]],
        Sqrt[b[[3]].b[[3]]]}

          2    2         2    2
{Sqrt[4 u  + 4 v ], Sqrt[4 u  + 4 v ], 1}
```

Unit Basis Vectors

```
e = {b[[1]]/h[[1]],
     b[[2]]/h[[2]],
     b[[3]]/h[[3]]};
(* Show the unit basis vectors as a column *)
TableForm[e]
```

$$\frac{2\,v}{\mathrm{Sqrt}[4\,u^2\,+\,4\,v^2\,]} \qquad \frac{2\,u}{\mathrm{Sqrt}[4\,u^2\,+\,4\,v^2\,]} \qquad 0$$

$$\frac{2\,u}{\mathrm{Sqrt}[4\,u^2\,+\,4\,v^2\,]} \qquad \frac{-2\,v}{\mathrm{Sqrt}[4\,u^2\,+\,4\,v^2\,]} \qquad 0$$

$$0 \qquad\qquad\qquad 0 \qquad\qquad\qquad 1$$

Test for Orthonormality

```
Simplify[Table[e[[j]].e[[k]],{j,1,3},{k,1,3}]];
MatrixForm[%]
```

```
1   0   0
0   1   0
0   0   1
```

Using the Vector Analysis Package

An Example: The Bipolar Coordinate System
The bipolar coordinate system has many useful applications, including: (1) the lines of flux and equipotential surfaces for an electric dipole, (2) the electric field between two spheres displaced by some amount; (3) parallel microwave transmission lines; and (4) a stripline transmission line (a wire parallel to an infinite conducting plane).

In order to use curvilinear coordinates it is necessary to load the Vector Analysis package. This was done at the beginning of this notebook. It is also useful to set the default coordinate system as is done in the next section. The user will find more information by opening and inspecting the Vector Analysis package distributed with the *Mathematica* software.

Set or Find Parameters

Set or Find Default Coordinate System
To set the default coordinate system

```
SetCoordinates[Bipolar]
```

```
Bipolar[u, v, z, 1]
```

To find the default coordinate system

```
CoordinateSystem
```

```
Bipolar
```

Find the Symbols (Parameters) for the Curvilinear Coordinates

```
Coordinates[Bipolar]
{u, v, z}
```

Find Transformation Functions

```
transcoord = CoordinatesToCartesian[{u,v,z}]

      Sinh[v]              Sin[u]
{----------------, ----------------, z}
 -Cos[u] + Cosh[v]  -Cos[u] + Cosh[v]

asdf = transcoord[[1]]
qwer = transcoord[[2]]


      Sinh[v]
----------------
-Cos[u] + Cosh[v]


      Sin[u]
----------------
-Cos[u] + Cosh[v]
```

plot11 = ContourPlot[asdf,{u,0.01,1},{v,0.01,1},
            AspectRatio->1,ContourShading->False]
plot22 = ContourPlot[qwer,{u,0.01,1},{v,0.01,1},
            AspectRatio->1,ContourShading->False]
Show[plot11,plot22]

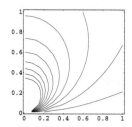

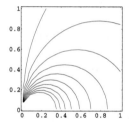

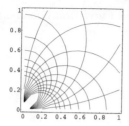

Gradient
Find the form of the gradient of a generic function.

```
f = ff[u,v,z]
Grad[f];
TableForm[%]
```

```
ff[u, v, z]
```

```
                                (1,0,0)
(-Cos[u] + Cosh[v]) ff          [u, v, z]
```

```
                                (0,1,0)
(-Cos[u] + Cosh[v]) ff          [u, v, z]
```

```
   (0,0,1)
ff         [u, v, z]
```

Find the form of the gradient of a specific function.

```
funct = u*v*z
Grad[funct];
TableForm[%]
```

```
u v z
v z (-Cos[u] + Cosh[v])
u z (-Cos[u] + Cosh[v])
u v
```

Divergence
Find the form of the divergence of a generic vector.

```
vvf = {vvfuu[u,v,z],vvfvv[u,v,z],vvfww[u,v,z]}
Div[vvf]
```

```
{vvfuu[u, v, z], vvfvv[u, v, z], vvfww[u, v, z]}
```

```
                    2    Sin[u] vvfuu[u, v, z]
(-Cos[u] + Cosh[v])  (-(---------------------) -
                                              2
                          (-Cos[u] + Cosh[v])
```

```
                                        (0,0,1)
    Sin[v] vvfvv[u, v, z]      vvfww          [u, v, z]
    -------------------- + --------------------- +
                      2                           2
    (-Cos[u] + Cosh[v])      (-Cos[u] + Cosh[v])
```

```
        (0,1,0)                    (1,0,0)
    vvfvv         [u, v, z]    vvfuu         [u, v, z]
    ------------------- + -------------------)
      -Cos[u] + Cosh[v]          -Cos[u] + Cosh[v]
```

Find the form of the divergence of a specific vector.

```
vvfunct = {v,0,u}
Div[vvfunct]

{v, 0, u}
-(v Sin[u])
```

Curl

```
vecCart = {fx[x,y,z],fy[x,y,z],fz[x,y,z]}

{fx[x, y, z], fy[x, y, z], fz[x, y, z]} 25_0}

Curl[vecCart]

      (0,0,1)                  (0,0,1)
  {-fy        [x, y, z], fx         [x, y, z],


                        2      fy[x, y, z] Sin[u]
  (-Cos[u] + Cosh[v])   (-(-------------------) +
                                        2
                              (-Cos[u] + Cosh[v])


       fx[x, y, z] Sinh[v]
       -------------------)}
                 2
       (-Cos[u] + Cosh[v])
```

Test to See if the Identities Hold in This Coordinate System

```
Div[Curl[vecCart]]

0

Curl[Grad[f]]

{0, 0, 0}
```

The Jacobian, Jacobian Matrix, and Scale Factors

Jacobian Matrix

```
jacMatrix = JacobianMatrix[Bipolar]

        Sin[u] Sinh[v]
  {{-(-------------------),
                   2
      (-Cos[u] + Cosh[v])


                                      2
        Cosh[v]              Sinh[v]
    ---------------- - -------------------, 0},
    -Cos[u] + Cosh[v]              2
                      (-Cos[u] + Cosh[v])
```

$$\{\frac{Cos[u]}{-Cos[u] + Cosh[v]} - \frac{Sin[u]^2}{(-Cos[u] + Cosh[v])^2},$$

$$-(\frac{Sin[u]\ Sinh[v]}{(-Cos[u] + Cosh[v])^2}),\ 0\},\ \{0,\ 0,\ 1\}\}$$

Note: You can pick off the tangent vectors from the components of the Jacobian matrix.

The Jacobian

The Jacobian is the determinant of the Jacobian matrix.

```
jac = Det[jacMatrix] 30}
```

$$-(\frac{Cos[u]\ Cosh[v]}{(-Cos[u] + Cosh[v])^2}) + \frac{Cosh[v]\ Sin[u]^2}{(-Cos[u] + Cosh[v])^3} +$$

$$\frac{Cos[u]\ Sinh[v]^2}{(-Cos[u] + Cosh[v])^3}$$

```
jacobyan = Simplify[jac] 31}
```

$$(Cos[u]\ -\ Cosh[v])^{-2}$$

Scale Factors

```
ScaleFactors[Bipolar] 32}
```

$$\{\frac{1}{-Cos[u] + Cosh[v]},\ \frac{1}{-Cos[u] + Cosh[v]},\ 1\}$$

Laplacian

```
Laplacian[f[u,v,z]]
```

$$(-Cos[u] + Cosh[v])^2\ (\frac{(ff[u,\ v,\ z])^{(0,0,2)}\ [u,\ v,\ z]}{(-Cos[u] + Cosh[v])^2} +$$

$$(ff[u,\ v,\ z])^{(0,2,0)}\ [u,\ v,\ z] + (ff[u,\ v,\ z])^{(2,0,0)}\ [u,\ v,\ z])$$

Test to see if Laplacian is div grad in this coordinate system.

`Div[Grad[f[u,v,z]]]`

$$(-Cos[u] + Cosh[v])^2 \left(\frac{(ff[u, v, z])^{(0,0,2)}[u, v, z]}{(-Cos[u] + Cosh[v])^2} + \right.$$

$$\left. (ff[u, v, z])^{(0,2,0)}[u, v, z] + (ff[u, v, z])^{(2,0,0)}[u, v, z] \right)$$

CHAPTER 2

Experimental Foundation

Reason and experiment have been indulged, and
error has fled before them.

Thomas Jefferson
Notes on the State of Virginia, Querry 6

. . . it behooves us to place the foundations of
knowledge in mathematics.

Roger Bacon
Opus Majus

In the following, we present an account of the experimental foundation of the classical theory of electromagnetism. It is not intended to
be in historical order, but merely a short compilation of the essential
empirical basis. Although electromagnetic interactions with matter are
of great interest to physics and are of great practical importance, nevertheless they are not *fundamental* to the topic—after all electromagnetic
fields do exist in complete vacuum. Therefore, in order to concentrate
on the essential aspects, we will treat fields in vacuum in this chapter, except for the isolated conductors necessary to hold charges and currents.
We will reserve our treatment of dielectric and magnetic materials for
later chapters, where they can be discussed in detail. In this chapter
we follow historical precedent by considering experiments performed
in the *steady state*, i.e., static charges and stationary currents.

2-1 Fields

The question of how forces are transmitted from one body to another was a serious problem for scientists even before Newton's time. The most familiar forces were contact forces: collisions, friction, etc.; but gravity, electrostatic, and magnetic forces operate without contact between the bodies. It was proposed that the void between bodies is filled with "vortices," or an intangible medium called aether. Then one body can exert a force on another, essentially by contact, through the aether. The concept of an aether, as a *substance* with various elastic and viscous properties, has been eliminated from modern science; but some of its traces remain in the fields we now associate with gravitational, electromagnetic, and nuclear forces.

As defined in physics, fields are quantities that depend on position and perhaps time. Examples include the temperature and density of a solid body like the Earth, which are scalar fields, and electric and magnetic fields, which are vector fields. Both scalar and vector fields occur in electrodynamics. The field concept is due to Michael Faraday, who introduced *lines of flux* to visualize vector fields. Although scientists were initially reluctant to accept Faraday's intuition, Maxwell put the concept into a mathematical form which is now universally accepted. Lines of flux (also called lines of force or LOF) are defined operationally by the measure-move procedure:

1. Start at some point ($r = (x, y, z)$) in the region and measure the vector field, $F(r)$, at that point. For example, you could measure the electric force field acting on a test charge.
2. Move the measuring device (e.g., the test charge) a small distance, Δr, in the direction of the field, and measure the field at the new point ($r + \Delta r$).
3. Continue this process as many times as is desired. The curve defined by the set of all the points reached in this way from the initial point is called a "line of flux." The field is tangent to the line of flux at each point.
4. Start at new initial points and repeat the previous procedure to determine a set of lines of flux. The equations of these lines of flux are:

$$d\mathbf{r} = K\mathbf{F}, \qquad (2\text{-}1.1)$$

or, eliminating the arbitrary constant, K, and using components:

$$\frac{dx}{F_x} = \frac{dy}{F_y} = \frac{dz}{F_z}. \qquad (2\text{-}1.2)$$

It is convenient to let the initial points of each line of flux lie on a surface called the *boundary surface*. Then the values of the field on the boundary surface are called the *boundary values* of the field. It is sometimes useful to define *flux tubes* which are small bundles of lines of flux.

2-2 COULOMB'S LAW

Publication of Newton's law of universal gravitation provided much stimulus for similar investigations of the electric and magnetic forces. The independent invention of the torsion balance by Cavendish and others made accurate measurement of the gravitational coupling constant possible. This same device also allowed Coulomb, and independently Cavendish (who declined publication), to determine the force law between two electrical charges. The result of the experiments, summarized by *Coulomb's law*, could not have been a great surprise, since the electrical and gravitational force laws are so similar. To express Coulomb's law in mathematical form, let $\mathbf{r}_s$ represent the location of the charge creating the electric field, called the *source* charge, and let $\mathbf{r}_f$ represent the location of the charge on which the force is measured, called the *field* charge. The displacement vector from the source charge to the field charge is:

$$\mathbf{r}_{fs} = \mathbf{r}_f - \mathbf{r}_s. \tag{2-2.1}$$

Figure 2-1 shows the configuration of charges and displacement vectors.

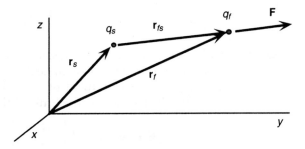

FIGURE 2-1. COULOMB'S LAW. This shows the position vectors for the source charge, q_s, and the test charge q_f. The direction of the force is along the line joining the charges. In contrast to gravitation, similar charges repel.

The force on a charge, q_f due to a charge q_s, separated by a distance, r_{fs}, where both charges are at rest in vacuum is, in Gaussian units:

$$\mathbf{F}_f = \frac{q_f q_s}{|r_{fs}|^3} \mathbf{r}_{fs}. \qquad (2\text{-}2.2)$$

Contrary to the gravitational force, like charges repel each other. Coulomb's law can be generalized to include multiple source charges by superposition. Then:

$$\mathbf{F}(\mathbf{r}_f) = q_f \sum_{s=1}^{N} \frac{q_s}{|r_{fs}|^3} \mathbf{r}_{fs}$$

$$\to q_f \iiint \rho(r_s) \frac{\mathbf{r}_{fs}}{|r_{fs}|^3} d^3 r_s. \qquad (2\text{-}2.3)$$

The summation is over all source charges (i.e., excluding the charge on which the force is measured). The source charge density, $\rho(\mathbf{r}_s)$, in the integral is a function of the variable, $\mathbf{r}_s$.

Coulomb's law, as given above, suggests direct interaction between charges. Faraday and Maxwell found it convenient, however, to propose that the interaction between two charges takes place with an intermediate step. In the first step, the source charge placed at some location modifies its neighborhood in some way. Specifically, we propose that an *electric field* is created around the source charge, which depends only on the source charge and its location. In the second step, the interaction between the electric field and a second charge creates a force on the second charge. Although the electric field can be defined consistent with Coulomb's law, the real justification for introducing the electric field lies in finding that it has other properties, beyond those of Coulomb's law. These other properties do exist and are expressed by Maxwell's laws, which we discuss below.

We operationally define the electric field to be the ratio of the force on a small charge (called the test charge) to the value of the charge:

$$\mathbf{E}(\mathbf{r}) \equiv \frac{\mathbf{F}(\mathbf{r})}{q}. \qquad (2\text{-}2.4)$$

In practical terms, the presence of a sizable test charge would cause the source charges on conductors to change their original distribution, thus preventing accurate measurement of the field. Consequently, a better definition of the electric field is the limit of the ratio:

$$\mathbf{E}(\mathbf{r}) \equiv \lim_{\Delta q \to 0} \left[\frac{\Delta \mathbf{F}(\mathbf{r})}{\Delta q} \right] = \frac{\partial \mathbf{F}}{\partial q}. \qquad (2\text{-}2.5)$$

Defined this way, the electric field created by the source charges depends on the point of measurement, but is completely independent of the value of the test charge. For a distribution of charges in vacuum, the electric field becomes an integral:

$$\mathbf{E}(\mathbf{r}_f) = \iiint \rho(\mathbf{r}_s) \frac{\mathbf{r}_{fs}}{|r_{fs}|^3} \, d^3 r_s. \tag{2-2.6}$$

The electric field can be expressed as the gradient of a scalar potential field. Since the source variables are constant with respect to the field variable gradient:

$$\frac{\mathbf{r}_{fs}}{|r_{fs}|^3} = -\nabla_f \frac{1}{|r_{fs}|} \tag{2-2.7}$$

we write:

$$\mathbf{E} = -\nabla\phi, \tag{2-2.8}$$

where the *electric potential*, ϕ, is a scalar field given by:

$$\phi(\mathbf{r}_f) = \iiint \rho(\mathbf{r}_s) \frac{1}{|r_{fs}|} \, d^3 r_s. \tag{2-2.9}$$

As we saw in Chapter 1, the differential properties of a field, i.e., the divergence and curl, are important in characterizing the field. They can be directly calculated for the static electric field from the properties of the potential:

$$\nabla \cdot \mathbf{E} = \iiint \nabla_f \cdot \left[\rho(\mathbf{r}_s) \nabla_f \left\{ \frac{-1}{|r_{fs}|} \right\} \right] d^3 r_s$$

$$= - \iiint \rho(\mathbf{r}_s) \left[\nabla_f^2 \frac{1}{|r_{fs}|} \right] d^3 r_s. \tag{2-2.10}$$

In Chapter 1 we found that the Laplacian of $1/r$ is proportional to a Dirac delta function with the spike at the origin. Then using the integral property of the delta function the divergence of the electric field becomes:

$$\nabla \cdot \mathbf{E} = 4\pi\rho(\mathbf{r}_f), \tag{2-2.11}$$

where the charge density has become a function of the field variable.

The curl of the (Coulomb) electric field can also be calculated from the scalar potential.

$$\nabla \times \mathbf{E} = -\nabla \times \nabla\phi \equiv 0. \tag{2-2.12}$$

The consequence of the vanishing curl is that the electrostatic field is *conservative* in the following sense: First, recall that the line integral defines a single-valued scalar function. That is, by equation (1-5.24), the change in electrical potential:

$$\Delta\phi(\mathbf{r}) = -\int_{\mathbf{r}_0}^{\mathbf{r}} \mathbf{E} \cdot d\mathbf{r} \qquad (2\text{-}2.13)$$

is independent of the path of integration. But this integral is just the work per unit charge done on a charge in moving a charge from r_0 to r. Then for a closed path, exactly zero work is done, and energy is conserved. Other vector fields with vanishing curl have similar conservation properties.

Now consider a second scalar quantity associated with electric fields. Consider a closed system containing movable charges and electric fields, and suppose that work is done on the system by an external force which is not necessarily electric. For this external force, we define an *electromotive force* (emf) to be the work per unit charge:

$$\Delta(\text{emf}) = \frac{\Delta W_{\text{external}}}{q} \qquad (2\text{-}2.14)$$

whose units are ergs/statcoulomb, i.e., statvolts. From this definition, we see that *emf is to work as potential is to potential energy*. There is, however, a significant difference: potential is always single valued, while emf is not single valued, since work is not.

An elementary calculation in mechanics shows that the external work done on an isolated system is proportional to the increase of internal energy of the system. That is:

$$\Delta W_{\text{external}} = \Delta U, \qquad (2\text{-}2.15)$$

where U is the internal energy of the system, i.e., the sum of its kinetic and potential energies. This equation is a limited case of the first law of thermodynamics represented by Figure 2-2. However, work done on the system depends on both the force and the path of integration. In thermodynamic terms, internal energy dU is called an exact differential, while work dW and heat dQ are called inexact differentials,

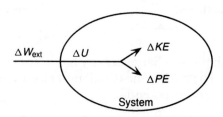

FIGURE 2-2. ENERGY BALANCE. There are two mechanisms for energy transport into a closed system, external work and heat transfer (ignored in this case). The external work done on the system ΔW increments the internal energy of the system ΔU, shared between kinetic energy ΔKE and potential energy ΔPE.

dependent on the path. The electromagnetic consequence of work being path dependent, is that emf is not a single-valued field as potential
is, and therefore not conserved. As a consequence, a charged particle
can be accelerated to very high energy, by circulating around a closed
loop containing an emf.

Example

The operation of the van de Graaff generator illustrates the relation
between external work and emf. This device consists essentially of a
hollow conducting sphere with a hole in it, and an insulating belt
passing into the sphere. The moving belt is charged externally and
thereby carries charge into the interior of the sphere where the charge
is transferred to the sphere by contact. Because the charge on the belt is
repelled by the charge on the sphere, work is required to move the belt.
This work may be supplied by any convenient external device (such as a
motor). Since the charges move relatively slowly, the work done on the
electrical charges takes the form of potential energy. Thus the effect of
the external work is to create a potential difference between the sphere
and ground. If the van de Graaff generator is inserted into a circuit, as
shown in Figure 2-3, it acts as a source of emf. The total energy given
to the circulating charge is proportional to the number of times the

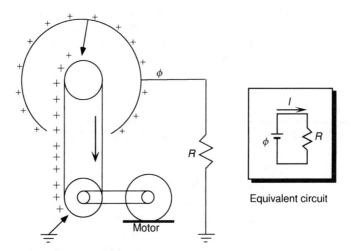

Equivalent circuit

Figure 2-3. Illustrating the relation between emf and potential. Work is done by the motor
in moving the charges on the belt toward the charged sphere. The emf is the work done per
unit charge transferred to the sphere; the potential on the sphere equals the emf. Although
potential is single valued, emf is multivalued since work is done on a circulating charge in
each circulation.

charge goes around the circuit. A resistor in an electric circuit acts as a negative source of emf since it removes energy from the system (in the form of heat which radiates away). In this circuit, the net effect of the two emfs is to transform mechanical work into heat.

Energy gained by the system from the emf may be stored as electrical potential energy associated with internal electric fields and charges. In that case we express the emf in a form similar to the expression for electric potential:

$$\Delta W = \int_{\text{path}} q\mathbf{E} \cdot d\mathbf{r}. \qquad (2\text{-}2.16)$$

Only the electric force contributes to the integral since, as we shall see in the next section, the velocity and magnetic force are orthogonal and therefore do no work.

The preceding discussion treats charges at rest, or moving as constant currents. Therefore, we cannot be certain from this discussion, that electrostatic fields are the most general electric fields. Time dependence may have a role to play, and the existence of emfs raises the distinct possibility of associated nonconservative electric fields, whose curl does not vanish. All this suggests a more general expression for the electric field:

$$\mathbf{E} = \mathbf{E}_{\text{coulomb}} + \mathbf{E}_{\text{nonconservative}}, \qquad (2\text{-}2.17)$$

where the second term is related to emf. When integrated around closed paths, these two fields are, respectively, vanishing and nonvanishing. In fact, as we will see in Section 2-4, where we discuss time-dependent magnetic fields, it *does* become necessary to add a nonconservative contribution to the electric field.

2-3 AMPÈRE'S LAWS FOR THE MAGNETIC FIELD

Both electric and magnetic forces have been known since antiquity, but it was not clear to the older investigators that the two forces were separate phenomena. The invention of the electric cell by Alessandro Volta in 1800 enabled detailed investigation of magnetism by Oersted, Ampère, and others, which produced evidence for distinguishing electricity from magnetism. Volta's discovery permitted steady currents, which previous experiments with electrical discharges did not.

In 1819 Hans Christian Oersted, led by intuition and philosophical considerations, discovered that a steady current in a wire exerts a torque on a compass needle. In his experiments, Oersted observed a torque on a permanent magnet exerted by an electric current in a nearby conductor. This observation implied a connection between electricity and magnetism and suggested further detailed investigation. André-Marie Ampère proceeded with a series of investigations measuring not only current–magnet forces but also current–current forces in various geometrical configurations. Among other results, it was found that the force exerted by one current loop on another could be given by the expression (in modern notation):

$$\mathbf{F}_2 = \frac{1}{c^2} I_2 I_1 \oint_2 \oint_1 \frac{d\mathbf{r}_2 \times (d\mathbf{r}_1 \times \mathbf{r}_{21})}{|r_{21}|^3}, \tag{2-3.1}$$

where the paths of integration are along the current loops. This is shown in an idealized form in Figure 2-4. We suppose the two current loops are in vertical planes. The moving loop is supported by a vertical fiber, which is in the plane of the loop, but offset horizontally. The fiber is the axis of rotation of the loop and provides the restoring torque. The force on the loop is determined from the offset (i.e., the lever arm) and the torque. Because the fiber's spring constant is very small, the torsion balance has great sensitivity.

Several observations can be made about this result. First, it is very similar to the Coulomb law for electric forces in that the force depends directly on the product of the two sources (the currents) and inversely on the square of the distance between them. Second, the direction of the force is more complicated. Unlike the Coulomb force which is parallel to the line connecting the two charges, the direction of the magnetic force for two elementary currents depends on the product:

$$d\mathbf{F}_2 \rightarrow d\mathbf{r}_2 \times (d\mathbf{r}_1 \times \mathbf{r}_{21}), \tag{2-3.2}$$

which is not, in general, parallel to the vector $\mathbf{r}_{21}$. This expression appears to contradict Newton's third law because of the asymmetric way in which $d\mathbf{r}_1$ and $d\mathbf{r}_2$ occur. That is, exchanging loop 1 and loop 2 does

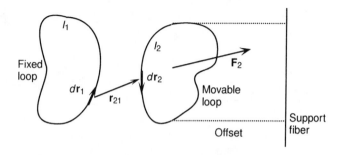

FIGURE 2-4. IDEALIZATION OF AMPÈRE'S EXPERIMENT. The movable loop is supported by a thin elastic fiber which provides a restoring torque. The supporting fiber is offset horizontally from the center of the loop. The amount of offset is the lever arm of the torque.

not necessarily make the expression negative. However, the double integral in equation (2-3.1) can be written as a sum of two terms, one of which is an exact integral and vanishes for a closed path. Thus:

$$
\begin{aligned}
\mathbf{F}_2 &= \frac{I_2 I_1}{c^2} \oint_2 \oint_1 \frac{d\mathbf{r}_2 \times (d\mathbf{r}_1 \times \mathbf{r}_{21})}{|r_{21}|^3} \\
&= \frac{I_2 I_1}{c^2} \oint_2 \oint_1 \frac{(d\mathbf{r}_2 \cdot \mathbf{r}_{21}) d\mathbf{r}_1 - (d\mathbf{r}_2 \cdot d\mathbf{r}_1)\mathbf{r}_{21}}{|r_{21}|^3} \\
&= \frac{I_2 I_1}{c^2} \oint_2 \oint_1 \left(d\mathbf{r}_2 \cdot \nabla \frac{-1}{|r_{21}|} \right) d\mathbf{r}_1 - \frac{(d\mathbf{r}_2 \cdot d\mathbf{r}_1)\mathbf{r}_{21}}{|r_{21}|^3} \\
&= -\frac{I_2 I_1}{c^2} \oint_2 \oint_1 \frac{(d\mathbf{r}_2 \cdot d\mathbf{r}_1)\mathbf{r}_{21}}{|r_{21}|^3} .
\end{aligned}
\tag{2-3.3}
$$

The remaining term does obey Newton's third law. Finally, the double integral expression can be factored into the product:

$$
\mathbf{F} = \frac{1}{c} I_f \oint d\mathbf{r}_f \times \mathbf{B}, \tag{2-3.4}
$$

where the subscripts have been changed to represent source and field quantities, and where the *magnetic induction* field, $\mathbf{B}(r_f)$, is defined:

$$
\mathbf{B}(r_f) = \frac{1}{c} I_s \oint \frac{d\mathbf{r}_s \times \mathbf{r}_{fs}}{|r_{fs}|^3} . \tag{2-3.5}
$$

This expression for the magnetic induction is called the *Biot–Savart* law.

The magnetic equations may be generalized to include distributions of currents by applying the *current density*, i.e., the current per unit area

$$
\mathbf{J} \equiv \frac{\Delta q}{\Delta t \, \Delta A} \frac{\mathbf{v}}{|v|} = \frac{\Delta q}{\Delta (\text{Vol})} \mathbf{v}. \tag{2-3.6}
$$

In Figure 2-5 an amount of charge, dq, passes a point in the charge flux

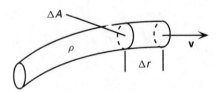

Figure 2-5. Current density.   This is the amount of charge passing through a unit area per unit time. Since the volume of the small cylinder is $\Delta r \, \Delta A$, the current density becomes $\mathbf{J} = \rho \mathbf{v}$.

tube per unit time; then the current density satisfies the useful equation:

$$\mathbf{J} = \rho\mathbf{v}. \tag{2-3.7}$$

The electric current through some surface is the integral of the current density:

$$I = \iint \mathbf{J} \cdot d\mathbf{S}. \tag{2-3.8}$$

It is well established experimentally that electric charge is conserved. This is not a trivial principle, considering that in experiments at high energies, charged particles are created and annihilated. For example, the creation of a single-charged particle in a reaction would violate conservation—but in experiments this never occurs. Instead, charged particles are always observed to be created in pairs or larger combinations in which the over-all charge remains constant. Charge conservation can be put into mathematical form. The only way that the amount of charge inside a closed region can increase is if it enters the region through the surface. That is, the rate of increase of charge inside the region equals the rate of flux through the surface:

$$\frac{dq}{dt} = - \oiint \mathbf{J} \cdot d\mathbf{S}. \tag{2-3.9}$$

The minus sign results from taking the outward direction for $d\mathbf{S}$. If the charge inside the surface has the charge density, ρ, this equation can be written in integral form:

$$\iiint \left(\frac{\partial \rho}{\partial t} + \nabla \cdot \mathbf{J} \right) d^3r = 0. \tag{2-3.10}$$

It can also be expressed in differential form. Although the integrand of a vanishing integral need not vanish, we require the integral to vanish regardless of the size or shape of the region of integration. In this case, only a vanishing integrand will suffice, giving the *equation of continuity*:

$$\frac{\partial \rho}{\partial t} + \nabla \cdot \mathbf{J} = 0. \tag{2-3.11}$$

Applying the current density, we may express the Biot–Savart law as a volume integral. Thus:

$$\mathbf{B}(r_f) = \frac{1}{c} \iiint \frac{\mathbf{J}(r_s) \times \mathbf{r}_{fs}}{|r_{fs}|^3} d^3r_s \tag{2-3.12}$$

It is interesting to compare this equation with equation (2-2.6) for the electric field—the current density and cross product replace the charge density and scalar product.

Equations (2-2.4) and (2-3.4) give expressions for the electric and magnetic forces on a charged body moving in an electric and magnetic field. The two forces can be combined into a single-volume integral by a short calculation:

$$\mathbf{F} = \iiint \rho \mathbf{E} \, d^3r + \frac{1}{c} \int I \, d\mathbf{r} \times \mathbf{B}$$

$$= \iiint \rho \mathbf{E} \, d^3r + \frac{1}{c} \int dq \mathbf{v} \times \mathbf{B}$$

$$= \iiint \rho (\mathbf{E} + \frac{1}{c} \mathbf{v} \times \mathbf{B}) \, d^3r. \tag{2-3.13a}$$

If the charged body is a point particle, its charge density is a delta function, and the force reduces to the expression:

$$\mathbf{F} = q(\mathbf{E} + \frac{1}{c} \mathbf{v} \times \mathbf{B}), \tag{2-3.13b}$$

which is called the *Lorentz force*. Because the velocity of the particle depends on the coordinate system of the observer, combining the two forces is not purely an ad hoc matter of convenience, but contains the seeds of a deeper symmetry. This matter will be discussed further in Chapter 4 as part of the treatment of special relativity.

Like the electric field, the magnetic field may be expressed in terms of a potential. Applying vector identity (1-3.28), equation (2-3.12) becomes:

$$\mathbf{B}(r_f) = \frac{1}{c} \iiint \mathbf{J}(r_s) \times \nabla_f \left[\frac{(-1)}{|r_{fs}|} \right] d^3r_s \tag{2-3.14}$$

$$= \nabla_f \times \mathbf{A},$$

where the vector field:

$$\mathbf{A}(r_f) = \frac{1}{c} \iiint \frac{\mathbf{J}(r_s)}{|r_{fs}|} d^3r_s \tag{2-3.15}$$

is the *vector potential*. Note its similarity to the electric scalar potential, equation (2-2.9).

Now, expressing the magnetic field in terms of the vector potential, we can calculate the differential properties of the magnetic field. The divergence vanishes due to the identity:

$$\nabla \cdot \mathbf{B} = \nabla \cdot \nabla \times \mathbf{A} = 0. \tag{2-3.16}$$

A field whose divergence vanishes is said to be *solenoidal*. It is possible to find a physical interpretation of equation (2-3.16) in terms of lines of flux. Define the *magnetic flux* through a surface by the integral:

$$\Phi_M = \iint \mathbf{B} \cdot d\mathbf{S}. \tag{2-3.17}$$

If **B** is represented by magnetic lines of flux, then Φ_M represents the number of lines that pass through the surface of integration. Thus, by the divergence theorem, the total magnetic flux through a closed boundary surface vanishes:

$$\oiint \mathbf{B} \cdot dS = \iiint \nabla \cdot \mathbf{B}\, d^3r = 0. \qquad (2\text{-}3.18)$$

That is, every line of flux that enters the region at some point must leave at some other point—lines of flux do not begin or end on any finite point. Disregarding lines of flux that might come from infinity and return to infinity, lines of flux form closed loops. This is illustrated in Figure 2-6.

The curl of the magnetic field is calculated as follows. Express **B** in terms of **A**:

$$\nabla \times \mathbf{B} = \nabla \times \nabla \times \mathbf{A} = \nabla(\nabla \cdot \mathbf{A}) - \nabla^2\mathbf{A}. \qquad (2\text{-}3.19)$$

However, the vector potential, **A**, is not uniquely determined, since adding the gradient of an arbitrary scalar function does not affect the field, **B**. Thus we could apply an auxiliary (gauge) condition on the potential, $\nabla \cdot \mathbf{A} = 0$, to let the first term vanish. Then the curl becomes:

$$\nabla \times \mathbf{B} = -\nabla^2\mathbf{A} = -\frac{1}{c}\iiint \mathbf{J}(r_s)\left[\nabla_f^2\frac{1}{|r_{fs}|}\right]d^3r_s. \qquad (2\text{-}3.20)$$

Using the Dirac delta function, we get for the curl of the magnetic field:

$$\nabla \times \mathbf{B} = \frac{4\pi}{c}\mathbf{J}(r_f). \qquad (2\text{-}3.21)$$

Although we eliminated it above, the divergence term contains some interesting consequences. Let us consider this matter further. Write the

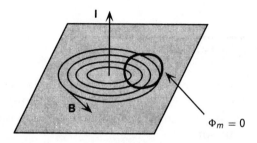

Figure 2-6. Solenoidal Lines of Flux.   The magnetic lines of flux are shown in two dimensions. The lines of flux are closed loops, therefore the total flux through the arbitrary, closed one-dimensional "surface" shown here vanishes.

divergence explicitly:

$$\nabla_f \cdot \mathbf{A} = \frac{1}{c} \iiint \nabla_f \cdot \frac{\mathbf{J}(r_s)}{|r_{fs}|} d^3 r_s. \qquad (2\text{-}3.22)$$

The del operator does not operate on the current density because it is a function of source variables only. Then, after using the symmetry:

$$\nabla_f \frac{1}{|r_{fs}|} = -\nabla_s \frac{1}{|r_{fs}|} \qquad (2\text{-}3.23)$$

and the vector identity (1-3.27), charge conservation, and the divergence theorem, the divergence becomes:

$$\nabla_f \cdot \mathbf{A} = \frac{1}{c} \iiint \mathbf{J}(r_s) \cdot \nabla_f \frac{1}{|r_{fs}|} d^3 r_s$$

$$= -\frac{1}{c} \iiint \nabla_s \cdot \frac{\mathbf{J}(r_s)}{|r_{fs}|} d^3 r_s + \frac{1}{c} \iiint \frac{\nabla_s \cdot \mathbf{J}(r_s)}{|r_{fs}|} d^3 r_s$$

$$= -\frac{1}{c} \oiint \frac{\mathbf{J}(r_s) \cdot d\mathbf{S}_s}{|r_{fs}|} - \frac{1}{c} \frac{\partial}{\partial t} \iiint \frac{\rho(r_s, t)}{|r_{fs}|} d^3 r_s. \qquad (2\text{-}3.24)$$

Take the integration over all space, but assume that the source current density is finite in extent (i.e., ignore cosmic size currents which would need experimental justification anyway). Then the surface integral vanishes, leaving:

$$\nabla(\nabla \cdot \mathbf{A}) = -\nabla \frac{\partial}{c \partial t} \iiint \frac{\rho(r_s, t)}{|r_{fs}|} d^3 r_s = \frac{1}{c} \frac{\partial \mathbf{E}_{\text{coulomb}}}{\partial t}, \qquad (2\text{-}3.25)$$

which vanishes for strictly steady-state conditions, in agreement with our gauge condition. However, the occurrence of this term is a strong hint that it may be needed in the general case. We will re-examine this matter, and the associated question of gauge conditions, in Section 2-5 and again in Chapter 4.

2-4   FARADAY'S INDUCTION LAW

If, as we have seen, electrical currents can cause a magnetic field, then by symmetry, shouldn't magnetism cause some electrical effect? Such symmetry arguments led Michael Faraday in 1831 to discover magnetic induction.

A representation of the Faraday experiments is shown in Figure 2-7. A loop of wire with variable area and orientation is present in a magnetic field. In general, the magnetic flux through the loop is a function

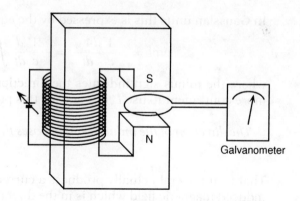

Figure 2-7. A modern representation of Faraday's experiment. The magnetic flux through the loop can be changed by changing the magnetic field (i.e., by changing the current through the winding), by rotating the loop, or by changing the area of the loop. Only one turn is indicated here for the loop.

of time since any of the three variables, the magnetic field, the area of the loop, or its orientation, may be functions of time. The loop need not lie in a plane normal to the field, since the flux through it is defined by the surface integral. The details of the loop are shown in Figure 2-8.

Then, an emf is produced around the loop, which is proportional to the rate of change of the magnetic flux through the loop with respect to time. The emf around the loop is determined by measuring the potential across any small gap in the loop with the galvanometer. Alternatively, one could measure the current with the loop closed, and determine the emf from the resistance of the loop.

The results of the Faraday experiment may be stated:

The emf measured around a closed loop is proportional to the rate of change of magnetic flux through the loop.

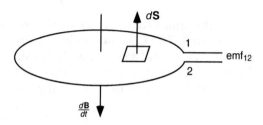

Figure 2-8. Details of the loop in Faraday's experiment. The emf induced in the loop is proportional to the rate of change of the magnetic flux through the loop. For the orientations shown here, the potential difference across the gap satisfies $\phi_2 > \phi_1$.

In Gaussian units this is expressed by the equation:

$$\text{emf} = -\frac{1}{c}\frac{d\Phi_M}{dt} = -\frac{1}{c}\frac{d}{dt}\iint \mathbf{B}\cdot d\mathbf{S}, \qquad (2\text{-}4.1)$$

where the minus sign indicates the direction of the emf. It is usually more convenient to use *Lenz's law* for this purpose:

The direction of the induced emf opposes the change of the inducing flux.

That is, if the emf actually produces a current, that current creates an induced magnetic field which is in the direction opposite to that of the change in the inducing magnetic field. The existence of the emf does not require the presence of a conductor, and if no actual current flows, the emf is the same as when a current does flow.

Since the induced emf in the Faraday experiment can create a steady current in a loop, there must be a corresponding electric field exerting a force on the charge. Applying equation (2-2.16), the Faraday emf measured around a closed loop must be:

$$\text{emf} = \oint \mathbf{E}_{\text{Faraday}}\cdot d\mathbf{r}, \qquad (2\text{-}4.2)$$

which is not a single-valued field. The total work done on the charge is proportional to the number of times the charge is carried around the loop. Therefore, if the loop consists of multiple turns, the emf measured around the loop is proportional to the number of turns. Consequently, the electric field associated with the emf must be nonconservative, and we write: $\mathbf{E}_{\text{nonconservative}} = \mathbf{E}_{\text{Faraday}}$. The positive sign comes from the definition; the emf is defined by *work done on the system* by the nonconservative field.

Compare the emf with the potential between two points. See equation (2-2.13):

$$\Delta\phi = -\int_a^b \mathbf{E}_{\text{Coulomb}}\cdot d\mathbf{r}, \qquad (2\text{-}4.3)$$

where the minus sign refers to *work done by the system* due to the change in potential. The Coulomb field is conservative, as we have seen, and its integral around a closed path vanishes. Consequently, there are two kinds of electric field, conservative and nonconservative. The general electric field must be the sum of the two:

$$\mathbf{E} = \mathbf{E}_{\text{Coulomb}} + \mathbf{E}_{\text{Faraday}}. \qquad (2\text{-}4.4)$$

Compare this equation with equation (2-2.17).

Now apply the preceding discussion to Faraday's law. We replace the nonconservative field, $\mathbf{E}_{\text{Faraday}}$, in equation (2-4.2) by the total electric field, since the closed path integral of the conservative field, E_{Coulomb} vanishes. Therefore Faraday's law may be expressed by the integro-differential equation:

$$\oint \mathbf{E} \cdot d\mathbf{r} = - \iint \frac{1}{c} \frac{\partial \mathbf{B}}{\partial t} \cdot d\mathbf{S}. \tag{2-4.5}$$

The sign has physical significance, but it also depends on the sign conventions of the integrals, i.e., a right-handed relationship between the direction of rotation of the loop integral and the direction of the surface element, $d\mathbf{S}$. This is indicated in Figure 2-8.

In order to express Faraday's law in differential form, apply Stokes's theorem and get:

$$\iint \nabla \times \mathbf{E} \cdot d\mathbf{S} + \iint \frac{1}{c} \frac{\partial \mathbf{B}}{\partial t} \cdot d\mathbf{S} = 0. \tag{2-4.6}$$

Because this equation must hold for an arbitrary surface regardless of size, shape, or orientation, it will be satisfied only if the integrands are equal. Therefore Faraday's law written in differential form must be:

$$\nabla \times \mathbf{E} + \frac{1}{c} \frac{\partial \mathbf{B}}{\partial t} = 0. \tag{2-4.7}$$

We emphasize that Faraday's law is in no way dependent on the material properties of the loop. The emf occurs even in vacuum, and any charge introduced into the region of the changing flux will be accelerated in the direction of the induced (Faraday) electric field; the betatron particle accelerator is based on this principle. The details of this calculation are left to the exercises.

2-5 MAXWELL'S EQUATIONS

Careful inspection of equations shows that equation (2-3.21) is inconsistent with charge conservation, equation (2-3.11)! Applying the divergence to equation (2-3.21) implies that the current density is solenoidal:

$$\nabla \cdot \mathbf{J} = 0, \tag{2-5.1}$$

which, *in general*, contradicts experiment. However, since the experiments are done under steady-state conditions, it is possible there are missing terms containing time derivatives. Consider what would be required to repair charge conservation. The charge density appears in equation (2-2.11); taking the time derivative gives:

$$4\pi \frac{\partial \rho}{\partial t} = \nabla \cdot \frac{\partial \mathbf{E}}{\partial t}. \tag{2-5.2}$$

From this we see that the expression, $(1/4\pi)(\partial \mathbf{E}/\partial t)$, has units of current density. This is called, after Maxwell, the *vacuum polarization current density*. Then if we add this new current density to the ordinary current density:

$$\nabla \times \mathbf{B} = \frac{4\pi}{c} \mathbf{J} + \frac{1}{c} \frac{\partial \mathbf{E}}{\partial t}, \tag{2-5.3}$$

charge conservation is satisfied. We saw some preliminary evidence of the vacuum polarization current density term in equation (2-3.25). We summarize our results in Table 2-1.

Introduction of the vacuum current density completes the symmetry between electricity and magnetism—in the absence of charges equations (2-5.6) and (2-5.7) are almost symmetric under exchange of $\mathbf{E}$ and $\mathbf{B}$. The symmetry introduced into the equations unifies electricity and magnetism into two aspects of a single interaction. There is, however, a sign difference in the two equations which cannot be eliminated by changing the sign of one of the fields. We shall see that this sign difference is essential in obtaining a wave equation for the fields, in which the speed of propagation is the speed of light. In this sense, a minus sign in one of the equations is necessary for the existence of light!

In the Gaussian system of units there are three useful mnemonic devices for Maxwell's equations. (1) Factors of 4π multiply electric charge (including the current density). (2) Factors of c are associated with time (including the current density). (3) Electric and magnetic fields have the same units. It is appropriate in relativistic treatments to have the velocity of light appear explicitly, especially when dealing with the wave equation, where it plays a central role.

TABLE 2-1.   SUMMARY OF THE ELECTROMAGNETIC EQUATIONS

Equations (2-5.4) through (2-5.7) are called *Maxwell's equations*.

$\nabla \cdot \mathbf{E} = 4\pi\rho$	(2-5.4)
$\nabla \cdot \mathbf{B} = 0$	(2-5.5)
$\nabla \times \mathbf{E} + \dfrac{1}{c} \dfrac{\partial \mathbf{B}}{\partial t} = 0$	(2-5.6)
$\nabla \times \mathbf{B} - \dfrac{1}{c} \dfrac{\partial \mathbf{E}}{\partial t} = \dfrac{4\pi}{c} \mathbf{J}$	(2-5.7)
$\nabla \cdot \mathbf{J} + \dfrac{\partial \rho}{\partial t} = 0$	(2-5.8)
$\mathbf{F} = q\left(\mathbf{E} + \dfrac{1}{c} \mathbf{v} \times \mathbf{B}\right)$	(2-5.9)

Equation (2-5.6) requires modification of the potentials associated with the electric field. Specifically, the scalar potential no longer determines the electric field, since curl $\mathbf{E}$ no longer vanishes. We must add another term to take induction effects into account. Equation (2-5.6) and the vector potential for the magnetic field, $\mathbf{B} = $ curl $\mathbf{A}$, enable us to determine the form of the additional potential for the electric field. That is, the expressions:

$$\mathbf{E} = -\nabla\phi - \frac{1}{c}\frac{\partial\mathbf{A}}{\partial t}, \tag{2-5.10}$$

$$\mathbf{B} = \nabla \times A, \tag{2-5.11}$$

are consistent with Maxwell's equations, since their substitution reduces equations (2-5.5) and (2-5.6) to identities, leaving only the two equations with sources. However, inspection of the potentials $\mathbf{A}$ and ϕ shows that they are not unique. That is, the *gauge transformation* of the potentials:

$$\mathbf{A} \rightarrow \mathbf{A} + \nabla\Gamma, \tag{2-5.12}$$

$$\varphi \rightarrow \varphi - \frac{1}{c}\partial_t\Gamma, \tag{2-5.13}$$

where Γ is an arbitrary *gauge function*, leaves $\mathbf{E}$ and $\mathbf{B}$ unchanged.

Substituting the potentials into the remaining Maxwell equations enables us to express them in very elegant and compact form. Thus substituting and rearranging gives:

$$\frac{1}{c^2}\partial_t^2\mathbf{A} - \nabla^2\mathbf{A} = \frac{4\pi}{c}\mathbf{J} - \nabla(\nabla \cdot \mathbf{A} + \frac{1}{c}\partial_t\varphi), \tag{2-5.14}$$

$$\frac{1}{c^2}\partial_t^2\varphi - \nabla^2\varphi = 4\pi\rho + \frac{1}{c}\partial_t(\Delta \cdot \mathbf{A} + \frac{1}{c}\partial_t\varphi), \tag{2-5.15}$$

which are coupled equations. However, they can be simplified because the existence of the gauge ambiguity allows us to add an auxiliary condition on the potentials. The most elegant choice for the gauge condition is:

$$\nabla \cdot \mathbf{A} + \frac{1}{c}\partial_t\varphi = 0 \tag{2-5.16}$$

called the *Lorentz gauge*. With this condition, equations (2-5.14) and (2-5.15) reduce to:

$$\frac{1}{c^2}\partial_t^2\mathbf{A} - \nabla^2\mathbf{A} = \frac{4\pi}{c}\mathbf{J}, \tag{2-5.17}$$

$$\frac{1}{c^2}\partial_t^2\varphi - \nabla^2\varphi = 4\pi\rho. \tag{2-5.18}$$

That is, the potentials **A** and ϕ satisfy the same equation. Other gauge conditions are possible, having utility in special cases; they will be discussed further in Chapter 4.

2-6   A Mechanical Model of the Electromagnetic Field

We consider here a mechanical model of the electromagnetic field, essentially that used by Maxwell in developing the field equations (it disappeared from the final formulation when it had served its usefulness). Although the analogy with mechanical systems cannot be carried very far, it is useful to visualize some features of the theory, and we present it here for pedagogical purposes only.

This type of model is sometimes called a granular model of space. Refer to Figure 2-9. First we consider a region of space which is free from electric and magnetic fields. Imagine that in all of space there are two uniform overlapping charge densities of equal magnitude but opposite sign. The positive charge has uniform spatial density and is fixed and unchanging. It is entirely passive; its only function is to insure that empty space is electrically neutral. We will call it the background charge.

Now, suppose that all space is also filled with two types of wheels or gears in a regular array and in mechanical contact with each other. (Since the model is three dimensional they must actually be spherical.) The first set of wheels, which we will call *cells*, are electrically neutral and have a dimension too small to be observable. They are free to move

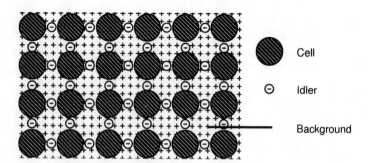

Figure 2-9. Mechanical model of empty space.   The wheels composed of infinitesmal cells and idlers are the active part of the model. The background charge keeps the total charge zero. The cells and idlers may translate and rotate on each other without slipping, and transmit their motion, but are otherwise frictionless.

laterally and to rotate. In some versions of the model they may be elastic and deformable.

The second set of wheels, called *idlers*, lie between the cells and mechanically connect them. The idler wheels are also free to move laterally and rotate but have a negative electric charge. The amount of charge on the idlers is just large enough to cancel the background charge, on a macroscopic scale. Therefore, space contains no observable charge, on this scale. Although the cells and idlers roll on each other without slipping, they are otherwise frictionless.

Figure 2-10 illustrates the electric field in this model. The cells and idlers have been displaced. The electric field is proportional to the amount of displacement of the cells (and idlers). In the uniform displacement shown in the diagram the electric charge density remains zero in the interior. To have consistent units the electric field is given by:

$$\mathbf{E} \approx \rho \mathbf{d}, \tag{2-6.1}$$

where $\mathbf{d}$ represents the local charge displacement and ρ is the charge density. The units of $\mathbf{E}$ are those of *electric dipole moment* per unit volume. Note that a uniform displacement in a finite volume creates surface charge densities, σ, on the boundaries. Electric lines of flux may be represented by lines of displacement. In the general case, where the displacement is not uniform, a volume charge density appears because charges do not locally cancel. This justifies equation (2-5.4). Thus:

$$\rho \approx \frac{\Delta E_x}{\Delta x} \; \to \; \nabla \cdot \mathbf{E}. \tag{2-6.2}$$

This is shown in Figure 2-11 (in somewhat exaggerated form where an actual gap appears between the wheels).

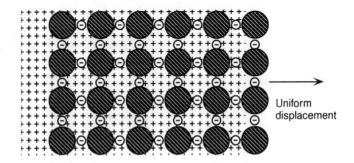

FIGURE 2-10. MECHANICAL MODEL OF THE ELECTRIC FIELD.   The electric field is represented by the charge density times the displacement. This has units of dipole moment per unit volume, as required. Lines of flux are represented by lines of displacement. An effective current flows while the displacement occurs.

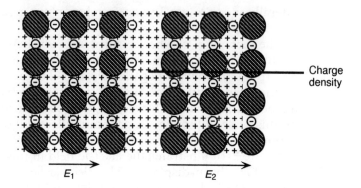

Charge density

FIGURE 2-11. CHARGE DENSITY.   The nonuniform electric field represented here produces a local charge. Discrete charge is shown here, but continuous charge density can occur as the wheels become vanishingly small.

This diagram also represents a model of the vacuum polarization current. When the displacement is a function of time, the current density is the time derivative. Thus the current density becomes proportional to the displacement velocity! Thus:

$$\mathbf{J} \approx \rho \frac{\Delta \mathbf{d}}{\Delta t} \approx \frac{d\mathbf{E}}{dt}. \tag{2-6.3}$$

Now consider the magnetic field in this model. Refer to Figures 2-12 and 2-13. As Figure 2-12 shows, a single spinning cell surrounded by motionless cells would cause its nearest idlers to rotate around it because of the differential velocities. This produces a minicurrent around it.

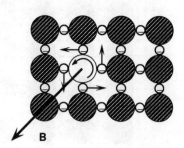

B

FIGURE 2-12. MAGNETIC FIELD MODEL.   The magnetic field at each point is represented by the angular velocity of the cell. A single spinning cell surrounded by motionless cells causes the nearest idlers to rotate around it, producing a circulating electric current.

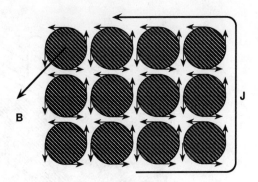

Figure 2-13. **Cancellation of internal currents for uniform B.**   This leaves only the external, i.e., boundary current. For simplicity the idlers are not shown.

When several contiguous cells rotate at the same rate, the interior tangential velocities cancel and there is no internal current. (The idlers spin, but we ignore this in the model.) However, there is a net rotation around the edges of that collection of cells, and this produces a surface current on the boundary. See Figure 2-13.

Now consider the magnetic field produced by a rotating charge distribution. Applying equation (2-5.7), integrating, and using Stokes's theorem, we get *Ampère's law*:

$$\oint \mathbf{B} \cdot d\mathbf{r} = \frac{4\pi}{c} I_{\text{linking}}, \qquad (2\text{-}6.4)$$

where I_{linking} is the total current passing through (linking) the loop of integration. Now, apply this to a cylinder of charge of length L, radius R, containing constant charge density ρ, and rotating with angular velocity ω. The current developed by this rotating charge density is calculated by integrating over the cylinder from the center to the radius:

$$I = \frac{\rho\omega L}{2} R^2. \qquad (2\text{-}6.5)$$

The magnetic field calculated by Ampère's law becomes:

$$B = \frac{4\pi}{cL} I = \frac{4\pi}{cL} \frac{\rho\omega L}{2} R^2 = \frac{\rho\omega}{c} 2\pi R^2$$

$$= \frac{\rho A}{c} \omega. \qquad (2\text{-}6.6)$$

Thus, in the model, a magnetic field is represented by the angular velocity of the local cell.

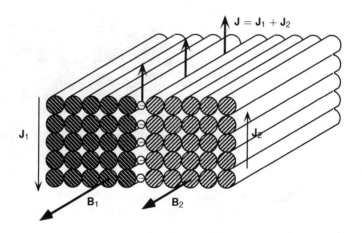

FIGURE 2-14. MODEL OF AMPÈRE'S LAW.   Current density occurs where there is a local change in the magnetic field (represented in the model by rows of rotating cells). There is a net current density at the interface due to the difference in the fields. This generalizes to the curl of the magnetic field.

Now consider vacuum polarization currents. Figure 2-14 shows a translating section of charges (idlers). These moving charges are a local (vacuum polarization) current. On each side of this section there are fixed sections of cells. The shear with the moving idlers causes the fixed cells on either side to rotate. In this way a vacuum polarization current creates magnetic fields. This is analogous to a sheet of paper in a typewriter. If you pull the paper, then the rollers turn. If you turn the rollers, then the paper moves. In our case, the rollers are rotating with different angular velocities. Then the velocity of the "paper" is proportional to the difference between the angular velocities of the two rollers. The relationship between the magnetic field and the surrounding currents corresponds to Maxwell's equation (2-5.7).

This completes our discussion of the mechanical model. Readers might like to try to interpret the two remaining equations: Faraday's law and the solenoidal property of the magnetic field. It might be useful to consider the (sourceless) analogs of Gauss's law and Ampère's circuit equation.

This model of the electromagnetic field in vacuum bears a slight resemblance to theories of polarization and magnetization based on atomic theory. One should remember, however, that Thomson's discovery of the electron and the Bohr model of the atom were still 30–40 years in the future. The model had at least one result of consequence, the notion that there needs to be a medium for the electromagnetic waves. This *aether theory* led to attempts to discover some of its observable traces. The most famous of these searches was the Michelson–Morley experiment, which we now discuss.

2-7 THE MICHELSON–MORLEY EXPERIMENT

From the point of view of nineteenth century physics, the observation of wave motion required the existence of some kind of medium in which the wave moves. For this reason, an electromagnetic aether was postulated, whose elastic properties are determined by Maxwell's equations. Michelson and Morley proposed that it should be possible to observe certain properties of the aether, among them the motion of the Earth through it. Navigators certainly measure their ship's speed through the water, why should Michelson and Morley not do the same for the Earth?

In classical, i.e., Newtonian, physics, the observed velocity of propagation should depend on the velocity of the observer. To take a simple illustration of this, consider a boat in water. To a boat at rest in water, the velocity of a wave in the water depends only on the properties of the medium (e.g., density). But if the boat moves through the water, the observed velocity of the wave will be the sum of the two velocities:

$$c_{observed} = c_{water} + v_{boat}. \tag{2-7.1}$$

Furthermore, the velocity does not depend on any properties of the source of the wave. These principles apply to any classical wave motion, e.g., sound waves.

If this classical model is applied to waves in the aether, it ought to be possible, in principle, to determine the motion of the Earth through the aether, i.e., the absolute velocity of the Earth through space. As a model for such an experiment, suppose the Earth moves through the aether with velocity, V_{Earth}, and a light pulse in the aether at rest moves with velocity, c_{aether}. Now suppose the light pulse travels a distance, D, in the direction of the Earth's velocity, and is then reflected backward and travels opposite the Earth's velocity the same distance, D. The total time for the return of the light pulse is:

$$t_{total} = \frac{D}{c_{aether} + v_{Earth}} + \frac{D}{c_{aether} - v_{Earth}}. \tag{2-7.2}$$

Then the expected velocity of the light pulse is:

$$c_{predicted} = \frac{2D}{t_{total}} = c_{aether} \left[1 - \left(\frac{v_{Earth}}{c_{aether}} \right)^2 \right], \tag{2-7.3}$$

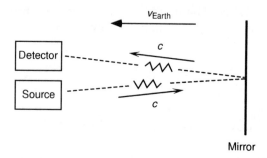

Figure 2-15. A possible experiment for measuring the earth's velocity through the aether. A light beam is reflected from the mirror. By Newtonian kinematics the observed velocity of light depends on the velocity of the earth. Consequently the time the signal takes to reach the detector should depend on the velocity of the earth. In this experiment time is measured as a phase shift.

which is smaller than the velocity of light in the rest frame of the aether. Figure 2-15 shows a possible experiment for performing this experiment.

The actual experiment was more sophisticated. It is necessary to measure the ratio of the velocity of the Earth to the speed of light to sufficient accuracy. In order to detect the effect of the Earth's orbital velocity (about 20 miles/s) it would take an instrument capable of measuring the velocity of light to within about one part in 10^8 to perform the experiment! Furthermore, the measurement should be repeated several times for at least a half year in order to test the velocity of the Earth through the aether in different directions. Although the experiment might be done in a half day using the rotational velocity of the Earth, that would require even more accuracy.

The Michelson interferometer is quite capable of such measurements by detecting small phase shifts in a split ray of light. A diagram of the interferometer is shown in Figure 2-16. The incident light ray is split into two secondary rays which then travel in orthogonal directions, return and form an interference pattern. It can be seen from the diagram that any change in either of the two orthogonal optical paths, whether length or velocity, will result in a change in the phase of the returning ray. This interference pattern is extremely sensitive to any changes in the relative velocity of light in the two orthogonal directions.

The aether drift measurement is made as follows: an interference pattern is observed with the interferometer in some initial orientation. Then the instrument is rotated and the observer looks for any change in the interference pattern. A shift in the pattern will indicate a relative phase shift between the two arms and therefore a corresponding relative change in the velocities in the two directions. The rotation can be

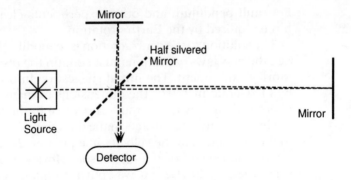

Figure 2-16. Michelson Interferometer.   The incident light ray is split by the half silvered mirror. The two secondary rays then travel in orthogonal directions, are reflected by the two front-surface mirrors, return to the half silvered mirror reflected again, recombine, and the resulting interference pattern is observed by the detector. The extra rays from the second reflection at the half silvered mirror are not shown.

done manually (on a frictionless bearing) or by waiting for the Earth to rotate during a day or revolve during a year. Although the aether drift experiment is the most celebrated application of the Michelson interferometer, the interferometer can be used for other measurements such as the determination of refractive indices.

The Michelson–Morley experiment was first performed in 1887. The experiment was repeated by Michelson and others many times in different configurations and with increasing accuracy. The result is null within experimental error, i.e., the observable velocity of the Earth through the aether is zero. Another way of stating the results is that *the velocity of light is the same in any reference frame, regardless of the velocity of the reference frame.*

The experiment was uncomplicated, reproducible, and extremely accurate—it could not be ignored. But it posed fundamental problems, by contradicting Galilean–Newtonian kinematics. Certain attempts were made to salvage the situation, but the cures tended to be worse than the disease. It was suggested that the Earth carries a local portion of the aether along with it. But that would require an aberration in the positions of the stars as the Earth rotates and revolves around the sun, i.e., the positions of the stars would appear to shift relative to each other as a function of time. This aberration is not observed.[1] The possibility that the Earth is at rest in the aether, and that the rest of the universe revolves around it through the aether is untenable, since the

---

[1] However, an unrelated aberration, due to the Earth's atmosphere, is well known to navigators.

Foucault pendulum and other experiments clearly show the Coriolis force required by the Earth's rotation.

The solution to the contradiction is to modify Newtonian kinematics, i.e., the new laws of motion must contain the results of the Michelson–Morley experiment. The initial suggestion in this direction was made by FitzGerald in 1892: "the dimensions of material bodies are slightly altered when they are in motion relative to the aether."[2] If the arm of the interferometer contracts in the direction of motion by a factor equal to the coefficient of the velocity in equation (2-7.3), this could account for the null result of the Michelson–Morley experiment. This *ad hoc* suggestion by FitzGerald was put into quantitative form by Lorentz and Poincaré and is an integral part of Einstein's theory of relativity.

2-8 Systems of Units

In spite of efforts at standardization, there are, unfortunately, three systems of electromagnetic units in use today: the Gaussian, SI, and Heaviside–Lorentz systems. Each system has advantages. For calculations involving Maxwell's equations, either the SI or Heaviside systems are preferred because they are rationalized, i.e., contain no factors of 4π. Theoretical literature is frequently written in Heaviside units, especially with c set to one. However, atomic and condensed matter literature mostly uses Gaussian units, and handbook values of dielectric constants, etc., are usually expressed in Gaussian or SI units. Engineering literature is mostly in SI units. Relativistic calculations are awkward in SI units since the speed of light is distributed between two of Maxwell's equations, and the electric and magnetic fields have different units. The SI system is recommended by international committees on units. Most elementary textbooks are in SI units, while few or none of them are in Heaviside units. For these reasons students cannot avoid at least occasional exposure to each of the three systems. The best compromise for our purposes is the Gaussian system since it is widely used and suitable for relativistic calculations. The following two tables may be useful to students who are mainly familiar with the SI system. Useful discussions of units are given in textbooks by Marion and Heald, and Jackson, both of which also use Gaussian units.

In some discussions it is desirable to refer to the vacuum permittivity and permeability defined in SI or Heaviside units. In these terms the constants K_E and K_M in Coulomb's law and Ampère's law become in the

---

[2] Whittaker, p. 404.

three systems of units:

$$\mathbf{E}(r_f) = K_E \iiint \rho(\mathbf{r}_s) \frac{\mathbf{r}_{fs}}{|r_{fs}|^3} \, d^3 r_s, \qquad (2\text{-}8.1)$$

$$\mathbf{B}(r_f) = K_M \iiint \frac{\mathbf{J}(r_s) \times \mathbf{r}_{fs}}{|r_{fs}|^3} \, d^3 r_s, \qquad (2\text{-}8.2)$$

where the coupling coefficients can be expressed by the equations:

$$K_E = \frac{1}{4\pi\varepsilon_0}, \qquad K_M = \frac{\mu_0}{4\pi}, \qquad (2\text{-}8.3 \text{ a,b})$$

with the values given in Table 2-2.

Charge and current are related by the equations:

$$q_H = \sqrt{4\pi} q_G, \qquad I_H = \sqrt{4\pi} I_G. \qquad (2\text{-}8.4 \text{ a,b})$$

There are no special names for Heaviside units of charge or current. Electric and magnetic fields have identical units in Gaussian and Heaviside systems. All four components of the potential have identical units:

$$\mathbf{E}_H = \frac{1}{\sqrt{4\pi}} \mathbf{E}_G, \qquad \mathbf{B}_H = \frac{1}{\sqrt{4\pi}} \mathbf{B}_G, \qquad (2\text{-}8.5 \text{ a,b})$$

$$\mathbf{A}_H = \frac{1}{\sqrt{4\pi}} \mathbf{A}_G, \qquad \phi_H = \frac{1}{\sqrt{4\pi}} \phi_G. \qquad (2\text{-}8.6 \text{ a,b})$$

Both Gaussian and Heaviside units are cgs systems. Mechanical quantities such as energy and force are unchanged.

$$\mathbf{F}_H = q_H \mathbf{E}_H = q_G \mathbf{E}_G = \mathbf{F}_G. \qquad (2\text{-}8.7)$$

Conversion of other electromagnetic quantities follow from these. Tables 2-3 and 2-4 give a summary of conversion factors and electromagnetic equations in the three systems of units.

Equations (2-8.3) and (2-8.4) enable us to find conversion constants between Gaussian and SI systems of units. An additional complication

TABLE 2-2.   COUPLING COEFFICIENTS

	ε_0	μ_0	K_E	K_M
Heaviside	1	$\dfrac{1}{c}$	$\dfrac{1}{4\pi}$	$\dfrac{1}{4\pi c}$
Gaussian	$\dfrac{1}{4\pi}$	$\dfrac{4\pi}{c}$	1	$\dfrac{1}{c}$
SI	$\dfrac{1}{\mu_0 c^2}$	$4\pi \times 10^{-7}$	$\dfrac{1}{4\pi\varepsilon_0}$	$\dfrac{\mu_0}{4\pi}$

Table 2-3.   Summary of Conversion Factors: $c = 2.997930 \times 10^8$ m/s

Quantity	Heaviside to Gaussian	SI to Gaussian
Charge, current	$q_G = \dfrac{1}{\sqrt{4\pi}} q_H$	q_G (statcoulomb) $= 10 c q_{SI}$ (Coulomb)
Electric field	$E_G = \sqrt{4\pi} E_H$	$E_G \left(\dfrac{\text{dyne}}{\text{statcoulomb}} \right) = 10^{-4} c E_{SI} \left(\dfrac{\text{Newton}}{\text{Coulomb}} \right)$
Magnetic field	$B_G = \sqrt{4\pi} B_H$	B_G (Gauss) $= 10^{-4} B_{SI}$ (Tesla)
Scalar potential	$\phi_G = \sqrt{4\pi} \phi_H$	ϕ_G (statvolt) $= 10^{-6} c \phi_{SI}$ (Volt)
Vector potential	$A_G = \sqrt{4\pi} A_H$	A_G (statvolt) $= 10^{-6} c A_{SI}$ (Volt)
Force	$F_G = F_H$	F_G (dyne) $= 10^5 F_{SI}$ (Newton)
Energy	$U_G = U_H$	U_G (erg) $= 10^7 U_{SI}$ (Joule)

arises since Gaussian is in cgs, and SI units are in MKS. For this, we use the conversion constant between units of energy:

$$\text{Joule} = 10^7 \text{ erg}. \tag{2-8.8}$$

To find the conversion constant for units of charge, express the potential energy of two equal static charges in the two systems of units. Thus:

$$\frac{q_G^2}{q_{SI}^2} = \frac{\text{PE (erg)} \, r \, (\text{cm})}{4\pi\varepsilon_0 \text{PE (joule)} \, R \, (\text{meter})} = \frac{1}{4\pi\varepsilon_0} 10^2 10^7.$$

$$= (10c)^2 \tag{2-8.9}$$

Then the conversion for electric charge is:

$$q_G \text{ (statcoulomb)} = (10c) q_{SI} \approx 3 \times 10^9 q_{SI} \text{ (coulomb)}. \tag{2-8.10}$$

The charge of the electron is 1.602×10^{-19} coulombs and 4.806×10^{-10} statcoulombs, respectively. The ratio of force between two identical charges in Gaussian and SI units is 10^5. Therefore the ratio of the electric field in the two systems is:

$$\frac{E_G}{E_{SI}} = \frac{q_{SI}}{q_G} \frac{F_G}{F_{SI}} = (10c) \left(10^{-5} \right)$$

$$= 10^{-4} c. \tag{2-8.11}$$

In Gaussian units **E** and **B** have the same units (Gauss). However, in SI units the magnetic field has different units from the electric field, as

TABLE 2-4.   COMPARISON OF EQUATIONS

Gaussian	Heaviside	SI
$\phi = \dfrac{q}{r}$	$\phi = \dfrac{1}{4\pi}\dfrac{q}{r}$	$\phi\dfrac{1}{4\pi\varepsilon_0}\dfrac{q}{r}$
$\mathbf{A} = \dfrac{1}{c}\dfrac{q\mathbf{v}}{r}$	$\mathbf{A} = \dfrac{1}{4\pi c}\dfrac{q\mathbf{v}}{r}$	$\mathbf{A} = \dfrac{\mu_0}{4\pi}\dfrac{q\mathbf{v}}{r}$
$\mathbf{E} = -\nabla\phi - \dfrac{1}{c}\dfrac{\partial\mathbf{A}}{\partial t}$	$\mathbf{E} = -\nabla\phi - \dfrac{1}{c}\dfrac{\partial\mathbf{A}}{\partial t}$	$\mathbf{E} = -\nabla\phi - \dfrac{\partial\mathbf{A}}{\partial t}$
$\mathbf{B} = \nabla\times\mathbf{A}$	$\mathbf{B} = \nabla\times\mathbf{A}$	$\mathbf{B} = \nabla\times\mathbf{A}$
$\mathbf{F} = q\left(\mathbf{E} + \dfrac{1}{c}\mathbf{v}\times\mathbf{B}\right)$	$\mathbf{F} = q\left(\mathbf{E} + \dfrac{1}{c}\mathbf{v}\times\mathbf{B}\right)$	$\mathbf{F} = q(\mathbf{E} + \mathbf{v}\times\mathbf{B})$
$\nabla\cdot\mathbf{E} = 4\pi\rho$	$\nabla\cdot\mathbf{E} = \rho$	$\nabla\cdot\mathbf{E} = \dfrac{1}{\varepsilon_0}\rho$
$\nabla\cdot\mathbf{B} = 0$	$\nabla\cdot\mathbf{B} = 0$	$\nabla\cdot\mathbf{B} = 0$
$\nabla\times\mathbf{E} + \dfrac{1}{c}\dfrac{\partial\mathbf{B}}{\partial t} = 0$	$\nabla\times\mathbf{E} + \dfrac{1}{c}\dfrac{\partial\mathbf{B}}{\partial t} = 0$	$\nabla\times\mathbf{E} + \dfrac{\partial\mathbf{B}}{\partial t} = 0$
$\nabla\times\mathbf{B} - \dfrac{1}{c}\dfrac{\partial\mathbf{E}}{\partial t} = \dfrac{4\pi}{c}\mathbf{J}$	$\nabla\times\mathbf{B} - \dfrac{1}{c}\dfrac{\partial\mathbf{E}}{\partial t} = \dfrac{1}{c}\mathbf{J}$	$\nabla\times\mathbf{B} - \dfrac{\partial\mathbf{E}}{\partial t} = \mu_0\mathbf{J}$

we can see from the Lorentz force equation:

$$\mathbf{F}_{\text{SI}} = q_{\text{SI}}(\mathbf{E}_{\text{SI}} + \mathbf{v}\times\mathbf{B}_{\text{SI}}). \tag{2-8.12}$$

Therefore the unit of magnetic field is smaller by a factor of c in SI units:

$$B_G \text{ (Gauss)} = 10^{-4}B_{\text{SI}} \text{ (Tesla)} \tag{2-8.13}$$

Similarly, the electrostatic potential has different units from the vector potential in SI units. Using $\mathbf{E} = -\nabla\phi$ and converting meters to centimeters gives the conversion:

$$\phi_G \text{ (statvolt)} = 10^{-6}c\phi_{\text{SI}} \approx 300\phi_{\text{SI}} \text{ (volt)}. \tag{2-8.14}$$

To give some feeling for these units, the potential in household circuits (110 volts) is a little more than 1/3 of a statvolt while a 10 ampere current is 3×10^{10} statamperes!

2-9 Selected Bibliography

J. Edminister, *Electromagnetics*, Schaum Outline Series, McGraw-Hill, 1979. A supplement to undergraduate texts.

A. French, *Special Relativity*, Norton, 1968.

D. Griffiths, *Introduction to Electrodynamics*, Prentice-Hall, 1981.

J. Jackson, *Classical Electrodynamics*, Wiley, 1967. A standard graduate level text.

L. Landau and E. Lifshitz, *The Classical Theory of Fields*, 4th ed., Pergamon, 1975. This is Volume 2 of the 10 volume Course of Theoretical Physics. It presents electromagnetism in both three- and four-dimensional notation, and general relativity.

J. Marion and M. Heald, *Classical Electromagnetic Radiation*, 2nd ed., Academic Press, 1980. A graduate level text emphasizing radiation.

R. Nelson, *SI: The International System of Units*, 2nd ed., American Association of Physics Teachers. Includes a short history of the systems of units.

W. Panofsky and M. Phillips, *Classical Electricity and Magnetism*, Addison-Wesley, 1955. Contains a useful summary of the Michelson–Morley experiments.

E. Purcell, *Electricity and Magnetism*, 2nd ed., McGraw-Hill, 1985. This is Volume 2 of the Berkeley Physics Course for undergraduates.

J. Reitz, F. Milford, and R. Christy, *Foundations of Electromagnetic Theory*, 3rd ed., Addison-Wesley, 1979.

E. Whittaker, *A History of the Theories of Aether and Electricity*, Dover, 1989.

2-10 Problems

1. The source of the electromagnetic field in Maxwell's equations is electric charge, i.e., magnetic fields are produced by moving electric charges. Single, isolated magnetic charges do not appear in Maxwell's equations.

A. Suppose that tomorrow a particle with a magnetic charge is discovered. Rewrite Maxwell's equations as if only magnetic charges existed.

B. Rewrite Maxwell's equations if both electric and magnetic charges exist.

C. Find appropriate potentials for the modified Maxwell's equations.

D. What gauge conditions must the potentials satisfy?

2. Discuss how to determine the magnetic lines of flux using the measure-move procedure.

3. Find the force of a proton on the electron in the ground state hydrogen atom. How does this compare in magnitude with the gravitational force between the two particles?

4. Maxwell's equations are normally expressed as partial differential equations. However, they can also be expressed as integro-differential equations. Derive the following from Maxwell's equations:

$$\oiint \mathbf{E} \cdot d\mathbf{S} = 4\pi Q_{inside}, \qquad \text{Gauss's law,}$$

$$\oint \mathbf{B} \cdot d\mathbf{r} = \frac{1}{c} \frac{d}{dt} \iint \mathbf{E} \cdot d\mathbf{S} + \frac{4\pi}{c} I_{linking}, \qquad \text{Ampère's law,}$$

$$\oint \mathbf{E} \cdot d\mathbf{r} = -\frac{1}{c} \frac{d}{dt} \iint \mathbf{B} \cdot d\mathbf{S}, \qquad \text{Faraday's law,}$$

$$\oiint \mathbf{B} \cdot d\mathbf{S} = 0, \qquad \text{Solenoidal law.}$$

5. The velocity of propagation of electromagnetic fields can be estimated by finding the ratio of magnetic to electric effects. Do this by finding E and B for a straight beam of charged particles moving with velocity, v.

6. Equation (2-7.3) gives a Newtonian theory calculation of the observed velocity of light.
   A. Calculate the fractional change in the velocity of light Δc due to the Earth's orbital velocity around the Sun.
   B. For a Michelson interferometer with arms one meter long at "rest," calculate the amount of FitzGerald contraction necessary to account for the observed speed of light?

7. Express Maxwell's equations in nonvector form, i.e., as eight equations (two equations are vector equations and two equations are scalar equations) in terms of the x, y, z components of E and B.

8. The equation for charge conservation is not independent of Maxwell's equations. Show that equation (2-5.8) can be derived from equations (2-5.4) and (2-5.7).

9. It was stated in the text that when the electric and magnetic fields are expressed in terms of their potentials

$$\mathbf{E} = -\nabla \phi - \frac{1}{c} \frac{\partial \mathbf{A}}{\partial t},$$

$$B = \nabla \times A,$$

makes Faraday's law an identity. Prove this.

10. Show that the expressions, $\mathbf{E} = \mathbf{F}(x - ct)$ and $\mathbf{B} = \mathbf{G}(x - ct)$, where $\mathbf{F}$ and $\mathbf{G}$ are vector functions, are solutions of the source-free Maxwell equations if certain assumptions are made about the relationship between $\mathbf{F}$ and $\mathbf{G}$. State the assumptions.

11. Resolve this paradox. Consider a magnetic field, directed parallel to the z-axis and constant everywhere in space, but changing in time. Now suppose there is a circular path of integration in the x–y plane. By Faraday's law:

$$\oint \mathbf{E} \cdot d\mathbf{r} = -\frac{1}{c} \frac{d}{dt} \iint \mathbf{B} \cdot d\mathbf{S},$$

there will be an electric field tangent to the path of integration. The electric field can be calculated by direct evaluation of these integrals:

$$E = \frac{r}{2C} \frac{dB}{dt}.$$

That is, E depends on the radius of the circle. Now consider two loops which are tangent to each other but which have different radii. At the point of tangency, the electric field has two different values. Since the point of tangency can be any point in the region, the electric field is ambiguous everywhere!

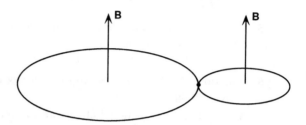

12. Resolve this paradox. Consider a magnetic field which is nonzero inside a toroidal region but zero outside. Now suppose there is a circular path of integration which links the toroid and lies in a plane which is orthogonal to the toroid as shown in the diagram. Calculating the electric field on this path gives:

$$\mathbf{E} = -\frac{1}{2\pi rc} \frac{d}{dt} \iint_{\text{toroid}} \mathbf{B} \cdot d\mathbf{S},$$

where the integral vanishes except on the cross section of the toroid. Now consider a second path of integration in the same plane as the first but with larger radius and displaced so that the two paths of integration are tangent at a point. The total magnetic flux through this loop is a constant zero, so that the electric field around it vanishes. Then at the point of tangency the electric field has two values! If you put an electron at rest at the tangent point, what will its acceleration be?

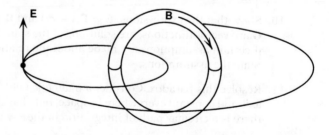

13. An electron is moving in a uniform magnetic field of B kilogauss which is increasing at a fixed rate of $\partial B/\partial t$ kilogauss per second.

   A. Find the radial component of force on it as a function of time.
   B. Find the tangential component of the force. (Hint: Use Faraday's law.)
   C. Find the radius of the orbit as a function of time.
   D. Find the energy of the electron as a function of time.

   Describe what the actual apparatus of this accelerator might look like.

14. Two circular current loops of radius, R_1 and R_2, lie in the x–y plane, and are separated by distance D along the z-axis. Each loop carries a current I in opposite directions. Calculate the force of one current loop on the other. (Hint: Use Ampère's formula.)

15. Assume the electric field is known in a region. Write a program in an appropriate computer language to trace a line of flux from an arbitrary initial location. Then use the program to trace some of the lines of flux for an electric dipole.

APPENDIX 2-1   FIELD PLOTTER IN 2-D

This notebook inputs a collection of charges, and calculates the resulting potential and electric field strength. It then plots the potential as a surface plot and a contour plot. The electric field is calculated from the gradient and can be plotted as a set of lines of flux. Any distribution of point charges can be used, but not continuous distributions of charge. Although a system of charged conductors can be simulated, the plots show the individual charges at high resolution.

Initialization

```
Remove["Global'@"]
```

Equipotential Surfaces

Enter the Values and Positions of Charges

Any number of charges may be entered in the form of a list of positions and charges. The example chosen here simulates two parallel lines of charge of opposite sign.

```
(* Enter the positions and values of the charges *)
(* to be plotted in the form: {x,y,q}.            *)

charges =  List[{-1,0,1},{1,0,-1},
                {-1,-1,1},{1,-1,-1},
                {-1,1,1},{1,1,-1},
                {-1,-1/2,1},{1,-1/2,-1},
                {-1,1/2,1},{1,1/2,-1}
                ];
```

Plot the Potential

The potential is plotted as both a 3-D surface and as contours. The contours are equipotential lines.

```
distance := Sqrt[(Part[charges,m,1]-x)^2+
                       (Part[charges,m,2]-y)^2]
potential = Sum[Part[charges,m,3]/
                    distance,
            {m,1,Length[charges]}];
Plot3D[potential,{x,-2,2},{y,-2,2}]
(* The number of contours, the range of values    *)
(* of the potential, and the contour shading       *)
(* can be changed by changing the options           *)
ContourPlot[potential,{x,-2,2},{y,-2,2},
     ContourShading -> False,
     Contours -> 20,
     PlotRange -> {-5,5}]
```

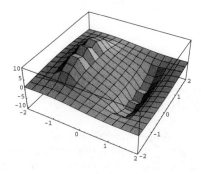

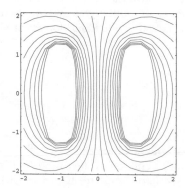

Electric Field Lines

The lines of flux are found by the measure-move procedure. A point in the region is chosen, and the electric field is calculated at that point. Next, a short line is drawn in the direction of the electric field. Then the electric field is calculated at the end of the line. This procedure is continued for a large number of points, giving the "electric line of flux." If this procedure is continued for a number of

initial points, the lines of flux map out the field in the region. At each point on
a given line, the electric field is tangent to the line of flux.

Input Starting Points for Electric Lines of Flux
Input a list of starting points and the path-length increment *ds* for the lines
of flux. There is no upper limit to the number, except that a large number of
points will require a long calculation time.

In order to make this section independent of the preceding, a new potential
function was chosen. The plot boundaries are chosen to be ±2.

```
(* Potential function *)
pot = x*y;


(* Plot boundaries *)
xLo = -2;
xHi = 2;
yLo = -2;
yHi = 2;


(* Maximum number of iterations in LOF calculations *)
nnMax = 40;


(* Starting points for Lines of Flux *)
(* The third value is the path-length increment *)
starts = {{-1.9,-1.9,.1},{-1.9,-1.7,.1},{-1.9,-1.5,.1},
          {-1.9,-1.3,.1},{-1.9,-1.1,.1},{-1.9,-0.9,.1},
          {-1.7,-1.9,.1},{1.7,1.9,.1},{1.9,1.7,.1}};
(* Calculate electric field *)
feld = -{D[pot,x],D[pot,y]};
magnit = Sqrt[feld[[1]]^2+feld[[2]]^2];
(* Unit vector in direction of electric field *)
u = feld/magnit;
```

Field Line Procedure
The points on the line of flux are calculated recursively, subject to conditions
on the length of the line of flux.

```
LoF :=
Block[{feldPtx,feldPty,n},
 feldPtx[n_] := feldPtx[n] = (u[[1]]/.x:>feldPtx[n-1]/.
                                    y:>feldPty[n-1])*ds+
                                    feldPtx[n-1];


  feldPty[n_] := feldPty[n] = (u[[2]]/.x:>feldPtx[n-1]/.
                                    y:>feldPty[n-1])*ds+
                                    feldPty[n-1];
        feldPtx[0] := xxStart;
        feldPty[0] := yyStart;


        abler = {};
        For[n=0,feldPtx[n]<xHi && feldPty[n]<yHi &&
                feldPtx[n]>xLo && feldPty[n]>yLo &&
        n<nnMax,++n,
        feldLineTable =
        AppendTo[abler,{feldPtx[n],feldPty[n]}]
```

```
    ];
    feldLine = Line[feldLineTable];
]
```

Make a Table of Points on the Lines of Flux
This can be a long calculation. Prepare to wait.

```
lineList = {};
For[m=1,m<Length[starts]+1,++m,
    xxStart := starts[[m,1]];
    yyStart := starts[[m,2]];
    ds := starts[[m,3]];
    LoF;
    lineList = {lineList,feldLine};
]
```

Plot the Lines of Flux and Equipotential Contours
The lines of flux (LOF) are drawn from the starting point in the clockwise or counterclockwise direction, depending on the quadrant of the starting point. To draw a LOF in the opposite direction, choose the opposite sign for *ds* for that curve.

Note some general properties of the equipotentials and the LOFs. (1) The equipotentials and LOF are mutually orthogonal when plotted on the same scale. (2) Neither the equipotentials nor the LOFs intersect themselves.

```
(* Plot Lines of Flux *)
FeldGraph = Show[Graphics[lineList],
                    Axes->True,
                    AspectRatio->1]
(* Plot equipotential contours *)
potPlot = ContourPlot[pot,{x,xLo,xHi},{y,yLo,yHi},
                ContourShading -> False]
(* Superimpose plots *)
Show[potPlot,FeldGraph]
```

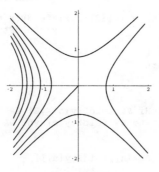

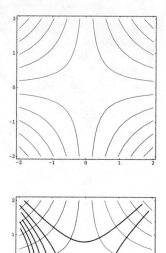

Using the Built-in Field Plotting Routines

For some purposes it is useful to use the built-in field plotting routine. The value and direction of the field at points in the plot are represented by arrows whose bases lie at the plot points. The options (e.g., HeadWidth) control the parameters of the arrow; they can be adjusted to give the best plot.

```
Needs["Graphics'PlotField'"]

PlotGradientField[pot, {x, -2, 2},{y, -2, 2},
      HeadWidth->0.01,
      HeadLength->0.01,
      HeadCenter->0]
```

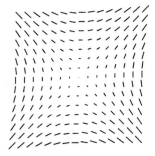

CHAPTER 3

Dielectric and Magnetic Materials and Boundary Conditions

I Sing the Body Electric.

Walt Whitman
Leaves of Grass

Nature cannot be ordered about, except by
obeying her.

Francis Bacon
Novum Organum (1620)

We have so far considered only electromagnetic fields in vacuum, in order not to obscure the most essential features of the theory by details associated with properties of matter. However, some interesting physical effects, and many of the most important practical applications result from the electromagnetic properties of matter. The interaction of matter with electric and magnetic fields derives from the fact that a material's behavior observed at the macroscopic level depends strongly on its microscopic structure, and that matter is composed of charged particles. For example, the periodic properties of the elements are a consequence of the atomic electrons residing in nested sets of orbital shells (resulting from certain quantum mechanical rules). Electrons in the outermost (valence) shell also make the largest contribution to the electrical properties of materials. Exact treatment of these topics requires detailed knowledge of condensed matter physics, including quantum theory, and is beyond the scope of this book. Instead, this chapter is limited to an empirical, mostly macroscopic, treatment of Maxwell's equations applied to dielectric and magnetic materials. We will restrict ourselves to low energies and low frequencies (adiabatic variations).

3-1   Dielectric Materials

Although materials exhibit an extremely wide range in electrical conductivity, it is conventional to distinguish conducting materials (like copper) from insulating materials (like sulfur). An electric field can be induced in an insulator, analogous to a magnetic field induced in a magnetic material. Thus, insulators are also called *dielectrics*. Consider an external electric field applied to a sample of dielectric. Electrical forces will be exerted on the charges of the sample's molecules, changing their charge distributions. The simplest of such changes is called *polarization*, a general lateral shift of the charges, producing a surplus of negative charge at one side of the molecule and leaving a corresponding surplus of positive charge at the opposite side. There are two common mechanisms for polarization:

1. Alignment of intrinsically polarized molecules that were initially randomly oriented.
2. Redistributing the charges in the molecules.

For a given molecule, we express its *dipole moment by*:

$$\mathbf{p} = q\,\Delta\mathbf{r}, \tag{3-1.1}$$

where q is the displaced charge, and $\Delta\mathbf{r}$ is the average displacement between the positive and negative charges. Then for a macroscopic sample of material the *polarization*, $\mathbf{P}$, is the dipole moment per unit volume:

$$\mathbf{p} = \iiint \mathbf{P}\,d^3r. \tag{3-1.2}$$

Applying an electric field to a dielectric sample produces a polarization by coordinated rotation of the molecules into a preferred direction. When an external field is applied to an unpolarized molecule, the electrons will tend to be displaced from their normal position (we may take a time average). Thus even if a molecule has no intrinsic dipole moment, the application of an external field will create one. Both of these effects occur in intrinsically polarized molecules.

EXAMPLE

The ethanol molecule in Figure 3-1 has an intrinsic polarization because of the strong electronegativity of its oxygen atom. A sample of ethanol in the absence of an external electric field has zero polarization, because the molecular dipoles are randomly oriented. When the molecules in such a sample are subjected to an external electric field, they rotate so as to align themselves parallel to the applied field, resulting in polarization of the sample.

EXAMPLE

The propane molecule in Figure 3-1 has no intrinsic polarization because of its composition and symmetry. An external electric field creates a dipole moment by shifting the electron orbits. The value of the dipole moment depends on the molecular electron mobilities and the orientation of the molecule relative to the field. The molecules align themselves parallel to the field since that configuration has the lowest energy.

Consider the electric field in the interior of a sample of dielectric material, located in an external electric field. When an isotropic medium is polarized, no local macroscopic charge density occurs in the sample interior, because for every charge that moves out of some small region, another one moves into the region to take its place. But, at an interface between two different dielectrics (or at a dielectric–vacuum boundary), there is a discontinuity in the displacements of the charges. In that case, more charge accumulates on one side of the discontinuity than leaves from the other. Consequently, charge accumulates on the discontinuity, producing a *surface charge density*, σ. Now suppose that the dielectric's nonuniformity is not discontinuous, but is instead a (macroscopically) smooth function of position. In that case, the charge associated with the nonuniformity smears out and becomes a *volume charge density*, ρ. Compare this polarization model for dielectric materials with Maxwell's

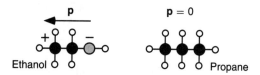

FIGURE 3-1. EXAMPLES OF POLAR AND NONPOLAR MOLECULES.   The ethanol molecule has an intrinsic dipole moment because of the strong attraction of the oxygen for electrons. A sample of ethanol is not normally polarized because the molecules are oriented randomly and moving randomly. The molecules rotate when placed in an external electric field giving a net polarization. Propane is not intrinsically polarized, but in an external electric field electrons spend more time at one end, thus exhibiting polarization.

mechanical model for the electric field in vacuum. In summary: a sharp discontinuity in the polarization creates a surface charge density, while a smooth change in the polarization creates a volume charge density. See Figure 3-2.

This may be expressed quantitatively. Consider a sample of uniformly polarized dielectric. The sample will have a surface charge on the vacuum interface; then a small volume on the vacuum interface, $dV = d\mathbf{S} \cdot d\mathbf{r}$, will contain a charge δq. Using equation (3-1.1) to find the surface charge density in terms of the polarization gives:

$$dq = \frac{d\mathbf{r} \cdot d\mathbf{S}}{dV} dq = \mathbf{P} \cdot d\mathbf{S}$$

$$= P_n \, dS. \tag{3-1.3}$$

Thus the surface charge density equals the normal component of the polarization:

$$\sigma = P_n. \tag{3-1.4}$$

Now consider an interface between two dielectric materials, each polarized in the x-direction. Then each dielectric will have its own surface charge, and the total surface charge density at the interface between the two will be the difference between the individual charge densities:

$$\sigma = \sigma_2 - \sigma_1 = P_2 - P_1. \tag{3-1.5}$$

Now suppose there is a transition length between the two media, Δx. Then in the transition region there will be a volume charge density:

$$\rho = \frac{P_2 - P_1}{\Delta x} \rightarrow -\frac{dP_x}{dx}. \tag{3-1.6}$$

In three dimensions this becomes:

$$\rho = -\nabla \cdot \mathbf{P}. \tag{3-1.7}$$

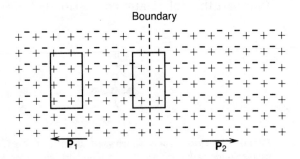

Figure 3-2. Polarization surface charge density. In the region of uniform polarization on the left, the box contains zero net charge. The polarization has a discontinuity at the boundary. Negative charge has moved away from the boundary, creating a positive surface density there.

The negative sign accounts for the sign of the charge relative to the direction of the change in polarization. The polarization charge is a quasi-macroscopic effect since, if the volume is made small enough, it can be made to contain a single charge, or no charge. See Figure 3-3.

The polarization charges create an electric field in the interior of the dielectric:

$$\mathbf{E}_{\text{interior}} = \mathbf{E}_{\text{applied}} + \mathbf{E}_{\text{polarization}}. \tag{3-1.8}$$

Due to the direction of the polarization, the polarization electric field $\mathbf{E}_{\text{polarization}}$ tends to be opposite in direction to the external field $\mathbf{E}_{\text{applied}}$ and partially cancel it:

$$\mathbf{E}_{\text{polarization}} \sim -\mathbf{P}. \tag{3-1.9}$$

Note the distinction between these two quantities: polarization represents the displacement of charges, while $\mathbf{E}_{\text{polarization}}$ is the electric field caused by the polarization.

An external field applied to a sample of dielectric material creates forces on the charges constituting the dielectric; then these forces cause the polarization. We may use a truncated Taylor series to write an empirical equation for the polarization:

$$P^i = P_0^i + \sum_k^3 \chi_k^j E^k + \sum_k^3 \sum_\ell^3 \chi_{k\ell}^j E^k E^\ell + \cdots. \tag{3-1.10}$$

The coefficient, χ_k^j, is called the *electric susceptibility*. Although for most materials, the residual polarization, $\mathbf{P}_0$ and the nonlinear coefficients vanish, some very interesting physics evolves from the exceptions to linearity. For example, materials exhibiting a residual or permanent dipole moment are called *electrets or ferroelectrics*, the phenomenon being the electric equivalent of permanent magnetism. Materials in which the quadratic term survives, produce nonlinear optical effects

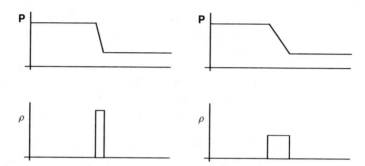

FIGURE 3-3. POLARIZATION AND VOLUME CHARGE DENSITY. In one dimension the charge density is proportional to the slope of the polarization.

which are both interesting and have practical applications. For the moment we restrict ourselves to the simplest case, where the polarization is proportional to the electric field. Then we write for *linear dielectrics*:

$$\mathbf{P} = \chi\mathbf{E}. \tag{3-1.11}$$

Except for low-symmetry crystals, the electric susceptibility is scalar and can often be treated as constant for a wide range of temperature and frequency. In equations (3-1.10) and (3-1.11), the electric field $\mathbf{E}$ represents the total electric field, including the polarization electric field, and not the externally applied field. This is sometimes a source of confusion.

If we write Maxwell's equation for the combined electric charges, we get:

$$\nabla \cdot \mathbf{E} = 4\pi\left(\rho_{\text{true}} + \rho_{\text{polarization}}\right) = 4\pi\rho_{\text{true}} - 4\pi\nabla \cdot \mathbf{P}. \tag{3-1.12}$$

Rewriting gives an equation in terms of the *true charge* only:

$$\nabla \cdot \mathbf{D} = 4\pi\rho_{\text{true}}, \tag{3-1.13}$$

where the field $\mathbf{D}$ is called the *displacement*.

$$\mathbf{D} = \mathbf{E} + 4\pi\mathbf{P} = \varepsilon\mathbf{E}. \tag{3-1.14}$$

It has the same units as $\mathbf{E}$. Equation (3-1.13) is the standard form for Maxwell's equation in a dielectric medium. We have distinguished true charge from other charge in the medium, which (in a specific energy range) is bound charge. In a purely macroscopic treatment, it is moot whether the bound charge is bound to an atom, molecule, or crystal; their displacements produce the observed polarization. True charge includes the familiar electrostatic charges produced by friction and the moving charges in electric circuits.

In equation (3-1.14) the quantity:

$$\varepsilon = 1 + 4\pi\chi \tag{3-1.15}$$

is the *permittivity* of the material. In Gaussian units it is identical to the dielectric constant. The permittivity of vacuum is one in Gaussian units. For typical insulating materials, it is less than 10 but for some materials it is greater than 100. Table 3-1 illustrates the range of values. Barium titanate is also interesting because it is ferroelectric.

Although nominally constant, the permittivity may actually be a function of temperature, frequency, and sample history. Because the electric field associated with the polarization charge partially cancels the applied field, the observed field inside a dielectric is smaller than the field would be in vacuum with the same applied field:

$$\mathbf{E}_{\text{dielectric}} = \frac{1}{\varepsilon}\mathbf{E}_{\text{vacuum}}. \tag{3-1.16}$$

TABLE 3-1. DIELECTRIC CONSTANT: From the
Handbook of Chemistry and Physics, 49th ed.

Material	ε
BaTiO$_3$	$\approx$ 12000.00
TiO$_2$ (parallel to axis)	170.00
TiO$_2$ (transverse)	86.00
Water (20° C)	80.37
SiO$_2$ (quartz)	4.30
Polyethelyene	2.30

At sufficiently low frequency, the electric field inside a conductor vanishes, because the charges are so mobile that the electric field due to the polarization completely cancels the applied field.

3-2 CURRENTS

As we have just seen, an electric field will induce a polarization field in a sample of dielectric. When the sample is also conductive, the electric field may also cause a current in the sample. Expressing this empirical relation in terms of the properties of the material:

$$\mathbf{J}^j = \sigma_k^j E^k. \qquad (3\text{-}2.1)$$

This is *Ohm's law*. Although the conductivity, σ_k^j, is in general a tensor, it is scalar for gases, liquids, high symmetry crystals, and polycrystalline solids. The resistivity is defined as the reciprocal of the conductivity. Table 3-2 gives resistivities for some elements; it can be seen that the range of values is enormous.

TABLE 3-2. RESISTIVITIES: In microhms/cm.

Material	Resistivity
Silver	1.59
Copper	1.67
Gold	2.35
Aluminum	2.65
Silicon	10.00
Germanium	46×10^6
Sulfur	2×10^{23}

The ohmic currents are due to the motion of true, i.e., unbound, charges. There are several other types of currents: convection currents, polarization currents, magnetization currents, and vacuum displacement currents. Each of these can contribute to the total effective current; we will treat each of these independently.

A nonconductive fluid having a volume charge density can produce *convective currents* due to the motion of the material. Examples are: thunderclouds, interstellar dust clouds, charged transformer oil, electronic space charges. Convective currents have much in common with the ohmic conduction discussed above, but are usually associated with the motion of charged particles. However, because the charged particles are actually free, i.e., not bound to the conductor, *convection charges accelerate in the presence of an electric field*. Thus convection currents do not obey Ohm's law.

An ideal dielectric material (insulator) would not carry a current in steady-state conditions. All the charge is bound charge. There is a way, nevertheless, for a dielectric to carry a current, if only momentarily or as an alternating current. When the electric field applied to a dielectric change, its bound charge moves in response and the polarization changes. While it is true that the displacement of these charges is microscopic, its rate of change with respect to time may still be appreciable. (A vibrating violin string does not move very far either, but it moves quickly.) Currents associated with the rate of change of polarization are called *polarization currents*.

The polarization current may be calculated as follows. Consider the length scale in which we must operate. As we mentioned above, the molecules of a sample material have random orientations, velocities, and other attributes. Generally, the observed properties are obtained from microscopic values averaged over a rather large number of molecules. On the other hand, we are treating the polarization as a function of position and time; therefore the volume should be smaller than any macroscopic probe. In a given sample some charges are more mobile than others, e.g., the negative electrons can move farther and quicker than the positive nuclei, and so contribute more to polarization currents.

From equation (3-1.7) find the time derivative of the polarization charge density:

$$\frac{\partial \rho}{\partial t} + \nabla \cdot \frac{\partial \mathbf{P}}{\partial t} = 0. \tag{3-2.2}$$

Comparing this equation with equation (2-3.11) shows the polarization current density must be:

$$\mathbf{J}_P = \frac{\partial \mathbf{P}}{\partial t}. \tag{3-2.3}$$

For a physical interpretation, refer to Figure 3-4. The cylindrical volume is small enough to consider the polarization as uniform in the region. Then during the polarization process, polarization charge is rigidly translated upward, creating a negative surface charge density on the bottom and a positive surface charge density on the top. Now define σ_P to be the charge per unit area that passes through the cross section δS during time δt in the polarization process. Use equation (3-1.4) to calculate the time derivative of the polarization:

$$\frac{\partial P}{\partial t} = \frac{\partial \sigma_P}{\partial t} = \frac{\partial^2 q}{\partial S\, \partial t} = J_P \tag{3-2.4}$$

in agreement with equation (3-2.3) in one dimension. Compare the polarization current in equation (3-2.3) with the vacuum polarization current in equation (2-5.2).

In the Bohr model of the atom, the orbital electrons form microscopic currents around the nucleus; the Schrödinger model alters this picture by replacing Bohr orbits by average orbits. In normal materials, in the absence of magnetic fields, the electronic currents will be oriented randomly. Then the average current density vanishes. In the presence of applied magnetic fields the currents will tend to become aligned, giving effective currents called *magnetization currents*.

We define the *magnetic dipole moment* for a charge in orbit:

$$\mathbf{m}(r_s) = \frac{1}{2c} \mathbf{r} \times q\mathbf{v} \bigg|_{\text{centered at } r_a} . \tag{3-2.5}$$

Since it is proportional to angular momentum, a charged particle in orbit with constant angular momentum will have constant magnetic moment. Define *magnetization* to be the magnetic dipole moment density:

$$\mathbf{M}(r_s) = \frac{1}{2c} \mathbf{r} \times \rho\mathbf{v} = \frac{1}{2c} \mathbf{r} \times \mathbf{J}. \tag{3-2.6}$$

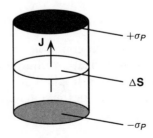

FIGURE 3-4. POLARIZATION CURRENT. The cylinder is small enough that the polarization, current density, and charge density are uniform in the region.

Magnetization is the magnetic analog of the electric dipole moment density (polarization):

$$\mathbf{P}(r_s) = r\rho\Big|_{\text{centered at } r_a} . \qquad (3\text{-}2.7)$$

Note that they have the same units. The magnetic moment of a general configuration of currents is the integral:

$$\mathbf{m} = \int \mathbf{M}\, d^3r. \qquad (3\text{-}2.8)$$

A simple calculation shows that the magnetic moment of a current carrying loop of arbitrary shape is:

$$\mathbf{m} = \frac{1}{c} I\mathbf{S}, \qquad (3\text{-}2.9)$$

where **S** is the area of the loop.

Now calculate the curl of the magnetization, averaged over a region larger than molecular size but smaller than a macroscopic probe. This gives:

$$\mathbf{M}_{\text{avg}}(r_f) = \frac{1}{2c} \frac{\iiint_{\Delta V} r_{fs} \times \mathbf{J}(r_s)\, d^3r_s}{\Delta V}, \qquad \Delta V \text{ small.} \qquad (3\text{-}2.10)$$

Using three-dimensional vector identities:

$$\nabla \times \mathbf{M}_{\text{avg}} = \frac{1}{2c}\Big[(\mathbf{r}_{fs} \cdot \nabla_f)\mathbf{J}(r_s) - \mathbf{r}_{fs}\left(\nabla_f \cdot \mathbf{J}(r_s)\right)$$

$$+ \mathbf{J}(\nabla_f \cdot \mathbf{r}_{fs}) - (\mathbf{J} \cdot \nabla_f)\mathbf{r}_{fs}\Big]_{\text{avg}}$$

$$= \frac{1}{2c}[2\mathbf{J}]_{\text{avg}}. \qquad (3\text{-}2.11)$$

The volume is small enough to be considered a point macroscopically. See Figure 3-5. Thus the magnetization current is related to the magnetization by:

$$\nabla \times \mathbf{M}(r_f) = \frac{1}{c}\mathbf{J}_M. \qquad (3\text{-}2.12)$$

Compare this with equation (3-1.7), which relates electrical polarization to the polarization charge density. The rotational relationship between **M** and **J** implied in equation (3-2.12) reminds us of the mechanical model of the magnetic field discussed in Chapter 2.

EXAMPLE

For a uniformly magnetized sample, the electronic currents cancel except on the boundaries of the sample, leaving a surface current

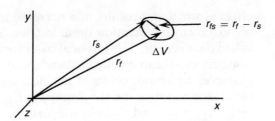

FIGURE 3-5. REGION OF INTEGRATION. The field point r_f is fixed for purposes of integration. For this small region, $r_{fs} \ll r_f$.

there. Now suppose that a sample has a magnetization with only an x-component, and that it changes only in the y-direction. Specifically, let:

$$\mathbf{M} = \left[-\frac{1}{c}\, by, 0, 0 \right],$$ (3-2.13)

where b is an unknown constant with units of current density. Now calculate the current density by taking the curl:

$$\nabla \times \mathbf{M} = \left[0, 0, \frac{1}{c}\frac{\partial[by]}{\partial y} \right] = \frac{1}{c}[0, 0, b],$$ (3-2.14)

and b must be the current density, i.e., $J = b$.

In general, Maxwell's equations include all currents:

$$\nabla \times \mathbf{B} - \frac{1}{c}\frac{\partial \mathbf{E}}{\partial t} = \frac{4\pi}{c}\frac{\partial \mathbf{P}}{\partial t} + 4\pi \nabla \times \mathbf{M} + \frac{4\pi}{c}\mathbf{J}_{\text{true}};$$ (3-2.15)

here $\mathbf{J}_{\text{true}}$ refers to the familiar currents carried by conductors (ohmic currents), and convection currents. Now define a new magnetic vector, the *magnetic field intensity*:

$$\mathbf{H} = \mathbf{B} - 4\pi\mathbf{M}$$ (3-2.16)

and use the displacement, $\mathbf{D}$. Then the magnetic Maxwell equation becomes:

$$\nabla \times \mathbf{H} - \frac{1}{c}\frac{\partial \mathbf{D}}{\partial t} = \frac{4\pi}{c}\mathbf{J}_{\text{true}}.$$ (3-2.17)

3-3 MAGNETIC MATERIALS

It is observed experimentally that a sample of material inserted into an external magnetic field modifies the field. For most materials, the effect is small and results in a decrease in the magnitude of the field, analogous to polarization in dielectrics. Such materials are therefore called *diamagnetic*. The physical basis for this effect lies in the interaction between the magnetic field and the atomic magnetic moments. For example, the atomic orbits of the electron currents in a sample of material tend to rotate (on the average in the quantum mechanical sense) and become aligned, thereby magnetizing the sample. The direction of the induced magnetization current is opposite to that of the external field, so the induced magnetic field diminishes the field observed in the sample. The magnitude and direction of magnetization depends on the nature of the sample. All materials exhibit a diamagnetic effect to some extent; however, some materials also have other magnetic interactions as well. Materials in which the magnetic field is increased slightly are called *paramagnetic* materials. Oxygen is a common paramagnetic material. If the intrinsic magnetic moments of the electrons in the material can be aligned, the sample will exhibit *ferromagnetism*. Ferromagnetic materials have extremely large permeabilities with strong dependence on sample history.

In treating magnetization mathematically one would like to find the functional dependence of the magnetization on the magnetic field; i.e., to write an equation of the form:

$$M = f(H). \tag{3-3.1}$$

When the dependence is small, and independent of the magnetic history of the sample, we may express the functional relation empirically as a truncated Taylor series:

$$M^j = M_0^j + \sum_k^3 \zeta_k^j H^k + \sum_k^3 \sum_\ell^3 \zeta_{k\ell}^j H^k H^\ell + \cdots. \tag{3-3.2}$$

Ferromagnetic materials satisfy neither of these conditions, and will be discussed separately below. For linear, non-ferromagnetic materials the magnetization takes the simple form:

$$\mathbf{M} = \zeta \mathbf{H} \tag{3-3.3}$$

analogous to the expression for the polarization. The symbol ζ is the *magnetic susceptibility* of the material. Then equation (3-2.16) becomes:

$$\mathbf{B} = \mathbf{H} + 4\pi \mathbf{M} = \mu \mathbf{H}. \tag{3-3.4}$$

The quantity μ is the *permeability* of the material. The permeability and the magnetic susceptibility are tensors in general, but for most materials may be taken as scalar. Then:

$$\mu = 1 + 4\pi\zeta. \tag{3-3.5}$$

The permeability of vacuum is one in Gaussian units.

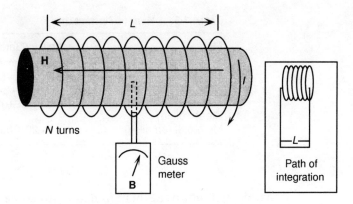

FIGURE 3-6. PERMEABILITY EXPERIMENT. The probe of the gaussmeter penetrates the interior of the sample.

Figure 3-6 illustrates an experiment to measure the permeability. The solenoid has N turns, length L, and a current I_{turn} in each turn. The magnetic field intensity, **H**, in the solenoid is calculated from equation (3-2.16):

$$\oint \mathbf{H} \cdot d\mathbf{r} = \iint \nabla \times \mathbf{H} \cdot d\mathbf{S} = \frac{4\pi}{c} \iint \mathbf{J} \cdot d\mathbf{S} = \frac{4\pi}{c} I_{\text{linking loop}}, \quad (3\text{-}3.6)$$

where we choose the path of integration through the rectangular loop shown in Figure 3-6. One side passes through the interior of the solenoid and the other is far from the solenoid where the magnetic field is vanishingly small. Then the path integral vanishes except inside the solenoid where the value is nearly constant. In the interior of the solenoid the H-field is:

$$H = \frac{4\pi N I_{\text{turn}}}{cL}. \quad (3\text{-}3.7)$$

That is, the H-field depends on the current and some measurable constants. Thus, when measuring permeabilities, measurements of the current can replace measurements of the magnetic intensity.

A magnetic sample is placed in the solenoid and the magnetic induction, **B**, is measured and the permeability determined. For diamagnetic and paramagnetic materials the permeability is nearly constant (although it may depend on temperature and frequency). Table 3-3 lists susceptibilities for some examples of magnetic materials. Those with negative susceptibilities are diamagnetic, those with small positive values are paramagnetic, and those with large values are ferromagnetic. The magnetic properties of materials are very complex and thorough understanding of them requires extensive application of the theory of

TABLE 3-3.   MAGNETIC SUSCEPTIBILITY

Material	Susceptibility (Gaussian units)
Copper	-5.46×10^{-6}
Oxygen (293° K)	$+3449 \times 10^{-6}$
Aluminum	$+16.5 \times 10^{-6}$
Steel (rolled)	$+180-2000$
Mu metal (an alloy)	$+20,000-100,000$

solids. We will not treat them in detail but some general discussion of the enormous range of values of the permeability is in order.

Now consider the *magnetization curve* for a ferromagnetic sample. See Figure 3-7. Assume the sample is initially unmagnetized. Then, as the applied H-field increases, the B-field initially rises roughly proportional to the applied field as would be expected for a diamagnetic sample, although it is much steeper. However as the applied field increases the slope decreases, and then becomes roughly linear again at very high fields (called the saturation region). For very small fields (far below the "knee"), the curve is roughly reversible during the initial magnetization, i.e., as the applied field is reduced to zero again, the same curve is followed in reverse.

After the sample is magnetized to saturation, the curve is no longer reversible, i.e., it does not follow the initial magnetization curve at all. Instead it follows the indicated curve until the applied field goes to zero. At this point the sample has a residual magnetization, B_r, called the *remanence*. If the applied field is now applied in the opposite direction, the measured induction **B** can be driven to zero and back to saturation in the opposite direction. The field necessary to drive the induction to zero, H_c, is called the *coercive force*. If the sample is driven back and forth from saturation to saturation, it follows the outer curve called

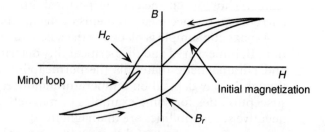

FIGURE 3-7. MAGNETIZATION CURVE FOR FERROMAGNETIC MATERIALS. This hysteresis loop shows the initial magnetization curve, the saturation hysteresis loop and a minor loop.

the *saturation hysteresis curve*. However, suppose the applied field is reversed at some other point than saturation. The sample then follows a *minor hysteresis curve* as shown in the diagram. The saturation hysteresis curve is, in fact, the envelope of all the minor curves including the initial curve. In general, a sample treated this way and removed from the inducing field will have a permanent magnetization. It is possible, by following one of the minor curves, to reduce the residual magnetization to zero, but one must know where to start on the saturation curve!

The shape of the hysteresis curve depends on the material. Two opposite extremes are the high silicon steel used in transformer cores, and the ferrites used in certain kinds of computer memory. The square shape of the ferrite curve ensures that the remanence is large and has a sharp transition between the high and low states. This makes them useful as an information storage mechanism. The very narrow curve (i.e., B is nearly a single-valued function of H) of the silicon steel is nearly reversible, but not linear. Since the energy converted to heat during a cycle is proportional to the area inside the loop, this material provides a high efficiency transformer core. The physical explanation for the heat loss is that hysteresis is intrinsically nonconservative. The energy balance in electromagnetic fields will be treated systematically in Chapter 6. The physical explanation of the hysteresis curve lies in the behavior of small regions of uniform magnetization called *magnetic domains*. However, the ultimate cause lies in the correlation of electron spins, which needs a quantum mechanical treatment, and will not be discussed further here. There is an analogous electric effect in certain materials called *ferroelectricity*. These effects are discussed in detail in texts on solid-state physics.

The preceding discussion of ferromagnetism shows that one cannot write an equation for the magnetization dependence like equation (3-3.1). In general, the permeability of ferromagnetic materials is multivalued and is also temperature and frequency dependent. On the saturation curve it may be useful to define an instantaneous value for the permeability:

$$\mu = \frac{dB}{dH}, \tag{3-3.8}$$

but even this cannot apply to the minor loops in the interior.

Although the four fields, **E**, **D**, **B**, and **H**, contain twelve independent quantities (their components), Maxwell's equations, when written in component form, provide only eight conditions. Therefore a unique determination of the fields requires additional information about the relationship between the four fields. Together with the Maxwell equations, the extra constraints, called *constitutive equations*, make a mathematically complete set of equations. The constitutive equations express the macroscopic properties of the medium which result from its microscopic structure: its molecular and crystal structure including impurities and imperfections, its temperature and frequency dependence, and sample history. On the atomic level, the electric and magnetic susceptibilities reduce to zero, making E and D identical and, likewise, B and H. At this level, ohmic currents also reduce to the motion of individual particles. Calculation and measurement of the susceptibilities and conductance is an important specialty of condensed matter physics.

Assume for the moment that the constitutive equations represent linear, but not necessarily isotropic materials. Then the complete set of *electromagnetic field equations* for the medium can be written:

$$\nabla \cdot \mathbf{D} = 4\pi \rho_{\text{true}}, \tag{3-3.9}$$

$$\nabla \cdot \mathbf{B} = 0, \tag{3-3.10}$$

$$\nabla \times \mathbf{E} + \frac{1}{c}\frac{\partial \mathbf{B}}{\partial t} = 0, \tag{3-3.11}$$

$$\nabla \times \mathbf{H} - \frac{1}{c}\frac{\partial \mathbf{D}}{\partial t} = \frac{4\pi}{c}\mathbf{J}_{\text{true}}, \tag{3-3.12}$$

$$\mathbf{D} = \varepsilon\mathbf{E}, \tag{3-3.13}$$

$$\mathbf{B} = \mu\mathbf{H}, \tag{3-3.14}$$

$$\mathbf{J}_{\text{ohmic}} = \sigma\mathbf{E}. \tag{3-3.15}$$

It is frequently useful to replace the electromagnetic fields by their potentials ϕ and $\mathbf{A}$ in the field equations. Chapters 5, 6, and 8 are devoted to solutions of the field equations.

3-4 Boundary Conditions

Solving the field equations includes applying appropriate boundary conditions. There are both mathematical and physical aspects to boundary conditions. We discuss the physical aspects here, and the mathematical aspects in the next section.

In many cases it may be convenient to break up the region of solution into a small number of subregions and find separate solutions in each. This is useful, for example, when ε and μ in the constitutive equations are piecewise constant. Similar methods of solution of the field equations may be used in each region, but the parameters change when going across the boundary, and one must consider how to relate the solution in one region to those in the adjacent region. Because the equations are of second order, two conditions apply at each boundary. Consider the displacement field first. See Figure 3-8.

Take a small imaginary cylinder (Gaussian Pillbox) bisected by the boundary surface between two dielectrics, ε_1 and ε_1. Make the whole pillbox small enough that the electric field is effectively constant over its top and bottom areas. Also make the thickness of the pillbox so small that its curved area (its side) is negligible compared to its top and bottom areas. Suppose also there is true surface charge σ distributed on the boundary surface. See Figure 3-8.

Integrating equation (3-3.9) gives:

$$\oiint \mathbf{D} \cdot d\mathbf{S} = \iint_{\text{top}} \mathbf{D} \cdot d\mathbf{S} + \iint_{\text{bottom}} \mathbf{D} \cdot d\mathbf{S} + \text{side}$$

$$\approx (\mathbf{D}_2 - \mathbf{D}_1) \cdot \Delta\mathbf{S}$$

$$= 4\pi \Delta q_{\text{true}} \tag{3-4.1}$$

supposing the flux through the side is negligible. Define the unit normal to the surface:

$$\mathbf{n} = \frac{d\mathbf{S}}{|d\mathbf{S}|}. \tag{3-4.2}$$

The displacement boundary condition becomes:

$$(\mathbf{D}_2 - \mathbf{D}_1) \cdot \mathbf{n} = 4\pi\sigma_{\text{true}}. \tag{3-4.3}$$

The interpretation of this equation is: the component of $\mathbf{D}$ normal to the boundary surface changes, when going across the surface, by an amount equal to the *true charge density* on the surface at that point.

The first magnetic boundary follows from similar arguments. Thus the divergence equation for the induction:

$$\nabla \cdot \mathbf{B} = 0 \tag{3-4.4}$$

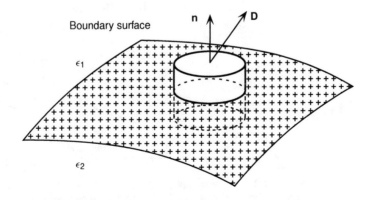

Boundary surface

ϵ_1

ϵ_2

FIGURE 3-8. BOUNDARY SURFACE AND DISPLACEMENT.

becomes:

$$(\mathbf{B}_2 - \mathbf{B}_1) \cdot \mathbf{n} = 0. \tag{3-4.5}$$

The equation for the magnetic intensity is a little more complicated due to the possible presence of true currents in the boundary surface. See Figure 3-9.

Take an imaginary rectangular path of integration, with two long sides parallel to the boundary surface, and two very short sides passing through it and normal to it. Let $\Delta\mathbf{r}$ be the length of the sides of the rectangle above and below the boundary surface, and let Δh be the length of the sides that penetrate the surface. We assume that Δr and Δh are small enough so that the magnetic field is effectively constant in the region. We also assume that $\Delta h \ll \Delta r$. Denote the unit normal to the rectangle by the unit vector $\mathbf{m}$ tangent to the boundary surface. Denote the unit normal to the boundary surface by $\mathbf{n}$. Then:

$$\frac{\Delta\mathbf{r}}{|\Delta\mathbf{r}|} = -\mathbf{n} \times \mathbf{m}. \tag{3-4.6}$$

Thus the three vectors $\mathbf{m}$, $\mathbf{n}$, and $\Delta\mathbf{r}$, are mutually orthogonal. Since $\mathbf{m}$ can point in any direction tangent to the surface, the rectangle has a rotational degree of freedom. The vector quantity $\Delta\mathbf{S}$, representing the area of the loop, is normal to the path of integration. Thus:

$$\Delta\mathbf{S} = \Delta\mathbf{r} \times \Delta h\mathbf{n} = \Delta r\, \Delta h\mathbf{m}. \tag{3-4.7}$$

Now suppose that there is current flowing in the boundary surface. Then define the vector *linear current density* $\mathbf{K}$, representing the current per unit length passing through the rectangle:

$$\mathbf{K} = \frac{\Delta I}{\Delta r}\frac{\mathbf{v}}{|v|} = \sigma\mathbf{v}, \tag{3-4.8}$$

where σ is the boundary surface charge density and $\mathbf{v}$ is its velocity in the surface. Compare this with equations (2-3.6), and (2-3.7), and

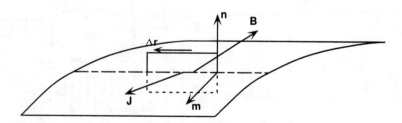

Figure 3-9. Loop of Integration Across Magnetic Boundary. The magnetic vectors are shown here but the integration loop is identical for the electric field. The vectors **m**, **n**, and Δ**r**, are mutually orthogonal, but **B** and **J** have arbitrary directions.

Figure 2-6. In some cases it might be convenient to write:

$$\mathbf{J}(x, y, z) = \rho(x, y, z)\mathbf{v} = \sigma(x, y)\,\delta(z)\mathbf{v} = \mathbf{K}(x, y)\,\delta(z) \tag{3-4.9}$$

for a current in the x–y plane. Recall that the delta function is a linear density.

Then, integrating equation (3-3.12) over this closed rectangular path gives:

$$(\mathbf{H}_2 - \mathbf{H}_1) \times \mathbf{n} \cdot \mathbf{m}\,\Delta r = -\frac{4\pi}{c} \iint_{\text{loop}} \mathbf{J}_{\text{true}} \cdot d\mathbf{S} - \frac{1}{c}\frac{\partial \Phi_D}{\partial t}$$

$$= -\frac{4\pi}{c}\mathbf{K}_{\text{true}} \cdot \mathbf{m}\,\Delta r. \tag{3-4.10}$$

The rate of change of displacement flux vanishes in the steady state, but we suppose it vanishes in general because the area of the rectangle can be made vanishingly small. Also, the contribution to the integral from the normal sides of the integration loop are negligible compared to the tangential sides, because they are so much shorter. Since the orientation of the loop is arbitrary, in that $\mathbf{m}$ may lie in any direction tangent to the boundary surface, the second magnetic boundary condition becomes:

$$(\mathbf{H}_2 - \mathbf{H}_1) \times \mathbf{n} = -\frac{4\pi}{c}\mathbf{K}_{\text{true}}. \tag{3-4.11}$$

This equation can be used to calculate the magnetic field for an infinite sheet of current.

We can use equation 3-3.11 to find a second boundary condition for dielectrics. Figure 3-9 indicates the path of integration again, but we now integrate the electric field. The integral becomes:

$$\oint \mathbf{E} \cdot d\mathbf{r} = \iint \nabla \times \mathbf{E} \cdot d\mathbf{S} = -\frac{1}{c}\iint \frac{\partial \mathbf{B}}{\partial t} \cdot d\mathbf{S} = -\frac{1}{c}\frac{\partial \Phi_m}{\partial t}. \tag{3-4.12}$$

The magnetic flux through the rectangle vanishes by arguments similar to those for the electric flux above. Then, reasoning as before:

$$(\mathbf{E}_2 - \mathbf{E}_1) \cdot \mathbf{n} \times \mathbf{m} = (\mathbf{E}_2 - \mathbf{E}_1) \times \mathbf{n} \cdot \mathbf{m} = 0 \tag{3-4.13}$$

and the second boundary condition for the electric field must be:

$$(\mathbf{E}_2 - \mathbf{E}_1) \times \mathbf{n} = 0. \tag{3-4.14}$$

That is, components of the electric field tangent to the boundary surface do not change across the boundary between two dielectrics.

It might be useful to compare the various fields for a very simple case: an infinite permanent polarized slab (an electret) in vacuum. See Figure 3-10. Note that whereas $\mathbf{P}$ and $\mathbf{E}$ vanish outside the electret, $\mathbf{D}$ is continuous across the boundary. We also show the analogous magnetic fields, where $\mathbf{M}$ and $\mathbf{H}$ vanish outside the magnet and $\mathbf{B}$ is continuous

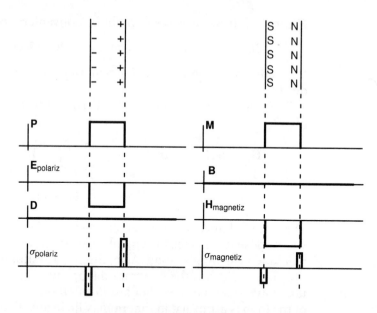

FIGURE 3-10. FIELDS ASSOCIATED WITH PERMANENT DIELECTRICS OR MAGNETS IN VACUUM. The polarization or magnetization vanishes outside the medium, while the displacement or induction is continuous across the boundary and thus vanish everywhere. The *E* and *H* fields shown are due to the permanent polarization or magnetization of the sample, not external fields.

across the boundary. The magnetic surface charge density is not real, of course, but is shown for comparison.

The boundary conditions can be expressed in terms of the potentials:

$$(\varepsilon_2 \nabla\phi_2 - \varepsilon_1 \nabla\phi_1) \cdot \mathbf{n} = 4\pi\sigma_{\text{true}}, \qquad (3\text{-}4.15)$$

$$(\nabla\phi_2 - \nabla\phi_1) \times \mathbf{n} = 0, \qquad (3\text{-}4.16)$$

$$(\nabla \times \mathbf{A}_2 - \nabla \times \mathbf{A}_1) \cdot \mathbf{n} = 0, \qquad (3\text{-}4.17)$$

$$\left(\frac{1}{\mu_2} \nabla \times \mathbf{A}_2 - \frac{1}{\mu_1} \nabla \times \mathbf{A}_1\right) \times \mathbf{n} = -\frac{4\pi}{c} \mathbf{K}_{\text{true}}. \qquad (3\text{-}4.18)$$

3-5 SOME MATHEMATICAL ASPECTS OF BOUNDARY CONDITIONS

Inspection shows that the electromagnetic equations are coupled, first-order, linear, partial differential equations. However, Maxwell's equations can be rewritten as uncoupled, second-order equations (by using the potentials, for example). As is the case for all differential equations, boundary values of the fields are required for unique solutions. Unfortunately, the treatment of boundary values for partial differential equations is considerably more difficult than for ordinary differential equations. The reason for this is, of course, that for partial differential equations, it is necessary to specify certain *functions* on a boundary *surface* rather than merely initial *values* at some initial *point*.

For our purposes, it is convenient to consider the second-order equations for the potentials. Since the scalar and vector potentials obey similar equations, we will consider only the scalar potential. The gauge condition adds an extra level of complexity to finding the vector potential, since the equations for the components are coupled equations. We will begin with the Laplace equation which is relatively simple and of great importance.

From familiarity with second-order ordinary differential equations, one might expect that both the potential and its first derivative (gradient) need to be specified on the boundary surface. Two arbitrary constants occur in the solution of a second-order ordinary differential equation. Then, for a specific problem, these constants are evaluated by applying auxiliary conditions to the solution. The usual conditions, called *Cauchy conditions*, are to specify the value of the function and its derivative at some initial point. Other alternatives exist: one can apply *Dirichlet conditions*, i.e., specify the value of the variable at the initial and end points on the range of the variable. A third possibility is to apply *Neumann conditions*, i.e., specify the values of the first derivative of the variable at the initial and end points. When considering partial differential equations, the boundaries become hypersurfaces and the boundary conditions become functions specified on the boundary surfaces. But the boundary surfaces may be closed, or open, or may extend to infinity, and it is not clear that one can choose boundary conditions arbitrarily. It is conceivable that Cauchy boundary conditions may overdetermine the solution. A mixed boundary condition is also possible, where the potential is given on part of the surface, and the normal component of its gradient on the rest.

For the Neumann boundary condition, *only the normal component of the gradient is given*, because choosing components of the gradient tangent to the surface is equivalent to giving the potential there by integration. We will denote the normal derivative by the symbol, $N(\phi)$:

$$N(\phi) = \nabla\phi \cdot \mathbf{n}. \tag{3-5.1}$$

We can prove that either the Dirichlet or Neumann boundary conditions gives a unique solution to the Laplace equation. Suppose there are two solutions, ϕ_1 and ϕ_2. Then form the quantity:

$$\mathbf{V} = (\phi_1 - \phi_2)\nabla(\phi_1 - \phi_2), \tag{3-5.2}$$

which does not vanish a priori. Applying Green's theorem gives:

$$\iiint \left[(\nabla(\phi_1 - \phi_2))^2 + (\phi_1 - \phi_2)\nabla^2(\phi_1 - \phi_2) \right] d^3r$$

$$= \oiint \left[(\phi_1 - \phi_2)\nabla(\phi_1 - \phi_2) \right] \cdot d\mathbf{S}. \tag{3-5.3}$$

Let the closed surface be the boundary surface. The surface integral vanishes when either the two potentials are equal on the surface, or the normal component of their gradients are equal on the surface. These are the Dirichlet and Neumann conditions, respectively. Because ϕ_1 and ϕ_2 are solutions, the second term in the volume integral vanishes, leaving just:

$$\iiint \left[\nabla(\phi_1 - \phi_2) \right]^2 d^3r = 0. \tag{3-5.4}$$

Since the integrand is positive or zero everywhere, and the region of integration is arbitrary, the integrand must vanish, implying that the two solutions can differ only by a constant. But a constant is a trivial solution and, since Laplace's equation is linear, we can subtract it off. Then ϕ_1 and ϕ_2 must be identical everywhere, including the boundary:

$$\phi_1 = \phi_2. \tag{3-5.5}$$

Thus the solution is unique.

This theorem has great practical importance. Any two solutions, no matter how they are found or how they are expressed, will be identical if they satisfy the boundary conditions. We are therefore free to apply many different mathematical techniques with confidence that the result will be unique.

A general solution can be found by applying Green's theorem:

$$\iiint (\Psi\nabla^2\Phi - \Phi\nabla^2\Psi) \, d^3r = \oiint (\Psi\nabla\Phi - \Phi\nabla\Psi) \cdot d\mathbf{S}. \tag{3-5.6}$$

Then let:

$$\Phi = \phi(r_s), \tag{3-5.7}$$

which is a function of the source variable, and:

$$\Psi = \frac{-1}{4\pi} \frac{1}{|\mathbf{r}_{\mathrm{fs}}|}. \tag{3-5.8}$$

Using Laplace's equation, the δ-function, and integrating over the source variable r_s gives:

$$\phi(r_f) = -\frac{1}{4\pi} \oiint \left[\phi(r_s)\nabla_s \frac{1}{|r_{fs}|} - \frac{1}{|r_{fs}|}\nabla_s\phi(r_s) \right] \cdot d\mathbf{S}_s. \qquad (3\text{-}5.9)$$

When the integration is over all space, the surface integral is at infinity where the integrand vanishes; then the potential vanishes everywhere.

This looks like a practical solution; all one needs to do is integrate. However, there are some problems with the boundary conditions. Because $1/r$ is a solution of the Laplace equation (except at $r = 0$), either term in the integral is a solution separately, with Dirichlet boundary conditions for the first term and Neumann boundary conditions for the second. Thus equation (3-5.9) actually implies a Cauchy boundary condition. But the Laplace equation actually only requires Dirichlet or Neumann conditions on closed boundaries. To see this, consider that an electric field is determined merely by placing charges on a set of conductors. The potential on the conductors completely determines the potential in the intervening space—the normal derivative is determined by specifying the potential. Thus, Cauchy conditions may overdetermine the solution. The following example illustrates some of the difficulties.

EXAMPLE

Find the potential in the interior of a split sphere where the top half of the sphere is at potential ϕ_1, and the bottom half is at potential ϕ_2. See Figure 3-11. The first term:

$$\phi(x_f) = -\oiint \phi(x_s)\nabla_s \frac{1}{|x_{fs}|} \cdot d\mathbf{S}_s$$

$$= \oint \phi \, d\Omega \qquad (3\text{-}5.10)$$

is a solution of the Laplace equation, as we have seen. Evaluating this integral over 4π steradians gives:

$$\phi(x_f) = \frac{1}{2}(\phi_1 + \phi_2) \qquad (3\text{-}5.11)$$

at the center of the sphere, which is reasonable. However, point A on the top edge of the sphere has potential ϕ_2. But the integral in equation (3-5.10) requires that the potential at point A is:

$$\phi(x_f) = a\phi_1 + b\phi_2, \qquad (3\text{-}5.12)$$

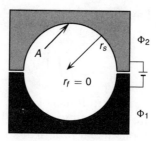

FIGURE 3-11. The potentials on the split sphere determine the potential in the interior. At the center, the potential is $\frac{1}{2}(\phi_1 + \phi_2)$ as expected. However, on a boundary point, such as point A, Cauchy boundary conditions present certain problems.

where neither a nor b vanish since point A sees some contribution from the lower solid angle. This is contradictory! The difficulty can be resolved only by including the second, gradient term, where $N(\phi)$ has values that give the correct potential on the boundary. The difficulty, for purposes of calculation, is that we are not free to choose the gradient, and we do not know what it is beforehand. Instead of being a solution to the problem, equation (3-5.9) turns out to be merely an integral equation for the problem. Rather disappointing!

For an ordinary differential equation, it is sometimes possible to replace Cauchy initial conditions by Dirichlet boundary conditions. For example, the motion of a stream of water from a hose can be expressed by a differential equation with Cauchy boundary conditions—its initial height and slope. Then, by adjusting the angle of the hose, a firefighter can spray water on a specific point on a wall. That is, the angle at the initial point determines the height at the end point, and Dirichlet conditions replace Cauchy conditions.

This firefighter technique can sometimes be generalized to partial differential equations, provided the expression for the solution contains arbitrary functions for the solution on the boundary, and its normal derivative. Consider Laplace's equation in two dimensions. First, assume we know a solution of the equation, ϕ, with adjustable functions ϕ_s and $N(\phi(s))$ on the boundary curve. Second, split the boundary curve into two parts. Third, adjust $N(\phi(s))$ *on the first boundary*, so that the solution ϕ *on the second boundary* equals the boundary function, i.e., $\phi = \phi(s)$. In the process of determining ϕ on the second boundary, $N(\phi)$ is determined on the first boundary. Then, on the entire boundary curve, the solution equals the boundary value (Dirichlet conditions), while $N(\phi)$ takes on *whatever value is necessary to achieve it*. Since ϕ is a solution, it is a unique solution, as we saw above.

From this point of view, we consider equation (3-5.9) to be an integral equation for the normal derivative. Assume that $\phi(s)$ is specified; then on the boundary:

$$\oiint \left[\frac{1}{|r_{fs}|} N(\phi(r_s)) \right] d\mathbf{S}_s = \phi(s) + \oiint \left[\phi(s) \nabla_s \frac{1}{|r_{fs}|} \right] \cdot d\mathbf{S}_s$$

$$= F(s), \tag{3-5.13}$$

where $F(s)$ is determined explicitly by performing the integral on the right. One can now solve this integral equation for $N(\phi)$ on the boundary.

The preceding discussion makes it clear that applying boundary conditions is not simple for partial differential equations. Some general discussion might be helpful to see what the possibilities are. The theory of characteristics can be applied to the linear, second-order, partial differential equations of electromagnetism. We give a standard treatment of the method. Mathews and Walker, and especially Myint-U, give more detailed treatments. We will restrict our discussion to two dimensions because the notation in three or more dimensions tends to obscure understanding. However, the generalization is straightforward. A linear second-order differential equation in two dimensions can be expressed by:

$$A\partial_x^2 \phi + 2B\partial_x\partial_y\phi + C\partial_y^2\phi = f(\partial_x\phi, \partial_y\phi, \phi, x, y), \tag{3-5.14}$$

where A, B, and C are functions of x and y. In Cartesian coordinates, they are constants for the equations discussed here, i.e., Laplace's equation and the D'Alembert wave equation.

A two-dimensional boundary curve can be expressed in parametric form:

$$\mathbf{r} = [x(s), y(s)]. \tag{3-5.15}$$

The pathlength is often a useful parameter:

$$ds^2 = dx^2 + dy^2. \tag{3-5.16}$$

The differential tangent and normal vectors for the boundary curve:

$$d\mathbf{t} = \frac{(dx, dy)}{\sqrt{dx^2 + dy^2}}, \tag{3-5.17}$$

$$d\mathbf{n} = \frac{(-dy, dx)}{\sqrt{dx^2 + dy^2}}, \tag{3-5.18}$$

are orthogonal to each other. On the boundary curve we calculate the derivatives:

$$d\partial_x\phi = \partial_x^2\phi \, dx + \partial_x\partial_y\phi \, dy, \tag{3-5.19a}$$

$$d\partial_y\phi = \partial_x\partial_y\phi\,dx + \partial_y^2\phi\,dy. \tag{3-5.19b}$$

Now let us treat the second derivatives as independent algebraic quantities on the boundary curve. Then we can solve equations (3-5.14) and (3-5.19) simultaneously for these second derivatives as functions of ϕ and $N(\phi)$, subject to the usual conditions on simultaneous linear equations. That is, if the discriminant:

$$D = \begin{vmatrix} dx & dy & 0 \\ 0 & dx & dy \\ A & 2B & C \end{vmatrix} = A(dy)^2 - 2B\,dx\,dy + C(dx)^2 \tag{3-5.20}$$

does not vanish there are solutions for the second derivatives. Similarly, the third derivatives can be found by differentiations, and ultimately all the derivatives can be expressed in terms of ϕ and $N(\phi)$ *on the boundary*. Then the function ϕ can be expanded as a Taylor series for all values of x and y, where the Taylor coefficients are proportional to the derivatives evaluated on the boundary. Therefore the solution depends on ϕ and $N(\phi)$ on the boundary, which is the Cauchy boundary condition.

But suppose the discriminant vanishes; then the equation:

$$A(dy)^2 - 2B\,dx\,dy + C(dx)^2 = 0 \tag{3-5.21}$$

determines a set of curves called *characteristics*. When A, B, and C are constants, the equation can be classified as elliptic, parabolic, or hyperbolic with corresponding characteristics. Examples of the three types of equation are the Laplace equation, the diffusion equation, and the D'Alembert wave equation, respectively. In order for Cauchy boundary conditions to hold, *the boundary must not be tangent to a characteristic*. Consequently, a consistent boundary condition depends on the classification of the equation and the choice of the boundary surface. Tables 3-4 and 3-5 classify the linear second-order equations and the allowed boundary conditions.

Table 3-4. **Classification of Differential Equations:** A, B, and C are constants

Classification	Condition	Example in two dimensions	
Elliptic:	$B^2 < AC$	Laplace:	$\partial_x^2\phi + \partial_y^2\phi = 0$
Parabolic:	$B^2 = AC$	Diffusion:	$\partial_x^2\phi - \partial_t\phi = 0$
Hyperbolic:	$B^2 > AC$	D'Alembert:	$\partial_t^2\phi - \partial_x^2\phi = 0$

TABLE 3-5. ALLOWED BOUNDARY CONDITIONS

Equation type	Example	Boundary type	Condition
Elliptic:	Laplace	Closed	Dirichlet or Neumann
Parabolic:	Diffusion	Open	Dirichlet or Neumann
Hyperbolic:	D'Alembert	Open	Cauchy

EXAMPLE

The wave equation characteristics are given by:

$$\left(\frac{dx}{dt}\right)^2 - c^2 = 0. \tag{3-5.22}$$

Thus the characteristics are straight lines, $x = \pm ct + x_0$, and it is possible to choose an open boundary which is nowhere tangent to any of these lines. See Figure 3.12. For example, an acceptable Cauchy boundary would be any straight line with any slope between, but not including $\pm c$. The solution to the wave equation is:

$$\phi = F(x - ct) + G(x + ct), \tag{3-5.23}$$

where F and G are arbitrary, independent functions. Since each function is defined on the boundary, there are two independent boundary conditions. This is equivalent to defining ϕ and its normal derivative $N(\phi)$ on the boundary, which are Cauchy boundary conditions. By contrast, a closed boundary with no discontinuities in its slope (i.e., no sharp corners) would be tangent to the characteristics at four points at

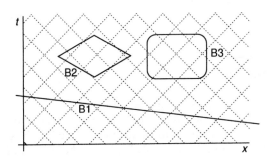

FIGURE 3-12. CHARACTERISTICS AND BOUNDARIES FOR THE WAVE EQUATION. The characteristics have slope ±1. Since boundary B1 has a smaller slope, it is an acceptable boundary. Boundary B2 has sharp corners and therefore undefined normal derivatives. Boundary B3 is tangent to a characteristic at each corner. A curved boundary would be allowed if its slope remains inside the limits of the slopes of the characteristics.

least. For a closed boundary with sharp corners, the normal derivative would not be well defined at the corners. Causality also poses a problem for a closed boundary, since the field would propagate back from the future. For these reasons, a closed hypersurface boundary is not appropriate for the wave equation.

For more information on the topic of boundary conditions, the reader can consult Mathews and Walker, Morse and Feshbach, or other advanced textbooks on mathematical methods of physics. Table 3-5 shows the allowed boundary conditions for the three kinds of linear second-order equations. We have not considered parabolic equations.

3-6 SELECTED BIBLIOGRAPHY

J. Jackson, *Classical Electrodynamics*, Wiley, 1962.

C. Kittel, *Introduction to Solid State Physics*, 3rd ed., Wiley, 1967.

L. Landau and E. Lifshitz, *The Electrodynamics of Continuous Media*, 2nd ed., Pergamon Press, 1975.

D. Langenberg, "Resource Letter OEPM-1 on the Ordinary Electronic Properties of Metals", Amer. J. Phys., **36**, 1, 1968.

J. Mathews and R. Walker, *Mathematical Methods of Physics*, 2nd ed., Addison-Wesley, 1970.

P. Morse and H. Feshbach, *Methods of Theoretical Physics*, McGraw-Hill, 1953.

T. Myint-U, *Partial Differential Equations of Mathematical Physics*, Elsevier North-Holland, 1980.

W. Smythe, *Static and Dynamic Electricity*, 2nd ed., McGraw-Hill, 1950.

A. Sommerfeld, *Partial Differential Equations in Physics*, Academic Press, 1967.

J. Stratton, *Electromagnetic Theory*, McGraw-Hill, 1941.

A. Webster, *Partial Differential Equations of Mathematical Physics*, 2nd ed., Dover, 1955.

3-7 PROBLEMS

1. Estimate the maximum magnetization of a crystal composed of hydrogen-like atoms arranged in a cubic array with distance B between atoms. Let the electrons move with velocity v in orbits in x–y planes of radius, R. What would the magnetic field be in the interior of this crystal if $B = 10$ Angstrom, $R = 1.0$ Angstrom, and $v = 0.01c$? Is this a realistic value?

2. Prove equation (3-2.9) for the magnetic dipole moment of a current loop.

3. Let the classical magnetic dipole moment of the electron be μ. Suppose the electron is a sphere of constant charge density ρ of radius R rotating with angular velocity ω.

A. Find the tangential velocity on the equator of the electron in terms of these parameters.

B. Assume that the mass of the electron is entirely due to the electrostatic energy of the charge confined to the sphere. Calculate the radius of the sphere in terms of the mass of the electron. (Hint: Use $E = mc^2$.)

C. From the known electron magnetic dipole moment, mass, charge, and spin, calculate the tangential velocity of the electronic sphere and compare this to the speed of light. Is this classical model reasonable?

4. Calculate the force on a particle with magnetic dipole moment μ in a nonuniform magnetic field. Let the field gradient in the z-direction be $G_z = dB/dz$. Discuss how this force can be used to separate particles with different magnetic moments in a particle beam experiment.

5. Suppose there is a uniform sheet of charge density σ lying in the x–y plane. The charged sheet lies in the x–y plane surrounded by vacuum. Use equation (3-4.3) to calculate the electric field at a point z meters above the sheet.

6. Suppose there is a uniform sheet of current density **K** directed parallel to the x–axis. The conducting sheet is in vacuum and lies in the x–y plane. Use equation (3-4.11) to calculate the magnetic field at a point z meters above the sheet.

CHAPTER 4

Electromagnetic Equations

I . . . keep my mind free so as to give up any
hypothesis as soon as the facts are shown to be
opposed to it.

Charles Darwin
Isaac Asimov's Book of Science and Nature Quotations

. . . it is more important to have beauty in
one's equations than to have them fit experiment.
[Since further developments may clear up the
discrepancy].

Paul Adrien Maurice Dirac
Scientific American, May 1963.

Einstein's paper, "On the Electrodynamics of Moving Bodies" published
in 1905, was the first of a series of papers revolutionizing our understanding of space and time and the kinematics of particles. It is
interesting that relativity, which is often treated primarily as a theory
of mechanics, has electromagnetism as its primary source.

In this chapter we introduce the relativistic form of Maxwell's equations. In our approach, we take a local Lorentz transformation of the
velocity vector, and show that this transformation is conformable with
the equation of motion of a charged particle in an electromagnetic
field. The use of local transformations has become a basic tool for the
treatment of the quantized fields of elementary particles; it is therefore
useful to see how it can be used in a classical application, free from the
impedimenta of quantization. We begin with a simple introduction to
tensor notation, sufficient to handle the field equations; for the most
part, the tensors used here may be treated just as 4 × 4 matrices. An

introduction to relativistic mechanics follows, and then the discussion of the electromagnetic fields, potentials, and the field equations.

4-1 TENSOR NOTATION

In Chapter 1 we defined a vector as an array of quantities with a single index, obeying a linear transformation. A matrix is a generalization of this with two or more indices. For example, if A^{jk} and B^{jk} are matrices, then T^{jk} is also a matrix:

$$T^{jk} = aA^{jk} + bB^{jk}. \tag{4-1.1}$$

As with a vector, multiplication by the scalar, a, means multiplying each element by a. Rank 2 matrices are usually displayed by a two-dimensional array. For example, a 2×3 matrix might be:

$$M^{jk} = \begin{bmatrix} m^{11} & m^{12} & m^{13} \\ m^{21} & m^{22} & m^{23} \end{bmatrix}, \tag{4-1.2}$$

where the j-index runs along columns and the k-index runs along rows. We will use column, row, or square matrices exclusively. A matrix can be formed by the *outer product* of two row matrices, two column matrices, or a row and a column matrix. Examples of row–column and row–row products are:

$$A^j \otimes B_k = \begin{bmatrix} a^1 \\ a^2 \end{bmatrix} \otimes [b_1 \quad b_2] = \begin{bmatrix} a^1 b_1 & a^1 b_2 \\ a^2 b_1 & a^2 b_2 \end{bmatrix}, \tag{4-1.3}$$

$$A_j \otimes B_k = [a_1 b_1, a_1 b_2, a_2 b_1, a_2 b_2]. \tag{4-1.4}$$

Column–column matrices are very awkward to illustrate, and will often be expressed in row–column form for *purposes of illustration*.

In the familiar three-dimensional Euclidean space, vectors may be expressed interchangeably as either row or column matrices. However, for relativity, we need a modification. In Section 1-4 we defined two types of vectors, covariant and contravariant, where by convention covariant vectors are represented by row matrices and contravariant vectors by column matrices. Further, for a column or row matrix to be a vector, it must obey a transformation condition in going from one coordinate system to a new coordinate system. Either its components transform by the linear combination:

$$\bar{V}^k = \sum_{j=1}^{4} \frac{\partial \bar{x}^k}{\partial x^j} V^j = \frac{\partial \bar{x}^k}{\partial x^j} V^j \tag{4-1.5}$$

for contravariant vectors, or:

$$\bar{V}_k = \sum_{j=1}^{4} \frac{\partial x^j}{\partial \bar{x}^k} V_j = \frac{\partial x^j}{\partial \bar{x}^k} V_j \tag{4-1.6}$$

for covariant vectors, where we use the Einstein summation convention on repeated indices. Then the transformation rule for the outer product of two contravariant vectors is:

$$\bar{A}^j \bar{B}^k = \frac{\partial \bar{x}^j}{\partial x^p} \frac{\partial \bar{x}^k}{\partial x^q} A^p B^q. \tag{4-1.7}$$

But the outer product of two vectors is a special case of a *second rank tensor*. A general second rank contravariant tensor obeys the same transformation rule:

$$\bar{T}^{jk} = \frac{\partial \bar{x}^j}{\partial x^p} \frac{\partial \bar{x}^k}{\partial x^q} T^{pq}. \tag{4-1.8}$$

Corresponding transformation rules hold for covariant and mixed tensors, with appropriate transformation coefficients and for higher rank tensors.

The inner product is the *trace* of a product tensor with mixed indices:

$$s = A^j B_j. \tag{4-1.9}$$

It is invariant under a coordinate transformation. Thus:

$$\bar{A}^j \bar{B}_j = \frac{\partial \bar{x}^j}{\partial x^k} \frac{\partial x^n}{\partial \bar{x}^j} A^k B_n = A^k B_k, \tag{4-1.10}$$

since:

$$\frac{\partial \bar{x}^j}{\partial x^k} \frac{\partial x^n}{\partial \bar{x}^j} = \frac{\partial x^n}{\partial x^k} = \delta_k^n. \tag{4-1.11}$$

The invariance only holds for mixed indices.

The two kinds of vectors, contravariant and covariant, have similar properties, differing only in their transformation. However, since their inner product is invariant, their transformations must be in some sense reciprocal—if one increments the other decrements to compensate. Thus there must be a transformation to convert a contravariant vector into a covariant vector, and vice versa. That is, an operator and its inverse:

$$V_j = g_{jk} V^k, \tag{4-1.12}$$

$$V^j = g^{jk} V_k, \tag{4-1.13}$$

obey the condition:

$$g^{jk} g_{kn} = \delta_n^j, \tag{4-1.14}$$

since raising and then lowering the index restores the original vector. The transformation is linear, as is appropriate for vectors. The tensor g_{jk} is called the *metric tensor*, and we shall call the transformed vector V_j the *associated vector*. Then, by transforming to the associated vector, an invariant inner product of either two covariant vectors or two contravariant vectors may be obtained. The metric tensor in mixed form is the Kronecker delta symbol, as can be seen by lowering an index and using equation (4-1.14) above:

$$g_k^j = g^{jm}g_{mk} = \delta_k^j. \tag{4-1.15}$$

This is not a trivial result since it holds in any coordinate system, including those in which the elements of the metric tensor are not constant. The Kronecker delta has the same form in any coordinate system. The contravariant and covariant metric tensors are usually displayed in row–column form for compactness. Similarly, the metric tensor raises or lowers an index of a tensor. Raising or lowering acts only on a specific index, not on all indices of the tensor. For example:

$$g_{pm}T^{jkmn} = T^{jk}{}_p{}^n. \tag{4-1.16}$$

Only the third index is lowered. The *metric*, defined as the magnitude of the displacement vector. That is:

$$ds^2 = dx^j\, dx_j$$

$$= g^{jk}\, dx_j\, dx_k$$

$$= g_{jk}\, dx^j\, dx^k. \tag{4-1.17}$$

Example

In the three-dimensional Euclidean space with which we are familiar, the metric tensor is just the Kronecker delta:

$$g_{jk} = \delta_{jk} = \begin{bmatrix} 1 & 0 & 0 \\ 0 & 1 & 0 \\ 0 & 0 & 1 \end{bmatrix}, \tag{4-1.18}$$

$$ds^2 = dx_1^2 + dx_2^2 + dx_3^2. \tag{4-1.19}$$

For this reason, in Euclidean spaces, covariant vectors are identical to contravariant vectors.

EXAMPLE

In the four-dimensional *Minkowski space* the metric tensor and the corresponding Minkowski metric are:

$$g_{jk} = g^{jk} = \begin{bmatrix} 1 & 0 & 0 & 0 \\ 0 & -1 & 0 & 0 \\ 0 & 0 & -1 & 0 \\ 0 & 0 & 0 & -1 \end{bmatrix} \quad \text{Minkowskian four dimensional,}$$

(4-1.20)

$$ds^2 = dx_0^2 - dx_1^2 - dx_2^2 - dx_3^2. \tag{4-1.21}$$

By contrast, in Minkowski space, covariant vectors differ from contravariant vectors in the sign of the spatial components. Placing the minus sign on the spatial terms is pure convention; other authors put the minus on the zeroth (time) coordinate. Although the two metric tensors appear very similar, the two spaces have fundamentally different properties. For example, the *Minkowski metric is not positive definite* for all pairs of points in the space.

Tensors can have symmetries in their indices. For example, symmetric and antisymmetric tensors are defined by the signs of their elements when the elements are transposed. The *symmetric tensor* is defined by the condition:

$$S_j^k = S_k^j. \tag{4-1.22}$$

Similarly, the *antisymmetric tensor* is defined by:

$$A_j^k = -A_k^j. \tag{4-1.23}$$

Symmetric and antisymmetric $N \times N$ matrices have $N(N + 1)/2$ and $N(N - 1)/2$ independent elements, respectively. Thus a 4×4 antisymmetric matrix has only six independent elements. Diagonal elements of antisymmetric matrices are zero. The metric tensor is symmetric; an example of an antisymmetric tensor is:

$$A_k^j = \begin{bmatrix} 0 & -A_2^1 & A_3^1 \\ A_2^1 & 0 & -A_3^2 \\ -A_3^1 & A_3^2 & 0 \end{bmatrix}. \tag{4-1.24}$$

It is possible to express any tensor as the sum of a symmetric and an antisymmetric tensor:

$$[M_k^j] = \frac{1}{2}[M_k^j + M_j^k] + \frac{1}{2}[M_k^j - M_j^k] = S_k^j + A_k^j. \tag{4-1.25}$$

The following identity is very useful for calculations:

$$A_k^j S_j^k = 0,$$ (4-1.26)

where we sum over both indices. To prove this, apply the transpose to the two matrices. Because we sum over both j and k indices, they are dummy indices, and we can use any symbol for them. Specifically, we may switch them: $j \leftrightarrow k$. Therefore, the sign of the product changes:

$$A_k^j S_j^k = -A_j^k S_k^j = -A_k^j S_j^k.$$ (4-1.27)

Since only the value zero is equal to its negative, the expression must vanish.

The Levi-Civita symbols, ε^{jk}, ε^{jkm}, ε^{jkmn}, ..., are the completely anti-symmetric quantities whose elements are 1, −1, or 0 depending on the permutation of the indices (in Cartesian coordinates). For example, the four index symbol in 4-space has the element:

$$\varepsilon^{0123} = 1$$ (4-1.28)

and the others take values according to the permutations:

$$\varepsilon^{jkmn} = \begin{cases} 1, & \text{even number of permutations of index values} \\ & jkmn = 0123, \\ -1, & \text{odd number of permutations of index values} \\ & jkmn = 0123, \\ 0, & \text{any two (or more) indices have equal values.} \end{cases}$$

(4-1.29)

Similar definitions hold for the other tensors; however, we will use the two tensors ε^{jkm} and ε^{jkmn}.

EXAMPLE

The 3-index Levi-Civita tensor in 3-space can be displayed as a 27 element array as follows:

$$\varepsilon^{jkm} = \begin{bmatrix} 0 & 0 & 0 \\ 0 & 0 & 1 \\ 0 & -1 & 0 \end{bmatrix} \begin{bmatrix} 0 & 0 & -1 \\ 0 & 0 & 0 \\ 1 & 0 & 0 \end{bmatrix} \begin{matrix} m=1 & m=2 & m=3 \\ \begin{bmatrix} 0 & 1 & 0 \\ -1 & 0 & 0 \\ 0 & 0 & 0 \end{bmatrix} \end{matrix} \begin{matrix} k=1 \\ k=2 \\ k=3 \end{matrix}$$

$$j=1 \qquad j=2 \qquad j=3$$

(4-1.30)

It is convenient to think of these three arrays as stacked into a cube.

The Levi-Civita tensors are important in defining the rotational properties of vectors and tensors. Every tensor may have a corresponding *dual tensor* which may be of different rank. For example, in four dimensions the duals for rank 1, 2, and 3 tensors are:

$$T^j = \varepsilon^{jkmn} T_{kmn},$$

$$T^{jk} = \varepsilon^{jkmn} T_{mn}, \qquad (4\text{-}1.31)$$

$$T^{jkm} = \varepsilon^{jkmn} T_n.$$

Only a second rank tensor has a dual of equal rank. In three-dimensional spaces, the dual of the vector has special significance. The dual of the outer product of two vectors is a rank one tensor. We may define a special kind of product of two vectors, the *vector product*:

$$C^j = \frac{1}{2} \varepsilon^{jkm} A_k B_m. \qquad (4\text{-}1.32)$$

The factor 1/2 compensates for summing over two indices; each pair of elements occurs twice. In three-dimensional notation this vector becomes:

$$\mathbf{C} = \mathbf{A} \times \mathbf{B} = \begin{bmatrix} A_y B_z - A_z B_y \\ A_z B_x - A_x B_z \\ A_x B_y - A_y B_z \end{bmatrix}. \qquad (4\text{-}1.33)$$

The product vector, **C**, behaves like a vector under linear transformations such as rotation. In a transformation in which the coordinates change sign (a parity transformation), a vector changes its sign since each of its components change sign:

$$V^j \rightarrow -V^j. \qquad (4\text{-}1.34)$$

However, if the two vectors, **A** and **B**, change sign, the product vector, **C**, does not change sign—it does not have the properties of a vector under parity transformations. It is a *pseudovector or axial vector*. Note that the cross product can only be defined in three-dimensional space (where the dual of a rank two tensor formed by the outer product of two vectors is a vector). Similarly, the inner product of a vector with a pseudovector changes sign under inversion of coordinates, whereas a scalar is invariant. Since it behaves like a scalar except under inversion of coordinates, this product is called a *pseudoscalar*.

Example

Although the prototypical tensor or matrix is a numerical quantity, i.e., each element is a number. However, we may also define various operators as tensors, provided their transformation satisfies the appropriate rules. In this spirit we define several tensor operators: The *gradient operator*:

$$\partial_j = \frac{\partial}{\partial x^j} \tag{4-1.35}$$

written here in Cartesian coordinates, increments the rank of the tensor on which it operates by one covariant index:

$$V_j = \partial_j \Phi = \frac{\partial \Phi}{\partial x^j},$$
$$T_{jk} = \partial_j V_k = \frac{\partial V_k}{\partial x^j}. \tag{4-1.36}$$

The gradient operator has a double nature: it is a differential operator, and it is a vector with an index and can form an inner product with other tensors. In general, the two operations need not act on the same quantity. For example, in the total differential operator:

$$dQ = dx^j \, \partial_j Q \tag{4-1.37}$$

the inner product acts to the left and the differential property acts to the right. Other operators are obtained from the gradient operator by additional operations. In the *divergence* of a vector:

$$\partial_j V^j = \frac{\partial V^j}{\partial x^j},$$
$$\partial_j T^{jk} = \frac{\partial T^{jk}}{\partial x^j}, \tag{4-1.38}$$

written here in Cartesian coordinates, the inner product and the differentiation act on the same quantity. The divergence decrements the rank of the tensor on which it operates by one contravariant index. The *curl* of a vector:

$$\frac{1}{2} \varepsilon^{jkm} \partial_m V_k = \frac{1}{2} \varepsilon^{jkm} \frac{\partial V_k}{\partial x^m} \tag{4-1.39}$$

is explicitly antisymmetric. In three-dimensions it is a pseudovector operator. Two very general identities involve the curl:

$$\operatorname{div} \operatorname{curl} V^j = \frac{1}{2} \left[\frac{\partial}{\partial x^j} \varepsilon^{jkm} \frac{\partial}{\partial x^m} \right] V_k \equiv 0, \tag{4-1.40}$$

$$\operatorname{curl} \operatorname{grad} \Phi = \frac{1}{2} \left[\varepsilon^{jkm} \frac{\partial}{\partial x^m} \frac{\partial}{\partial x^j} \right] \Phi \equiv 0. \tag{4-1.41}$$

These expressions vanish because they involve the double inner product of a symmetric tensor and an antisymmetric tensor (see equation (4-1.26)). The identities also apply to tensors of higher rank.

The scalar operator called the *D' Alembertian*, expressed here in Cartesian coordinates:

$$\partial_j \partial^j = \frac{\partial^2}{\partial x_0^2} - \frac{\partial^2}{\partial x_1^2} - \frac{\partial^2}{\partial x_2^2} - \frac{\partial^2}{\partial x_3^2}$$

$$= \frac{\partial^2}{c^2 \partial t^2} - \nabla^2 \qquad (4\text{-}1.42)$$

is the divergence of the gradient in four-dimensional Minkowski space. In our sign convention, the minus sign is associated with the spatial variables.

4-2   INTEGRAL THEOREMS

There are several integral theorems of great importance to electromagnetism, including the divergence theorem and Stokes's theorem. In order to express them in four-dimensional space, it is necessary to consider the differentials. Whereas in three-dimensional space there are line, surface, and volume integrals, in four-dimensional space there are line, surface, hypersurface, and hypervolume integrals.

The line integral is unchanged in four dimensions except for the additional $(j = 0)$ term. It includes ordinary three-dimensional line integration when the path of integration is at some fixed time, i.e., when $dt = 0$. We may also express this integral in terms of the invariant pathlength ds. The line integral of a vector in Minkowski space is:

$$\int V_j \, dx^j = \int V_0 \, dx_0 - V_1 \, dx_1 - V_2 \, dx_2 - V_3 \, dx_3. \qquad (4\text{-}2.1)$$

The surface element in three-dimensional integration, is taken to be a vector whose magnitude is equal to the area of the surface element, and whose direction is orthogonal to the surface (normally outward for a closed surface). The question here is how to express the surface element in four dimensions, i.e., without using the cross product. What is needed is some quantity whose value is equal to the surface area element and which is orthogonal to it.

In four dimensions, there are two dimensions orthogonal to a two-dimensional surface, and there is no unique orthogonal vector. The dual tensor:

$$dS^{jk} = \frac{1}{2} \varepsilon^{jkmn} \, d\sigma_m \, d\tau_n \qquad (4\text{-}2.2)$$

is the surface element where $d\sigma$ and $d\tau$ are tangent to the surface. This is shown in Figure 4-1. Using (4-1.26), we have:

$$dS^{ik}\, d\sigma_j = d\sigma_j \frac{1}{2}\, \varepsilon^{jkmn}\, d\sigma_m\, d\tau_n = 0, \qquad (4\text{-}2.3)$$

since the tensor, $d\sigma_j\, d\sigma_m\, d\tau_n$, is symmetric and the Levi-Civita tensor is antisymmetric. This shows that the quantities dS^{ik} and $d\sigma_j$ are orthogonal. Then the 2-surface integral in 4-space is:

$$\iint V_j\, dS^{jk} = W^k, \qquad (4\text{-}2.4)$$

which is a vector, in contrast to three-dimensions.

The volume of a parallelopiped in three dimensions, is the triple scalar product of the differential vectors defining its sides:

$$
\begin{aligned}
d^3 r &= \frac{1}{2}\, \varepsilon^{kmn}\, d\sigma_k\, d\tau_m\, d\zeta_n \\
&= d\sigma \cdot d\tau \times d\zeta.
\end{aligned}
\qquad (4\text{-}2.5)
$$

In four dimensions, this becomes the dual of the rank 3 tensor, $d\sigma_k\, d\tau_m\, d\zeta_n$, which is the hypersurface element:

$$dH^i = \frac{1}{6}\, \varepsilon^{ikmn} d\sigma_k\, d\tau_m\, d\zeta_n. \qquad (4\text{-}2.6)$$

The factor 1/6 compensates for the triple sum.

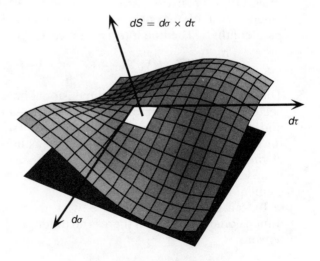

Figure 4-1.   Surface element for curved surface.

EXAMPLE

The vector dH^j is orthogonal to each of its component vectors since:

$$\varepsilon^{jkmn} \, d\sigma_j \, d\sigma_k \, d\tau_m \, d\zeta_n = 0, \tag{4-2.7}$$

analogous to $\mathbf{A} \cdot (\mathbf{A} \times \mathbf{B}) = 0$ in three dimensions. Let the covariant vectors $d\sigma$, $d\tau$, and $d\zeta$ lie in the spatial directions x, y, and z. Then:

$$[d\sigma_k, d\tau_m, d\zeta_n] = [dx, dy, dz] \tag{4-2.8}$$

and dH^0 is the volume element in three dimensions and lies in the t-direction, orthogonal to x, y, z:

$$dH^0 = dx \, dy \, dz = d^3 r. \tag{4-2.9}$$

A *relative tensor* is a generalized tensor defined by the transformation:

$$\bar{T}^{jk} = J^w \frac{\partial \bar{x}^j}{\partial x^p} \frac{\partial \bar{x}^k}{\partial x^q} T^{pq}, \tag{4-2.10}$$

where the *Jacobian* is the determinant formed from the transformation coefficients:

$$J = \left| \frac{\partial x}{\partial \bar{x}} \right| \tag{4-2.11}$$

and w is the *weight* of the relative tensor. By analogy with equation (4-2.5), the four-dimensional invariant hypervolume element is:

$$
\begin{aligned}
d^4 x &= dv_j \, dH^j \\
&= \frac{1}{6} \varepsilon^{jkmn} \, dv_j \, d\tau_k \, d\sigma_m \, d\zeta_n \\
&= J \, d\xi^0 \, d\xi^1 \, d\xi^2 \, d\xi^3,
\end{aligned}
\tag{4-2.12}
$$

where dv_j is the fourth differential vector.

EXAMPLE

The Jacobian is c in Cartesian and $cr^2 \sin \theta$ in spherical coordinates:

$$
\begin{aligned}
d^4 x &= c \, dt \, dx \, dy \, dz && \text{Cartesian,} \\
&= cr^2 \sin \theta \, dt \, dr \, d\theta \, d\phi, && \text{spherical.}
\end{aligned}
\tag{4-1.13}
$$

The integral theorems discussed in Section 1-5 have generalizations to four dimensions. We will not attempt to prove them but the reader can find details of the proofs (which are not especially difficult) in the references. In four-dimensional notation, *Stokes's theorem* becomes:

$$\oint V_j \, dx^j = \frac{1}{2} \iint (\partial_j V_k - \partial_k V_j) \, dS^{jk} \qquad (4\text{-}2.14)$$

The factor of 1/2 normalizes the double sum.

The second identity, the three-dimensional *divergence theorem*, takes the form in four-dimensional tensor notation:

$$\oiint V_j \, dS^{jk} = \iiint \partial^j V_j \, dH^k. \qquad (4\text{-}2.15)$$

Since no sum is performed over the index, k, the integral is a vector; the zeroth component corresponds to the three-dimensional expression where $dH^0 = d^3r$.

The integral identity:

$$\oiiint V_j \, dH^j = \iiiint \partial^j V_j \, d^4r \qquad (4\text{-}2.16)$$

is the four-dimensional generalization of the three-dimensional divergence theorem. Thus, the hypervolume of integration is bounded by the closed three-dimensional hypersurface. Each of these identities also applies to higher rank tensors; indices not summed over are unaffected.

4-3   Relativity: A New Kinematics

We have seen some evidence for the necessity of revising Galilean–Newtonian kinematics; in particular, the Michelson–Morley experiment cannot be ignored. In this section we present an introduction to relativistic kinematics, emphasizing those aspects required for understanding electromagnetism. The references provide more comprehensive treatments.

In Galilean–Newtonian kinematics, particle motion is described by a three-dimensional vector representing the position of the object, and the associated velocity and acceleration. The position of a particle is a parametric function of time, and the laws of motion are differential equations in this parameter. The following equations summarize Newtonian mechanics:

$$F^j(x^k, v^k, t) = ma^j(t),$$

$$a^j(t) = \frac{d^2 x^j(t)}{dt^2}. \qquad (4\text{-}3.1)$$

In principle, it is possible to rewrite these equations with the pathlength as parameter:

$$ds^2 = dx^2 + dy^2 + dz^2. \qquad (4\text{-}3.2)$$

In these terms, Newtonian kinematics *could* be summarized by the equations:

$$F^j(x^k, v^k, s) = ma^j(s),$$

$$a^j(s) = \frac{d^2x^j(s)}{ds^2},$$

(4-3.3)

with the relation between the two parameters:

$$s = s(t).$$

(4-3.4)

Not much is gained by this in three dimensions. However in four dimensions, where time itself becomes a coordinate, choosing s to be the parameter allows all four coordinates to occur symmetrically. This becomes a virtual necessity in relativity.

A salient feature of Newtonian kinematics is what is sometimes called Galilean relativity:

The laws of motion are invariant to a global translation or rotation of the coordinate system or any uniform motion of the coordinate system.

Expressing this in terms of the velocities:

$$\mathbf{V}_2 = \mathbf{V}_1 + \mathbf{V}_{\text{relative}}.$$

(4-3.5)

Specifically, Newton's laws are invariant to adding an arbitrary constant to the velocity since the acceleration is independent of constant displacement or velocity increments. We have encountered this equation previously in the context of the velocity of light seen from a moving reference frame attached to the Earth. The Michelson–Morley experiment shows that the velocity of light in vacuum is independent of the motion of the observer. This specifically contradicts Galilean relativity since equation (4-3.5) implies that the velocity of the light wave *must* depend on the velocity of the reference frame of the observer.

Einstein is famous for using *Gedankenexperimente* or thought experiments as a method of framing questions and testing the logic of hypotheses. Let us follow his method to discover how to modify the kinematics of Galileo and Newton to satisfy the Michelson–Morley experiment. In this spirit, suppose a pulse of light is emitted by a star (say a pulsar) and that two identical space laboratories (determined to be identical when both are at rest) are to measure the velocity of the pulse. The only difference between the observers is that one is moving with velocity v_1, and the other with velocity v_2 (say with respect to the star). The two observers measure the speed by using the length of their space labs and cesium clocks. See Figure 4-2. Then each reports:

$$L_1 = c_1 t_1,$$

(4-3.6)

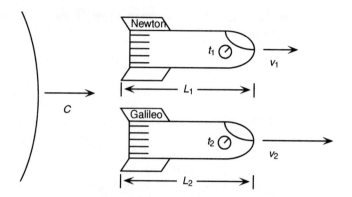

FIGURE 4-2. THOUGHT EXPERIMENT OF THE SPEED OF LIGHT.   Except for their velocities, the two observers are identical. The velocity of light is measured in each case by finding the ratio of the length to the time of passage of the light front.

$$L_2 = c_2 t_2. \tag{4-3.7}$$

If one observer has velocity $0.9999c$ toward the star, Galilean velocity implies that its observed light velocity ought to be almost $2c$. If its velocity is $0.9999c$, away from the star, the light pulse ought to be almost standing still! In fact, in Galilean–Newtonian kinematics the observer could recede from the star at greater than light speed and the light pulse would appear to go backward! However, by Michelson's experiment, they actually find that the two measurements of the speed of light are identical, regardless of their velocities. That is,

$$c_1 = c_2 = c. \tag{4-3.8}$$

(Incidentally, they would also find that they can *not* exceed the speed of light relative to the star or to each other!) It was suggested by FitzGerald in 1892, that the Michelson–Morley results could be explained by assuming that the arm of the interferometer contracts in the direction of motion of the Earth. This idea was adopted and developed by Lorentz to include a modification in the rate of passage of time. Thus one is led to consider some radical ideas about the nature of space and time: observed length and time duration depend on the velocity of the observer. These principles are integrated into Einstein's theory of special relativity.

The result of the thought experiment may be summarized by rewriting equations (4-3.6) and (4-3.7). That is,

$$c^2 t_1^2 - L_1^2 = 0,$$
$$c^2 t_2^2 - L_2^2 = 0. \tag{4-3.9}$$

Let us consider a more general quantity, $M^2 = c^2 t^2 - L^2$, which is not necessarily zero, generalize to three spatial dimensions, and ask what transformations satisfy invariance of the expression. Let:

$$M^2 = L_0^2 - L_1^2 - L_2^2 - L_3^2. \tag{4-3.10}$$

This is the magnitude of a vector in Minkowski space with the metric:

$$ds^2 = dx^j \, dx_j$$

$$= c^2 \, dt^2 - dx^2 - dy^2 - dz^2. \tag{4-3.11}$$

Since the metric defines the separation of any two points in space–time, it applies to all displacements, not merely that of light in vacuum. It can serve as the parameter in the kinematic equations of particles like equations (4-3.3).

Now consider linear transformations of the components that leave the magnitude of this vector invariant. With four components there are six possible linear combinations taken two at a time: $x - y, y - z, z - x$ and $x - t, y - t, z - t$. The rotations therefore have six degrees of freedom. The first three are simply rotations, but the second three are less familiar since they involve time; they are the *Lorentz transformations*. First, consider a rotation in the x–y plane shown in Figure 4-3.

The angles of rotation are related to the sides of a triangle inscribed inside a circle:

$$\tan \theta = y/x, \tag{4-3.12}$$

$$\sin \theta = \frac{y}{r} = \frac{y}{\sqrt{x^2 + y^2}} = \frac{y/x}{\sqrt{1 + (y/x)^2}}, \tag{4-3.13}$$

$$\cos \theta = \frac{x}{r} = \frac{x}{\sqrt{x^2 + y^2}} = \frac{1}{\sqrt{1 + (y/x)^2}}. \tag{4-3.14}$$

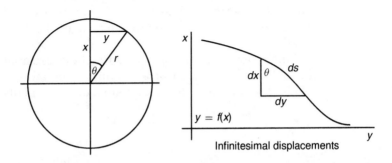

FIGURE 4-3. ORDINARY ROTATIONS. Transformations based on the circle are related to the sine, cosine, and tangent functions. The infinitesmal displacements satisfy $ds^2 = dx^2 + dy^2$, which is the equation of a circle.

The rotation in the x–y plane is:

$$\begin{bmatrix} c\bar{t} \\ \bar{x} \\ \bar{y} \\ \bar{z} \end{bmatrix} = \begin{bmatrix} 1 & 0 & 0 & 0 \\ 0 & \cos\theta & \sin\theta & 0 \\ 0 & -\sin\theta & \cos\theta & 0 \\ 0 & 0 & 0 & 1 \end{bmatrix} \begin{bmatrix} ct \\ x \\ y \\ z \end{bmatrix} = \begin{bmatrix} ct \\ x\cos\theta + y\sin\theta \\ y\cos\theta - x\sin\theta \\ z \end{bmatrix}$$

(4-3.15)

with corresponding matrices for rotations in x–z and y–z planes.

In this discussion the angle θ refers to a finite rotation for a circle of radius r. Now consider infinitesmal displacements along the path given by the equation $y = f(x)$, and the pathlength is given by $ds^2 = dx^2 + dy^2$. Since this is the equation of a circle of radius ds, we apply the preceding equations to the infinitesmal displacements. The slope of the curve at any point is the derivative y', which is the tangent. Then we write the circular functions for the infinitesmal displacements:

$$\tan\theta = y' = \frac{dy}{dx},$$

(4-3.16)

$$\sin\theta = \frac{y'}{\sqrt{1 + [y']^2}},$$

(4-3.17)

$$\cos\theta = \frac{1}{\sqrt{1 + [y']^2}},$$

(4-3.18)

where the derivative $y' = dy/dx$ replaces the ratio y/x. Then equation (4-3.15) refers to the circular functions as above for differential displacements along the curve $y = f(x)$.

Now consider Lorentz transformations. Terms involving space and time in (4-3.11) take opposite signs and the invariant, $s^2 = c^2t^2 - x^2$ for fixed s, describes a hyperbola rather than a circle. Figure 4-4 shows a triangle inscribed inside a hyperbola analogous to that in the circle in Figure 4-3. Note that the hypotenuse of the triangle does not have length s.

Calculations for the hyperbolic functions are analogous to those for the circular functions. Corresponding to equations (4-3.16), (4-3.17), and (4-3.18), we have for infinitesmal displacements in the x–t plane:

$$\tanh\theta = v/c,$$

(4-3.19)

$$\sinh\theta = \frac{v/c}{\sqrt{1 - (v/c)^2}},$$

(4-3.20)

$$\cosh\theta = \frac{1}{\sqrt{1 - (v/c)^2}},$$

(4-3.21)

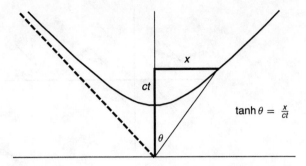

FIGURE 4-4. HYPERBOLIC ROTATIONS. The hyperbolic tangent is the ratio of the sides of the triangle inscribed in the hyperbola $s^2 = c^2t^2 - x^2$, where s is constant everywhere on the hyperbola. The asymptote, $s = 0$, corresponds to the speed of light.

in which the derivative is the velocity, $v = dx/dt$. It is customary to use the notation:

$$\beta = \frac{v}{c} = \tanh\theta, \tag{4-3.22}$$

$$\gamma = \frac{1}{\sqrt{1 - \beta^2}} = \cosh\theta. \tag{4-3.23}$$

In these terms, the Lorentz transformation in the x–t plane is:

$$\begin{bmatrix} c\bar{t} \\ \bar{x} \\ \bar{y} \\ \bar{z} \end{bmatrix} = \begin{bmatrix} \gamma & \beta\gamma & 0 & 0 \\ \beta\gamma & \gamma & 0 & 0 \\ 0 & 0 & 1 & 0 \\ 0 & 0 & 0 & 1 \end{bmatrix} \begin{bmatrix} ct \\ x \\ y \\ z \end{bmatrix} = \begin{bmatrix} \gamma(ct + \beta x) \\ \gamma(x + \beta ct) \\ y \\ z \end{bmatrix}. \tag{4-3.24}$$

Similar expressions hold for Lorentz transformations in $y - t$ and $z - t$. Note that all the nonvanishing coefficients are positive, unlike equation (4-3.15). All the rotations and Lorentz transformations taken together are called *Poincaré transformations*. It can be seen directly that the magnitude of the vector in equation (4-3.10) is invariant under these transformations.

If the transformation applies uniformly to the entire system, the transformation is said to be *global*, but if the angle of rotation is a function of position then the transformation is called *local*. These are shown in Figure 4-5.

The preceding discussion of the Michelson–Morley experiment and the Poincaré transformation group is merely a plausibility argument. The Michelson experiment does not prove that space is four dimensional, or that the Minkowski metric applies to the actual world.

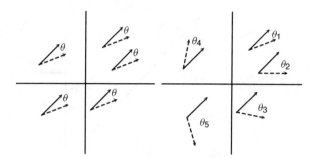

FIGURE 4-5. GLOBAL AND LOCAL ROTATIONS.   The angle in local rotations is position dependent.

Moreover, our thought experiment is just that: thought. Equations (4-3.9) only suggest the more general equation (4-3.10). What the discussion does do, however, is to suggest that the hypothesis might be reasonable. Thus we are justified in further investigation of the consequences of the hypothesis to see if they are internally consistent and predict results in agreement with experiment. In this spirit, we consider: the FitzGerald contraction and time dilation, the addition of velocities, causality, simultaneity, and particle kinematics.

Consider a time interval seen by two observers, one at rest and the other moving with a velocity comparable to the speed of light. The Lorentz transformation of time is:

$$\bar{ct} = \gamma(ct + \beta x). \tag{4-3.25}$$

Suppose we measure a time interval such as the decay time of a radioactive particle at the location $x = 0$. The decay time of the moving particle is longer than it would be at rest by the factor γ:

$$\bar{t} = \gamma t. \tag{4-3.26}$$

EXAMPLE

The lifetime of a K^- particle is 1.24×10^{-8} s at rest. By Newtonian kinematics, a beam of these particles should travel a few meters before decaying. In fact, at the large accelerators, a particle beam may travel hundreds of meters between the accelerator and the detector. The reality of time dilation is observed every day by experimental high-energy physicists.

The FitzGerald contraction is a little more subtle, but just as real. A measurement of the length of an object moving near light speed is difficult, but consider the particle beam above. The Lorentz transformation of length is:

$$\bar{x} = \gamma(x + \beta ct). \tag{4-3.27}$$

Setting $t = 0$, the distance in the lab frame becomes:

$$x = \bar{x}\sqrt{1 - \beta^2}. \qquad (4\text{-}3.28)$$

That is, for particles near high speed, the distance from the accelerator to the detector is short enough that particles can reach it before decaying. Using these relativistic length and time intervals together, one can understand the null result of the Michelson–Morley experiment. Thus, even though the length and time intervals are different in the two reference frames, their ratio, velocity, is constant. Equations (4-3.26) and (4-3.28) give:

$$\frac{\bar{x}}{\bar{t}} = \frac{x}{t}. \qquad (4\text{-}3.29)$$

The speed of light is an upper limit to velocity. This can be seen by calculating the addition of velocities using the Lorentz transformation, equations (4-3.25) and (4-3.27). Take the ratio of distance to time; then a little manipulation gives:

$$\bar{v} = \frac{\bar{x}}{\bar{t}} = \frac{\gamma(x + \beta_{rel}ct)}{\gamma(t + \beta_{rel}x/c)} = \frac{v_{rel} + v}{1 + vv_{rel}/c^2} \leq c, \qquad (4\text{-}3.30)$$

where v_{rel} is the relative velocity between the two reference frames. At low speed (i.e., where $v \ll c, v_{rel} \ll c$), velocities just add, but the upper limit of this expression is just c. In Figure 4-4 the asymptotic angle is 45°, corresponding to $v_{asymp} = c$.

Refer to Figure 4-6. Suppose that at some instant a particle is at the origin of coordinates. If the particle emits a light pulse at this time, then the set of all possible paths that the pulse can take forms a cone with the origin of coordinates at the vertex of the cone. Now suppose the particle is moving. Because particles cannot exceed the velocity of light, its future position in 4-space must lie inside the cone (or on it if the particle is massless). Two intervals are shown, the first is called *timelike* and the second is called *spacelike*. Two particles separated by a spacelike interval cannot have a *causal relationship* because a signal from one to the other would have to move faster than light.

Because of the Lorentz transformation, events occurring simultaneously in one reference frame may not appear simultaneous in another frame, which is moving relative to the first. This very counterintuitive feature is illustrated by Figure 4-7. The relativity of simultaneity of two events is analogous to the relative parallax of two objects seen by two observers at different angles. In relativity time is one of the coordinates, and the angle of observation corresponds to the relative velocity of the two observers. Thus two events (corresponding to two objects in 3-space) do not necessarily coincide for both observers.

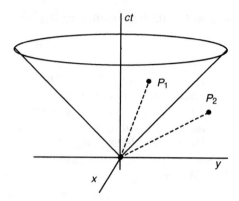

FIGURE 4-6. THE LIGHT CONE SEPARATES POINTS IN FOUR-DIMENSIONAL SPACE INTO TWO REGIONS. The third spatial dimension is not shown. Because no massive particle can exceed the velocity of light, any point P_1 inside the light cone may be causally connected to a particle at the vertex. No point outside the light cone P_2 can be. Points on the light cone can receive signals traveling at light speed—either light itself, gravity, or neutrinos which also travel at light speed.

EXAMPLE

A famous thought experiment of Einstein's illustrates this relativity of simultaneity. Suppose lightening strikes a moving train at both its front and rear ends. An observer standing alongside the train observes the strikes to be simultaneous. A second observer, inside the train, passes the first observer just as the strikes occur. The second observer observes the strike at the front of the train to occur before the strike at the rear of the train. What is simultaneous to one observer is not to the other.

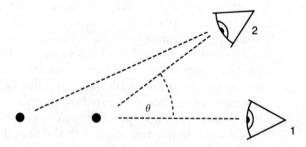

FIGURE 4-7. SIMULTANEITY ANALOG.   An observer at one angle may see two objects coincide; to another observer at a different angle they will have a definite parallax, i.e., they are separated. An analogy with simultaneous events occurs in relativity: the angle, θ, between the two observers is related to the relative velocity, v, of the two observers. Events that appear simultaneous to the "stationary" observer will appear nonsimultaneous to the "moving" observer.

A nonrelativistic explanation for this is that the velocity of light from the strikes adds to the velocity from the front of the train, whereas it subtracts from the rear. But which is "really" moving, the first or the second observer? If, as relativity requires, one cannot distinguish, it is necessary to relinquish the idea that time passes at the same rate for the two observers.

Now consider the relativistic kinematics of a particle. In four dimensions, time itself is one of the coordinates, and is therefore not the best choice for the parameter of motion. However, as in the three-dimensional case, it may be replaced by the pathlength. The parametric equations for the trajectory then take the form:

$$x^j = x^j(s) \tag{4-3.31}$$

and velocity and acceleration are also functions of s. Nevertheless, when it is useful, there is a simple relation between pathlength and time:

$$ds = \sqrt{dx^j\, dx_j} = \sqrt{1 - \beta^2}\, c\, dt = \frac{1}{\gamma} c\, dt. \tag{4-3.32}$$

The advantage of using s as the parameter is that it is invariant—it has the same value in any coordinate system. In three spatial dimensions, $v = c\beta$, where the velocity is $v^2 = v_x^2 + v_y^2 + v_z^2$. At low velocity ds and dt become equivalent:

$$ds \to c\, dt. \tag{4-3.33}$$

Thus, motion parametrized either by time or by pathlength is essentially similar at low velocity.

Now consider the four-dimensional equivalent of velocity and momentum. We define the 4-velocity to be proportional to the rate of change of displacement with respect to the pathlength:

$$U^j = c \frac{dx^j}{ds} = \frac{v^j}{\sqrt{1 - \beta^2}}, \qquad j = 1, 2, 3. \tag{4-3.34}$$

The factor, c, gives it units of velocity. Its limit as $v \ll c$ is the 3-velocity:

$$U^j \to v^j, \qquad j = 1, 2, 3. \tag{4-3.35}$$

There is also a zeroth component:

$$U^0 = c \frac{dx^0}{ds} = \frac{c}{\sqrt{1 - \beta^2}} \to c. \tag{4-3.36}$$

Each component of the velocity is the rate of displacement along that coordinate. Thus the time component, U^0, must be the rate of displacement along the time axis, i.e., the rate at which the moving body moves in time. At small velocities the rate is the same for all observers.

The relativistic 4-velocity has constant magnitude, unlike its 3-velocity counterpart. This can be seen directly: take equation (4-3.11) and divide by ds^2. This gives:

$$U^j U_j = c\frac{dx^j}{ds} c\frac{dx_j}{ds} = c^2. \qquad (4\text{-}3.37)$$

The Poincaré transformations preserve the length of the 4-vector; one consequence of this is that the change in the velocity, i.e., the acceleration, is orthogonal to the velocity. The derivative is:

$$\frac{dU^j}{ds} U_j = 0, \qquad (4\text{-}3.38)$$

which is the condition for orthogonality. Figure 4-8 shows particle trajectories in Minkowski space.

In Newtonian mechanics, the momentum is defined to be the product of the mass of an object and its velocity. But we have seen above that the three spatial components of the 4-velocity have the ordinary velocity as their limit. It is reasonable therefore, to define a relativistic momentum analogous to the Newtonian momentum:

$$P^j = mU^j = \frac{mv^j}{\sqrt{1-\beta^2}}, \qquad j = 1, 2, 3. \qquad (4\text{-}3.39)$$

At small velocities the 4-momentum becomes the Newtonian momentum. But there is also a zeroth component:

$$P^0 = mU^0 = \frac{mc}{\sqrt{1-\beta^2}}, \qquad (4\text{-}3.40)$$

which must be interpreted. First, the quantity cP^j has units of energy. Now write the Taylor expansion in the variable β. This is to first order:

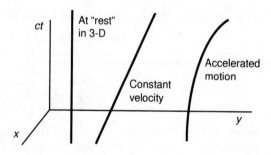

Figure 4-8. Particle trajectory in Minkowski coordinates. Note the time coordinate is placed vertically. The direction of the velocity must be less than 45 degrees from the ct-axis. The velocity of a particle "at rest" lies parallel to the ct-axis, i.e., time passes even for a particle at rest.

$$cP^0 = \frac{mc^2}{\sqrt{1 - \beta^2}}$$

$$= mc^2 + \frac{1}{2}mv^2 + O(\beta^2). \tag{4-3.41}$$

The first-order term is the Newtonian kinetic energy; we will call the velocity-dependent terms, $cP^0 - mc^2$, the relativistic kinetic energy. Second-order and higher terms are negligible at small velocity.

The zeroth order (constant) term is unprecedented in Newtonian mechanics—it represents a constant energy for a particle at rest. If we let $E = cP^0$ be the total energy, and let p be the spatial components of the momentum, then equations (4-3.39) and (4-3.41) give a useful equation for the energy:

$$E^2 - c^2p^2 = m^2c^4. \tag{4-3.42}$$

We also get this equation directly by multiplying equation (4-3.37) by m^2c^4. The equivalence of energy and mass implies that a given force exerted on a particle at different velocities results in different accelerations. It also replaces the separate principles of conservation of mass and of energy with a combined conservation principle. The energy of an object is plotted as a function of its velocity in Figure 4-9. It agrees with Newtonian energy at small velocities, but approaches infinity as the velocity of the particle approaches light speed. It would take an infinite amount of energy to accelerate a particle up to light speed.

Observing this effect at low speed is not easy. Energy and mass differences associated with ordinary physical or chemical processes are too small to be detected easily, although they presumably exist. Because

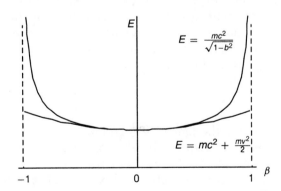

FIGURE 4-9. RELATIVISTIC ENERGY. This shows that the upper limit for relativistic velocity is the speed of light. At small velocities, Newtonian and relativistic expressions agree except for the constant rest energy.

the Newtonian laws of motion deal with the change of momentum and energy; adding a constant to the energy has no effect on the kinematics of a particle, *provided it is indeed constant*. However, at energies associated with nuclear and elementary particle reactions, the total mass in the reaction is not constant, and conversion between mass and energy can be observed.

Example

In the fusion reaction between two deuterons giving ^{3}He and a neutron:

$$D + D \rightarrow {}^3\text{He} + n, \tag{4-3.43}$$

about 0.1% of the mass of the particles is lost during the reaction—and reappears as the kinetic energy of the ^{3}He and neutron. For every kilogram of deuterium fused in this reaction, about 10^{14} Joules of energy are produced.

4-4 The Electromagnetic Field

We have seen that the laws of mechanics can be expressed in four-dimensional relativistic form. But Maxwell's equations are not obviously relativistic as they are written in Chapter 2. For consistency, we must find a relativistic representation of the electromagnetic equations. The approach taken here is to show that a local Poincaré transformation of the velocity is equivalent to the equation of motion of a charged particle in an electromagnetic field.[1] From this we can find a field tensor, whose elements we can interpret as the components of the electromagnetic field. Relativistic electromagnetic field equations can then be found by comparison with the experiments of Chapter 2. This method is systematic, efficient, and perhaps, even elegant.

The differential Poincaré transformation for the 4-velocity may be written:

$$dU^j = dP^j_k U^k = dP^{jk} U_k, \tag{4-4.1}$$

where the tensor dP^{jk} combines the Lorentz and spatial rotation operators. Expressed as a mixed tensor, the rotation and the Lorentz components have symmetric and antisymmetric matrix representations, respectively. It thus has no particular symmetry. However,

---

[1] For further discussion of the local gauge approach to fields, see Ramond (Ch. 6) and especially Utiyama. These treatments go far beyond the scope of this book, but may be lightly read as general background.

the contravariant matrix is completely antisymmetric. Expressed as infinitesmal angles of rotation, the operator becomes:

$$dP^{jk} = d\theta^{jk} = -d\theta^{kj}. \tag{4-4.2}$$

For example, $d\theta^{01}$ refers to an infinitesmal Lorentz rotation in the 0–1 plane, and $d\theta^{12}$ refers to a rotation in the 1–2 plane.

Due to the antisymmetry, there are six independent elements in this matrix—which just equals the number of elements of the electric and magnetic field vectors in three-dimensional space. Written in explicit matrix form, it is:

$$dP^{jk} = \begin{bmatrix} 0 & -dP^{01} & -dP^{02} & -dP^{03} \\ dP^{01} & 0 & -dP^{12} & dP^{13} \\ dP^{02} & dP^{12} & 0 & -dP^{23} \\ dP^{03} & -dP^{13} & dP^{23} & 0 \end{bmatrix}. \tag{4-4.3}$$

In general, rotation operators do not commute. The reader can test this by laying a book on a table and rotating it by 90^0 on a vertical axis; then rotate it by 90^0 on a North–South axis; then note the orientation of the book. Next, reverse the order of these rotations. The book will be in a different orientation this time. For this reason, it might seem impossible to find an operator to express arbitrary rotations without considering the order of rotations around the axes. In that case the matrix is multivalued. However, infinitesimal operations do commute. To see this, consider the unitary operation for a finite angle:

$$U^j = \Theta^{jk} U_{k0} = e^{\theta^{jk}} U_{k0}, \tag{4-4.4}$$

where U_{k0} is the unrotated value. These finite operators do not commute for $j, k \neq m, n$:

$$\Theta^{jk} \Theta^{mn} \neq \Theta^{mn} \Theta^{jk}. \tag{4-4.5}$$

Now let the angles of rotation be infinitesmal, and express the operators as a Taylor series truncated after first order. Then the operator product:

$$\begin{aligned} \Theta^{jk} \Theta^{mn} &= \left(1 + d\theta^{jk}\right)\left(1 + d\theta^{mn}\right) \\ &= 1 + d\theta^{jk} + d\theta^{mn} + d\theta^{jk} d\theta^{mn} \\ &= 1 + d\theta^{jk} + d\theta^{mn} \qquad \text{to first order} \\ &= \Theta^{mn} \Theta^{jk} \end{aligned} \tag{4-4.6}$$

does commute to first order, since addition is commutative. The differential matrix is therefore single valued, as are its derivatives such as dP^{jk}/ds, for example.

There is still ambiguity about which components are to be positive in the antisymmetric matrix. We will use this sign convention: the components $d\theta^{0j}$ ($j = 123$) are all negative, and components $d\theta^{jk}$ ($j, k = 123$) are negative if jk is in cyclical order, or positive if in anticyclical order. Then the upper right half of the matrix is negative except for the θ^{21} term. This convention is consistent with the right-hand rule for the cross product of two three-dimensional vectors. The mnemonic given in Figure 4-10 might be helpful in remembering the signs.

Equation (4-4.1) expresses the change in the velocity. It shows, for example, how to find the new velocity after some Lorentz transformation (increase in velocity). But now suppose that the Poincaré transformation depends on the location of the particle, i.e., the transformation is a *local transformation*, where the matrix elements are functions of the position:

$$d\theta^{jk} = d\theta^{jk}(xyzt). \tag{4-4.7}$$

Now suppose we want to know how the velocity of the particle changes as it moves from one point in space–time to another. Then equation (4-4.1) gives the differential change of velocity with respect to s (the relativistic acceleration) along the four-dimensional path. Dividing by ds gives the rate of change:

$$\frac{dU^j}{ds} = \frac{dP^{jk}}{ds} U_k = T^{jk}(x^r)U_k, \tag{4-4.8}$$

which is an *equation of motion* for a particle in a relativistic tensor field $T^{jk}(x^r)$ antisymmetric in the indices j, k. Let us determine whether this equation agrees with the equation of motion of a charged particle in an electromagnetic field. From the Lorentz force, equation (2-5.9), we see that the magnetic force is proportional to the velocity and to the magnetic field. We also know that the tensor, being antisymmetric, has six independent components which can be identified with the six independent components of the electric and magnetic fields. The equation

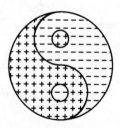

Figure 4-10. A mnemonic for the contravariant form of the Poincaré transformation tensor. The Yin-Yang symbol has a little − in the + region and a little + in the − region.

of motion is consistent with fixed magnitude for the velocity. To see this, multiply by U_j. Then:

$$\frac{dU^i}{ds} U_j = T^{ik} U_k U_j = 0,   \qquad (4-4.9)$$

which vanishes because T^{ik} is antisymmetric and $U_j U_k$ is symmetric. This agrees with equations (4-3.37) and (4-3.38).

Now suppose T^{ik} contains a multiplicative constant consistent with the dependence on mass and charge in the necessary units. Then equation (4-4.8) may be written:

$$\frac{dU^i}{ds} = \frac{q}{mc^2} F^{ik} U_k,   \qquad (4-4.10)$$

where the field tensor takes the form (displayed as row–column):

$$F^{ik} = \begin{bmatrix} 0 & -E^1 & -E^2 & -E^3 \\ E^1 & 0 & -B^3 & B^2 \\ E^2 & B^3 & 0 & -B^1 \\ E^3 & -B^2 & B^1 & 0 \end{bmatrix}. \qquad (4-4.11)$$

Compare the equation with the known equation of motion in an electromagnetic field. At small velocities equation (4-4.10) becomes:

$$m\frac{dv^i}{dt} = -\frac{q}{c} F^{ik} v_k, \qquad j = 1, 2, 3. \qquad (4-4.12)$$

The minus sign comes from expressing the spatial components of v_k as a covariant vector. After explicitly summing on the index k, equation (4-4.10) gives the Lorentz force for the spatial components ($j = 1, 2, 3$):

$$m\frac{d\mathbf{v}}{dt} = q(\mathbf{E} + \frac{\mathbf{v}}{c} \times \mathbf{B}). \qquad (4-4.13)$$

The zeroth component of this equation needs some discussion since there is no equivalent in three-dimensions. For small velocity it becomes:

$$\frac{dW}{dt} = qF^{0k} v_k = q\mathbf{E} \cdot \mathbf{v}. \qquad (4-4.14)$$

The term on the right is the work done per unit time on the particle by the electromagnetic field (no work is done by magnetic forces because they are orthogonal to the velocity) and we have seen that the zeroth component of the momentum is proportional to energy. Thus the equation represents energy balance—the increase of energy of a particle equals the work done on it.

We conclude that the Lorentz force may be interpreted as a local Poincaré transformation in which the transformation tensor is proportional to the electromagnetic field tensor.

We now want to express the electromagnetic field equations in terms of the electromagnetic field tensor. Before we can do that, we must redefine the current density as a 4-vector. The three-dimensional form was defined as the charge that passes through a cross sectional area per unit time, with a direction parallel to the direction of the velocity:

$$\mathbf{J} \equiv \frac{\Delta q}{\Delta t \, \Delta A} \frac{\mathbf{v}}{|v|} \tag{4-4.15}$$

We can use equation (2-3.11) to find its four-dimensional form. Integrating with respect to time gives the charge that passes through a closed surface in a given time interval:

$$\Delta q = \int \oiint \mathbf{J} \cdot d\mathbf{S} \, dt. \tag{4-4.16}$$

In four-dimensional space, the equivalent of a surface is a hypersurface of three dimensions. Generalizing to four-dimensions, the two integrals become a closed hypersurface integral:

$$\Delta q = \frac{1}{c} \oiiint J_k \, dH^k, \tag{4-4.17}$$

where dH^k is the hypersurface element; see equation (4-2.9). Then the current density J_k is the derivative:

$$J_k = \frac{dq}{dH^k} c. \tag{4-4.18}$$

That is, in four-dimensions the current density is the charge transmitted through a unit hypersurface. The three spatial components agree with equation (4-4.15); the zeroth component is:

$$J_0 = \frac{dq}{d^3 r} c. \tag{4-4.19}$$

An argument similar to that given in three-dimensions gives the useful formula:

$$J_k = \rho U_k. \tag{4-4.20}$$

The spatial components at small velocities are the same as the three-dimensional case, but the time component is proportional to the charge density:

$$J_0 \to \rho c. \tag{4-4.21}$$

For charge at "rest," the current density is entirely in the time direction, the hypersurface is the volume ($dx \, dy \, dz$) containing the charge, and the

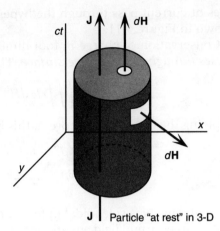

Particle "at rest" in 3-D

FIGURE 4-11A.

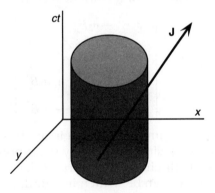

Moving particle leaves 3-volume

FIGURE 4-11B.

FIGURE 4.11. CURRENT DENSITIES FOR PARTICLES AT REST AND MOVING.   The hypersurface elements, $d\mathbf{H}$, are shown. The hypervolume is the volume of the cylinder, and the 3-volumes at particular times are represented by circular cross sections. Current density is defined as the amount of charge passing through a given hypersurface. Particles at rest in three dimensions remain in the finite 3-volume. The z-dimension is not shown.

lines of current pass through the hypersurface with velocity c. This is shown in Figure 4-11.

Conservation of charge in four-dimensions becomes: *no net charge passes through a closed hypersurface.* That is:

$$\oiint J_k \, dH^k = 0. \tag{4-4.22}$$

Applying the divergence theorem, this becomes in differential form:

$$\partial_k J^k = \partial_0 J^0 + \partial_1 J^1 + \partial_2 J^2 + \partial_3 J^3$$

$$= \frac{\partial \rho}{\partial t} + \nabla \cdot \mathbf{J} = 0. \tag{4-4.23}$$

Current density is solenoidal in four-dimensions.

We can now find field equations for the electromagnetic field tensor. Using Maxwell's equations as the model, we want a first-order vector equation.[2] Applying the divergence operator ∂_j to the field tensor F^{jk} produces the contravariant vector:

$$\partial_j F^{jk} = \begin{bmatrix} \partial_1 E^1 + \partial_2 E^2 + \partial_3 E^3 \\ -\partial_0 E^1 + \partial_2 B^3 - \partial_3 B^2 \\ -\partial_0 E^2 - \partial_1 B^3 + \partial_3 B^1 \\ -\partial_0 E^3 + \partial_1 B^2 - \partial_2 B^1 \end{bmatrix} = \begin{bmatrix} \nabla \cdot \mathbf{E} \\ \nabla \times \mathbf{B} - \dfrac{1}{c} \dfrac{\partial \mathbf{E}}{\partial t} \end{bmatrix}, \tag{4-4.24}$$

where we sum over the first index; this accounts for the signs. But these two expressions are exactly the expressions in the two Maxwell equations with charge and current density. Therefore these two Maxwell equations can be written in four-dimensional form:

$$\partial_j F^{jk} = \frac{4\pi}{c} J^k. \tag{4-4.25}$$

It is less obvious how to obtain the second (source-free) set of Maxwell's equations in relativistically invariant form. However the same principles apply: they must be first-order linear equations involving the field tensor. These conditions are satisfied by the equation:

$$\partial^i F^{jk} + \partial^j F^{ki} + \partial^k F^{ij} = 0. \tag{4-4.26}$$

Note the raised indices on the gradient operators. This equation is invariant under cyclical permutations of the indices ($i, j, k \rightarrow k, i, j \rightarrow j, k, i$). It is also highly redundant; with three free indices it gives 64 relations between the components.

---

[2] The reason for expressing the laws in tensor notation is to express physical laws in a form which is independent of the frame of reference of the observer.

Example

Illustrate equation (4-4.26) for indices $i, j, k = 123$ and $i, j, k = 120$:

$$\partial^1 F^{23} + \partial^2 F^{31} + \partial^3 F^{12} = \partial_1 B^1 + \partial_2 B^2 + \partial_3 B^3$$

$$= \nabla \cdot \mathbf{B} = 0, \tag{4-4.27}$$

$$\partial^1 F^{20} + \partial^2 F^{01} + \partial^0 F^{12} = -\partial_1 E^2 + \partial_2 E^1 + \partial_0 B^3$$

$$= -\left[\nabla \times \mathbf{E} + \frac{1}{c} \frac{\partial \mathbf{B}}{\partial t} \right]_3 = 0. \tag{4-4.28}$$

Equation (4-4.28) gives the third component of the equation; the other two components are obtained by cyclic permutation of the indices.

Although equation (4-4.26) gives satisfactory agreement with Maxwell's equations, it is perhaps not so pleasing esthetically. Moreover, it is not a vector equation, and its interpretation as a Poincaré transformation is unclear. However, it can be converted to vector form by multiplying by the Levi-Civita tensor and summing over three indices. To see this most elegantly, define the dual tensor:

$$D^{ik} = \frac{1}{2} \varepsilon^{iknn} F_{mn} = \begin{bmatrix} 0 & B^1 & B^2 & B^3 \\ -B^1 & 0 & -E^3 & E^2 \\ -B^2 & E^3 & 0 & -E^1 \\ -B^3 & -E^2 & E^1 & 0 \end{bmatrix}. \tag{4-4.29}$$

In terms of the dual field tensor, Maxwell's source free equations become:

$$\partial_i D^{ik} = 0, \tag{4-4.30}$$

which has the same form as the other equation and can be interpreted in terms of Poincaré transformations. Together, equations (4-4.25) and (4-4.30) put eight conditions on the six field components, as do the Maxwell equations. The extra conditions are reduced by charge conservation which is implicit in equation (4-4.25). Take the divergence:

$$\partial_k \partial_j F^{jk} = \frac{4\pi}{c} \partial_k J^k. \tag{4-4.31}$$

The left side of this equation vanishes because the two gradient operators are symmetric in j, k and the field is antisymmetric in j, k. This leaves:

$$\partial_k J^k = 0. \tag{4-4.32}$$

which is the differential form of charge conservation. Compare it with equation (2-3.11).

Comparing equations (4-4.25) and (4-4.30) suggests a way to incorporate magnetic charges, called magnetic monopoles, into the theory (Amaldi and Cabibbo, 1972). In fact, though careful searches have been made for these particles, no evidence for them has been found.

Maxwell's equations are a set of coupled, first-order, linear, differential equations for the elements of the field tensor. It is possible to uncouple the equations at the price of getting second-order equations. Take the gradient of equation (4-4.25) and commute:

$$\partial_j \partial^j F^{ik} = \frac{4\pi}{c} \partial^i J^k. \qquad (4\text{-}4.33)$$

Apply equation (4-4.26) to get:

$$\partial_j \left[-\partial^j F^{ki} - \partial^k F^{ij} \right] = \frac{4\pi}{c} \partial^i J^k. \qquad (4\text{-}4.34)$$

Then, using (4-4.25) again, we get an equation in which the elements of the electromagnetic field tensor are uncoupled:

$$\partial_j \partial^j F^{ki} = \frac{4\pi}{c} \left[\partial^k J^i - \partial^i J^k \right]. \qquad (4\text{-}4.35)$$

Both sides are antisymmetric as one might expect. By writing F^{ik} and J^k as components, it becomes possible to group the electric and magnetic equations:

$$\left[\frac{1}{c^2} \frac{\partial}{\partial t^2} - \nabla^2 \right] \mathbf{B} = \frac{4\pi}{c} \nabla \times \mathbf{J}, \qquad k = 1, 2, 3, \qquad (4\text{-}4.36)$$

$$\left[\frac{1}{c^2} \frac{\partial}{\partial t^2} - \nabla^2 \right] \mathbf{E} = -4\pi\nabla\rho - \frac{4\pi}{c^2} \frac{\partial \mathbf{J}}{\partial t}, \qquad k = 0. \quad (4\text{-}4.37)$$

In regions with no sources they satisfy the D'Alembert wave equation. They can also be derived directly from Maxwell's equations (2-5.4) through (2-5.7).

The electric and magnetic components are separated in these equations. Although it is pleasing to see the electric and magnetic field equations displayed separately, it obscures a very important aspect of the relativistic formulation. By observing the fields from a new moving reference frame (i.e., applying a Lorentz transformation to F^{ik}), the electric and magnetic components become linear combinations of the original ones. But which frame of reference is real? It is the essence of relativity that all inertial frames are equally valid. Thus there is only a *unified electromagnetic field*, not separate electric and magnetic fields. We have seen that the current density and charge density are components of a 4-vector, whereas the fields $\mathbf{E}$ and $\mathbf{B}$ are only 3-vectors. Thus equations (4-4.36) and (4-4.37) must not be interpreted to imply an absolute separation of electric and magnetic fields.

4-5 Electromagnetic Potentials and Gauge Conditions

When discussing the differential properties of the electric and magnetic fields in Sections 2-2 and 2-3, we found it convenient to express the fields in terms of scalar and vector potentials. However, those potentials are not expressed relativistically. In order to generalize the potentials, consider the properties of the field tensor, i.e., it is antisymmetric, and the magnetic field is expressed as the curl of the vector potential. Now generalize the curl to four dimensions:

$$F^{ik} = \partial^i A^k - \partial^k A^i,$$ (4-5.1)

where the vector potential A^j is a 4-vector. Note that the gradient operator in this expression is contravariant. Then, lowering the gradient indices, the spatial components give the magnetic field:

$$\mathbf{B} = \nabla \times \mathbf{A}$$ (4-5.2)

in agreement with equation (2-3.10). Choosing $j = 0$, the electric field is:

$$E^k = F^{k0} = -\partial_k A^0 - \partial_0 A^k \rightarrow -\nabla\phi - \frac{1}{c}\frac{\partial \mathbf{A}}{\partial t},$$ (4-5.3)

which, for the steady state, agrees with an *electrostatic potential*:

$$\mathbf{E} = -\nabla\phi \quad \text{(steady state)},$$ (4-5.4)

$$A^0 = \phi.$$ (4-5.5)

This identifies the zeroth component of A^k, which is not necessarily constant in time.

The vector potential, introduced specifically to express the electromagnetic field tensor, is not uniquely defined. The field tensor is unchanged by transforming the potential:

$$A^j \rightarrow A^j + \partial^j \Gamma.$$ (4-5.6)

This transformation is called a *gauge transformation*, where the arbitrary scalar function Γ is called a *gauge function*. Because of this ambiguity in the potential, an auxiliary condition may be put on it. A relativistic condition, called the *Lorentz gauge condition* is:

$$\partial_j A^j = 0,$$ (4-5.7)

which in three-dimensional notation is:

$$\nabla \cdot \mathbf{A} + \frac{1}{c}\frac{\partial \phi}{\partial t} = 0,$$ (4-5.8)

in agreement with equation (2-5.16). Requiring this gauge condition to hold in both the original and transformed coordinate systems puts a condition on the gauge function. Thus applying equations (4-5.6) and (4-5.7) together gives:

$$\partial^j \partial_j \Gamma = 0, \tag{4-5.9}$$

i.e., the D'Alembert wave equation:

$$\nabla^2 \Gamma - \frac{1}{c^2} \frac{\partial^2 \Gamma}{\partial t^2} = 0. \tag{4-5.10}$$

Therefore the Lorentz gauge function is an arbitrary function of the form:

$$\Gamma = \Gamma(\mathbf{k} \cdot \mathbf{x} - \omega t),$$
$$\omega = c|k|. \tag{4-5.11}$$

Obviously other gauge conditions are possible and may be useful for certain purposes, but they are not all relativistically invariant. Other gauge conditions are:

$$\nabla \cdot \mathbf{A} = 0 \qquad \text{radiation gauge,} \tag{4-5.12}$$

$$\partial_t \phi = 0 \qquad \text{Coulomb gauge,} \tag{4-5.13}$$

$$A^3 = 0 \qquad \text{axial gauge,} \tag{4-5.14}$$

$$\partial_i A^i = \partial_i \partial^i \Gamma \qquad \text{'t Hooft gauge.} \tag{4-5.15}$$

This derivation of the electromagnetic field is an application of a classical (nonquantum) gauge field theory. In gauge theories, the equations of motion are required to be invariant under some transformation (Poincaré in this case) and then allow the transformation to become local. The result of this is to subject the particle to new forces, or equivalently, to introduce new fields. Gravity may also be treated as a gauge field. Quantized gauge fields play a central role in the modern theories of elementary particles. Specifically, the *t' Hooft gauge* was useful in the unification of electromagnetism and weak interactions.

By applying the potentials to the electromagnetic field equations, (4-4.25), we get equations for the potentials:

$$\partial_i F^{ik} = \partial_i \partial^i A^k - \partial_i \partial^k A^i = \frac{4\pi}{c} J^k. \tag{4-5.16}$$

The Lorentz gauge condition makes the second term vanish, leaving the potential field equation:

$$\partial_i \partial^i A^k = \frac{4\pi}{c} J^k \tag{4-5.17}$$

which in three-dimensional notation is:

$$\left[\frac{1}{c^2} \frac{\partial^2}{\partial t^2} - \nabla^2 \right] A^k = \frac{4\pi}{c} J^k. \tag{4-5.18}$$

This equation is obviously relativistically invariant. Different gauge conditions give field equations which may not be invariant, e.g., the radiation gauge. The zeroth component in steady state, $A^0 = \phi$, is *Poisson's equation:*

$$\nabla^2\phi = -4\pi\rho. \qquad (4\text{-}5.19)$$

The potential, ϕ, is associated with the potential energy of a charged particle. Because relativistic kinetic energy and momentum are components of the same 4-vector, the spatial components of the vector potential must be associated with *potential momentum.*[3] Energy and momentum carried by the electromagnetic field will be discussed in detail in Chapter 6.

The source-free Maxwell equation becomes an identity when expressed in terms of the potential. Thus:

$$\partial^i(\partial^j A^k - \partial^k A^j) + \partial^j(\partial^k A^i - \partial^i A^k) + \partial^k(\partial^i A^j - \partial^j A^i) \equiv 0 \qquad (4\text{-}5.20)$$

vanishes identically. There are other types of potentials. We will discuss the Hertz potentials, which are useful for representing radiation in Chapter 8.

4-6 LORENTZ TRANSFORMED FIELDS

We have seen that the electromagnetic field can be represented by an antisymmetric tensor. However, the actual values of the components (i.e., the electric and magnetic fields) depend on the coordinate system in which they are observed. If an observer goes from a rest frame to a moving frame, the observed values of the electric and magnetic fields will change. Consider how the fields change for a new coordinate system. For a Lorentz transformation in the x–t plane, the transformation matrix is:

$$L_k^j = \begin{bmatrix} \gamma & \gamma\beta^1 & 0 & 0 \\ \gamma\beta^1 & \gamma & 0 & 0 \\ 0 & 0 & 1 & 0 \\ 0 & 0 & 0 & 1 \end{bmatrix}. \qquad (4\text{-}6.1)$$

Then the vector potential and gradient transform:

$$\bar{A}^j = L_k^J A^k, \qquad (4\text{-}6.2)$$

---

[3] See Konopinski in the Selected Bibliography.

$$\bar{\partial}^j = L^j_k \partial^k. \tag{4-6.3}$$

Because the field tensor is the outer product of two vectors:

$$F^{jk} = \partial^j A^k - \partial^k A^j \tag{4-6.4}$$

it transforms by applying the transformation to each index:

$$\overline{F^{mn}} = L^m_j L^n_k F^{jk}. \tag{4-6.5}$$

Transformations in other planes take similar form. The field tensor is, in general:

$$F^{jk} = \begin{bmatrix} 0 & -E^1 & -E^2 & -E^3 \\ E^1 & 0 & -B^3 & B^2 \\ E^2 & B^3 & 0 & -B^1 \\ E^3 & -B^2 & B^1 & 0 \end{bmatrix}. \tag{4-6.6}$$

Transforming it as above gives:

$$\overline{F^{mn}} = \begin{bmatrix} 0 & -E^1 & -\gamma(\beta B^3 + E^2) & \gamma(\beta B^2 - E^3) \\ E^1 & 0 & -\gamma(B^3 + \beta E^2) & \gamma(B^2 - \beta E^3) \\ \gamma(\beta B^3 + E^2) & \gamma(B^3 + \beta E^2) & 0 & -B^1 \\ \gamma(-\beta B^2 + E^3) & \gamma(-B^2 + \beta E^3) & B^1 & 0 \end{bmatrix} \tag{4-6.7}$$

where, as usual:

$$\gamma^2 = \frac{1}{1 - \beta^2}. \tag{4-6.8}$$

The transverse components of the fields are rotated by the transformation, while the longitudinal (i.e., parallel to the direction of the velocity) are not. Although this calculation was for a transformation in the x-direction, these properties of the electromagnetic fields are actually quite general since the choice of the x-direction for the velocity in the Lorentz transformation is arbitrary.

The striking feature of this transformation is the mixing of transverse electric and magnetic components. That is, *it is not possible to distinguish absolutely between electric and magnetic fields*. What appears to be an electric field in one coordinate system appears to be both electric and magnetic fields in a coordinate system moving with respect to the first.

EXAMPLE

Consider a long, linear distribution of charge at rest. A short segment is illustrated in Figure 4-12 on the left. The electric field has zero longitudinal component, and the lines of flux are as indicated. The equipotential surfaces are concentric cylinders, and there is no current, vector potential, or magnetic field.

Now consider the fields produced by these charges when they move longitudinally toward the right, as shown by Figure 4-12 on the right. That is, perform a Lorentz transformation on the system. As at rest, there is no longitudinal component of the electric field. The transverse component is increased by the increase of the linear charge density due to the FitzGerald contraction. Since the field lines originate at the charges and retain their shape, they may be thought of as moving also. The moving charges constitute a current, which creates a vector potential and magnetic field. This is shown in Figure 4-13 which illustrates the contribution of just a small element of charge. The vector potential is parallel to the velocity whose value is determined by the Lorentz transformation of the scalar potential:

$$\bar{A}^j = L_k^j A^k = [\gamma\phi, \beta\gamma\phi, 0, 0]$$

$$= \left[\frac{\lambda}{2\pi}\ln r, \frac{\lambda v}{2\pi c}\ln r, 0, 0\right], \qquad v \ll c, \tag{4-6.9}$$

where the scalar potential, $\phi = (\lambda/2\pi)\ln r$, is calculated for the system at rest, and λ is the charge per unit length. The transformed electric (and magnetic!) fields can be calculated either from these potentials or from the transformation equations (4-6.5).

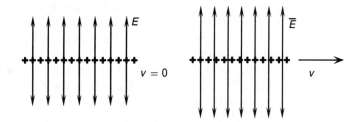

FIGURE 4-12. ELECTRIC FIELD DUE TO UNIFORMLY MOVING LINE OF CHARGE. The lines of flux in the moving system travel synchronously with the charges. Note the contraction along the direction of the motion and the magnification of the transverse fields, both due to the longitudinal FitzGerald contraction of the charge density.

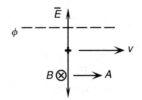

FIGURE 4-13. DETAILS FROM THE PREVIOUS DIAGRAM.   The equipotential surfaces form concentric cylinders which are represented here just by a single line. The positions of the equipotentials are unchanged by the motion of the charges. The vector potential lies in the equipotential cylinder, paralleled to the motion of the charge. Its occurrence here is due to Lorentz transforming the scalar potential from rest. Then a corresponding magnetic field occurs directed away from the viewer, as indicated.

4-7   THE LAGRANGIAN METHOD

The preceding derivation of the electromagnetic field assumed a local transformation of the velocity vector; the transformation matrix is then interpreted as an electromagnetic field tensor. We then showed that the field tensor satisfies Maxwell's equations. There is an alternative approach in which the path of a particle corresponds to the minimum of a scalar functional, the action. Galileo observed that an object not subject to external forces moves in a straight line—this has become known to us as Newton's first law. But a straight line is a special object in Euclidean space. It is a geodesic, i.e., the shortest interval between its end points. Least values are the subject of a branch of mathematics known as the calculus of variations. The original problem of this sort is Euler's brachystochrone problem: find the shape of the curve (the trajectory) on which a bead slides (under the influence of gravity but without friction) in the least time from its initial point to its final point. In this problem the time of flight depends on the position of the bead, and its derivative, velocity. Solution requires taking the joint variation of the two variables. Its generalization to Lagrangian mechanics provides a very useful way of finding equations of motion in new or complex situations.

Consider all the allowed trajectories of a particle between fixed initial and end points. Three such trajectories are shown in Figure 4-14. An arbitrary trajectory from a given initial point to a given final point can be specified by adding arbitrary increments, δx^j, to the path at each point, except at the initial and final points.

Now define action to be the integral of the Lagrangian functional between the boundary points. In order to keep the equations relativistically invariant the Lagrangian should be composed of invariant

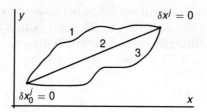

FIGURE 4-14. POSSIBLE TRAJECTORIES BETWEEN FIXED BOUNDARY POINTS. By taking an arbitrary variation of the path at all points except at the end points, an arbitrary path between the end points is created. The actual path is the one in which the action is a minimum.

quantities. This gives the candidates: $m, c, U^j U_j, ds$, products and sums. The free-particle trajectory must give constant velocity; we will try:

$$S = \int_{s_0}^{s} mc\, ds, \qquad (4\text{-}7.1)$$

or its equivalent, since $U^j U_j = c^2$:

$$S = \int_{s_0}^{s} \frac{m}{2} U^i U_j \frac{ds}{c}. \qquad (4\text{-}7.2)$$

Minimizing this quantity (which is proportional to the pathlength) gives a straight-line trajectory for the free particle. At small velocities the integrand contains the Newtonian kinetic energy, in agreement with the classical Lagrangian. The general form of the action must give equations that depend on the particle position in order to account for accelerated motion. Therefore the action becomes:

$$S = \int_{s_0}^{s} L\left(x^j, U^j\right) \frac{ds}{c}. \qquad (4\text{-}7.3)$$

Taking the least value of the action for fixed boundary points gives the *Lagrange equations*. Set the variation of the action equal to zero:

$$\delta S = \int \frac{\partial L}{\partial U^j} \delta U^j + \frac{\partial L}{\partial x^j} \delta x^j \frac{ds}{c} = 0. \qquad (4\text{-}7.4)$$

Now use the definition of velocity to express the variation in terms of δx^j only. Thus applying:

$$\delta U^j = c \frac{d\delta x^j}{ds} \qquad (4\text{-}7.5)$$

and integration of the first term by parts allows us to write:

$$\delta S = \int_{s_0}^{s} \left[-c \frac{d}{ds} \left(\frac{\partial L}{\partial U^j} \right) + \frac{\partial L}{\partial x^j} \right] \delta x^j \frac{ds}{c} + \left[\frac{\partial L}{\partial U^j} \delta x^j \right]_{s_0}^{s}. \qquad (4\text{-}7.6)$$

The last term vanishes because the variation at the end points is zero. But the variation at all other points is arbitrary. Since we require that the integral vanish for the path *whatever it is*, the integrand itself must vanish. This gives the *Lagrangian equation of motion*:

$$\frac{d}{ds}\left(\frac{\partial L}{\partial U^i}\right) - \frac{1}{c}\frac{\partial L}{\partial x^j} = 0. \qquad (4\text{-}7.7)$$

We define the *canonical momentum*:

$$P_j \equiv \frac{\partial L}{\partial U^j}. \qquad (4\text{-}7.8)$$

For free particles the equations of motion become:

$$\frac{dmU^j}{ds} = 0 \qquad (4\text{-}7.9)$$

which gives constant velocity as expected.

Generalize the Lagrangian to include the vector potentials. The form of the additional term is determined by taking the corresponding *potential energy*:

$$S = \int\left[\frac{m}{2}U_j U^j + \frac{q}{c}A_j U^j\right]\frac{ds}{c}. \qquad (4\text{-}7.10)$$

The canonical momentum includes an extra term because the interaction Lagrangian is velocity dependent.

$$P_j = mU_j + \frac{q}{c}A_j. \qquad (4\text{-}7.11)$$

The equations of motion become:

$$m\frac{dU_k}{ds} + \frac{q}{c}\frac{dA_k}{ds} - \frac{q}{c^2}\frac{\partial A_j}{\partial x^k}U^j = 0. \qquad (4\text{-}7.12)$$

Evaluating the second term by the chain rule of differentiation, this equation may be written in terms of the electromagnetic field tensor:

$$m\frac{dU_k}{ds} = \frac{q}{c^2}F_{kj}U^j \qquad (4\text{-}7.13)$$

which agrees with equation (4-4.10).

The electromagnetic field equations can also be found from a Lagrangian. For this purpose let the components of the vector potential A^j be the independent variables, and find the variation of the field Lagrangian with respect to them. Assuming a Lagrangian whose variables are functions of space–time, we define a *Lagrangian density* including both the particle and field terms. Many combinations of the potentials and its derivatives will yield a scalar; but we will show that the field equations can be derived from the invariant expression:

$$S = \iiiint \rho\left[\frac{m}{2}U_j U^j + \frac{q}{c}A_j U^j\right] + \frac{1}{16\pi}F^{jk}F_{jk}\,d^4r, \qquad (4\text{-}7.14)$$

where ρ is the *particle density* and $U^j(x)$ is the velocity field. In these terms, $J^k = q\rho U^k$. The last term can be written:

$$\frac{1}{16\pi} F^{jk} F_{jk} = \frac{1}{8\pi} (\mathbf{E}^2 - \mathbf{B}^2). \tag{4-7.15}$$

Now find the variation of the action with respect to the vector potential A_k:

$$\delta S = \iiiint \frac{1}{c} J^k \delta A_k + \frac{1}{8\pi} F^{jk} \delta F_{jk} \, d^4 x$$

$$= \iiiint \frac{1}{c} J^k \delta A_k + \frac{1}{8\pi} F^{jk} \left[\frac{\partial \delta A_j}{\partial x^k} - \frac{\partial \delta A_k}{\partial x^j} \right] d^4 x$$

$$= \iiiint \left[\frac{1}{c} J^k - \frac{1}{4\pi} \frac{\partial F^{jk}}{\partial x^j} \right] \delta A_k \, d^4 r + \text{surface term.} \tag{4-7.16}$$

As with the particle calculations above, the variations are arbitrary except for the boundary points. The boundary surface term, obtained by integration by parts, vanishes because the variation, δA_k, vanishes on the boundary surface. Then, setting the integral to zero, the integrand vanishes. This gives the expected field equations (4-4.25).

4-8 Selected Bibliography

Amaldi and Cabibbo, *Aspects of Quantum Theory* (Salam and Wigner, eds.), Cambridge University Press, 1972.

A. Barut, *Electrodynamics and Classical Theory of Fields and Particles*, Macmillan, 1965.

P. Bergmann, *Introduction to the Theory of Relativity*, Prentice-Hall, 1959.

A. French, *Special Relativity*, Norton, 1968.

B. Hoffmann, *Einstein Creator and Rebel*, New American Library, 1972.

J. Jackson, *Classical Electrodynamics*, Wiley, 1967.

E. Konopinski, "What the Electromagnetic Vector Potential Describes," Amer. J. Phys., **46**(5), May 1978.

L. Landau and E. Lifshitz, *The Classical Theory of Fields*, 4th ed., Pergamon, 1975.

C. Misner, K. Thorne, and J. Wheeler, *Gravitation*, Freeman, 1973.

W. Panofsky and M. Phillips, *Classical Electricity and Magnetism*, Addison-Wesley, 1955.

P. Ramond, *Field Theory: A Modern Primer*, 2nd ed., Addison-Wesley, 1990.

J. Reitz, F. Milford, and R. Christy, *Foundations of Electromagnetic Theory*, 3rd ed., Addison-Wesley, 1979.

W. Rindler, *Essential Relativity*, Van Nostrand, Reinhold, 1969.

F. Rohrlich, *Classical Charged Particles*, Addison-Wesley, 1965.

J. Stachel, Ed., *The Collected Papers of Albert Einstein*, Vols. I and II, Princeton University Press, 1987 (Vol. I) and 1989 (Vol. II).

J. Stratton, *Electromagnetic Theory*, McGraw-Hill, 1941.

R. Utiyama, "Invariant Theoretical Interpretation of Interactions," Phys. Rev. **101**, 1597, 1956.

4-9   PROBLEMS

1. For the coordinate transformation:

$$x = 2uv,$$

$$y = u^2 - v^2,$$

$$z = w.$$

   A. Evaluate the metric

$$ds^2 = dx^2 + dy^2 + dz^2 = g_{11}(uvw)\, du^2 + \cdots.$$

   B. Express the vector $\mathbf{r} = [x, y, z]$ in the uvw coordinate system.

   C. What is it in covariant form?

   D. Find the inverse transformation.

   E. Find a unit vector tangent to the curve $x = 2uv$ in uvw coordinates.

   F. Do the same for the other two, and show they are mutually orthogonal.

2. Write the following matrix as a sum of symmetric and antisymmetric vectors.

$$M = \begin{bmatrix} 1 & 2 & 3 \\ 4 & 5 & 6 \\ 7 & 8 & 9 \end{bmatrix}.$$

3. Define the antisymmetric, three-dimensional matrix B^{jk} by the equation:

$$F^j = B^{jk} v_k,$$

   where F^j and v_k are known vectors. Show that all three components of B^{jk} cannot be determined uniquely. With this in mind, how can the magnetic field in a region be measured?

4. Discuss why the metric tensor must be symmetric.

5. The hypersurface element in a five-dimensional space is defined by:

$$dH^i = K\varepsilon^{ijkmn} d\xi_j\, d\xi_k\, d\xi_m\, d\xi_n.$$

   Evaluate the constant K.

6. Find the Jacobian and the volume element for the paraboloidal system:

$$x = 2uv \cos \psi,$$
$$y = 2uv \sin \psi,$$
$$z = u^2 - v^2.$$

7. The Earth's orbital velocity is about 20 miles/s. For this velocity calculate β and γ. For the Michelson–Morley experiment, express this as a phase shift for light of wavelength. $\lambda = 550$ nm.

8. The Σ^0 particle decays via electromagnetic interactions:

$$\Sigma^0 \rightarrow \Lambda^0 + \gamma,$$

with a lifetime at rest of about 1.0×10^{-14} s. Calculate its average distance before decay at energy 1 Gev (1 ev $= 1.602 \times 10^{-19}$ Joule). At what energy would it travel 1 cm?

9. A pion decays into a muon and a massless neutrino:

$$\pi^+ \rightarrow \mu^+ + v.$$

Suppose the pion velocity is half the speed of light and that the muon velocity is parallel to the pion velocity. What is the magnitude of the muon velocity?

10. A particle of mass m_1 and velocity v_0 collides with another of mass m_2 at rest at relativistic energies. If the collision takes place in one dimension, calculate the final velocities of the two particles after the collision.

11. Derive the equations:

$$\oiint F^{ik} \, dH_k = \frac{4\pi}{c} \iiiint J^i \, d^4x = \frac{4\pi}{c} P^i,$$
$$\oiint D^{ik} \, dH_k = 0,$$

analogous to Gauss's law. Give a physical interpretation to P^i.

12. Use equation (4-5.17) to derive equations (4-4.36), (4-4.37).

13. Express equation (4-4.25) in terms of the dual tensor D^{ik}. What is the source term in this equation?

14. Discuss the status of Newton's third law in a relativistic context, i.e., when simultaneity is relative.

15. Verify that equation (4-4.12) becomes the equation of motion (4-4.13) in three dimensions.

APPENDIX 4-1 LORENTZ TRANSFORMATION OF THE FIELD TENSOR

Define the Lorentz transformation matrix for the x–t plane. The symbols b and g are used for β and γ.

```
lorentz  = {{g,g*b,0,0},
            {g*b,g,0,0},
            {0,0,1,0},
            {0,0,0,1}};
```

Display it as a matrix.

```
MatrixForm[lorentz]
```

```
g     b g   0    0
b g   g     0    0
0     0     1    0
0     0     0    1
```

Define the electromagnetic field tensor.

```
ftensor  = {{0,-E1,-E2,-E3},
            {E1,0,-B3,B2},
            {E2,B3,0,-B1},
            {E3,-B2,B1,0}};
```

Display it as a matrix.

```
MatrixForm[ftensor]
```

```
0     -E1    -E2    -E3
E1    0      -B3    B2
E2    B3     0      -B1
E3    -B2    B1     0
```

Find the transformed electromagnetic field tensor by operating on both indices with the Lorentz transformation matrix.

```
ftransformed = lorentz.ftensor.lorentz
```

```
                2        2       2
{{0, -(E1 g ) + b  E1 g , -(b B3 g) - E2 g, b B2 g - E3 g},
        2    2      2
   {E1 g  - b  E1 g , 0, -(B3 g) - b E2 g, B2 g - b E3 g},
   {b B3 g + E2 g, B3 g + b E2 g, 0, -B1},
   {-(b B2 g) + E3 g, -(B2 g) + b E3 g, B1, 0}}
```

Apply some algebraic simplification.

```
Simplify[ftransformed/.g^2->1/(1-b^2)]
```

```
{{0, -E1, -((b B3 + E2) g), (b B2 - E3) g},
   {E1, 0, -((B3 + b E2) g), (B2 - b E3) g},
   {(b B3 + E2) g, (B3 + b E2) g, 0, -B1},
   {(-(b B2) + E3) g, (-B2 + b E3) g, B1, 0}}
```

Display it as a matrix.

```
MatrixForm[%]
```

```
0                 -E1              -((b B3 + E2) g)   (b B2 - E3)
g
E1                0                -((B3 + b E2) g)   (B2 - b E3)
g
(b B3 + E2) g     (B3 + b E2) g    0                  -B1
(-(b B2) + E3) g  (-B2 + b E3) g   B1                 0
```

CHAPTER 5

Electromagnetic Fields in Steady States

Since 'tis Nature's law to change,
Constancy alone is strange.

Earl of Rochester
A Satire Against Mankind

Reality must take precedence over public
relations, for Nature cannot be fooled.

Richard Feynman
Appendix to the Commission Report on the Challenger Disaster

In Chapter 4 we derived the electromagnetic field equations; in this and following chapters, we find solutions to the field equations for special circumstances. This chapter deals with the case of essentially static charges or currents. Later chapters deal with various aspects of radiation and the motion of charges in electromagnetic fields.

5-1 STEADY-STATE EQUATIONS

In this chapter we are concerned with the steady-state solutions, i.e., when neither the field nor the source depends on time. The terms *static* and *time independent* are also used, but since currents, i.e., moving charges, are the sources of magnetic fields, the term steady-state best expresses the condition. First we must consider the conditions for the steady state. In general, when a conducting body has charge placed on it, the charge will distribute itself over the body, and come to rest after some time interval. Let us express the time for this redistribution in precise terms. Suppose some charge is placed into a sample of material having conductivity, σ, and dielectric constant, ε. Then electric fields will be set up in the material which will cause charge to flow until the minimum energy configuration is reached. At any given time conservation of charge holds at each point in the material.

$$\nabla \cdot J + \frac{\partial \rho}{\partial t} = 0. \tag{5-1.1}$$

Then applying Ohm's law, (3-2.1), and equation, (3-3.9), gives:

$$\sigma J + \frac{\varepsilon}{4\pi} \frac{\partial J}{\partial t} = 0. \tag{5-1.2}$$

The solution is the exponential function:

$$J = J_0 \, \exp(-t/\tau), \tag{5-1.3}$$

where the constant:

$$\tau = \frac{\varepsilon}{4\pi\sigma}, \tag{5-1.4}$$

is the *relaxation time* for the material. After the passage of several relaxation times, the current in the sample will have vanished and the charge will be at rest.

The mathematical properties of linear equations requires the total electromagnetic field at any point to be the sum of superimposed fields from various sources; this agrees with experience. For example, imagine a cavity with magnetic walls bearing a static charge, with a laser beam passing through, and containing a radiating plasma. The electromagnetic field at any point in the cavity is the sum of the individual fields. The electromagnetic field equations (4-4.25), (4-4.36), and (4-4.37) have the form:

$$\partial_i \partial^i F = s, \tag{5-1.5}$$

where F represents any of the fields and s its corresponding source. The general solution of this linear equation is the sum of solutions of several homogeneous and inhomogeneous equations. If we write:

$$\partial_i \partial^i F_1 = 0, \quad \text{D'Alembert wave equation}, \tag{5-1.6}$$

$$\partial_i \partial^i F_2 = s_2, \quad \text{inhomogeneous wave equation}, \tag{5-1.7}$$

$$\nabla^2 F_3 = 0, \quad \text{Laplace equation}, \tag{5-1.8}$$

$$\nabla^2 F_4 = s_4, \quad \text{Poisson equation}, \tag{5-1.9}$$

where F_2 and F_4 are particular solutions of inhomogeneous equations, then the general solution of equation (5-1.5) is:

$$F = F_1 + F_2 + F_3 + F_4. \tag{5-1.10}$$

The *principle of superposition* is therefore a result of the linearity of the electromagnetic field equations.

Poisson's equation for electrostatic potential in vacuum is a special application of equation (5-1.9):

$$\nabla^2 \phi = -4\pi\rho. \tag{5-1.11}$$

The particular solution can be found by using Green's theorem, equation (1-3.40):

$$\iiint (\Psi\nabla^2\Phi - \Phi\nabla^2\Psi)\, d^3x = \oiint (\Psi\nabla\Phi - \Phi\nabla\Psi)\cdot d\mathbf{S}. \tag{5-1.12}$$

Then choose the function:

$$\Phi = \phi(r_s), \tag{5-1.13}$$

which is a function of the source variable, and:

$$\Psi = \frac{-1}{4\pi}\frac{1}{|\mathbf{r}_f - \mathbf{r}_s|} = \frac{-1}{4\pi}\frac{1}{r_{fs}}, \tag{5-1.14}$$

which is a function of source and field variables. Now substitute equations (5-1.11), Φ, and Ψ, and integrate over the source variable r_s. *Use of the delta function gives:*

$$\int \left(\frac{-1}{4\pi r_{fs}}\nabla^2\phi - \phi\nabla^2\frac{-1}{4\pi r_{fs}}\right) d^3r_s = \int \left(\frac{\rho(r_s)}{r_{fs}} - \phi(r_s)\delta^3(r_{fs})\right) d^3r_s$$

$$= \int \left(\frac{\rho(r_s)}{r_{fs}}\right) d^3r_s - \phi(r_f). \tag{5-1.15}$$

Then, using Green's theorem and rearranging:

$$\phi(r_f) = \iiint \frac{\rho(r_s)}{r_{fs}}\, d^3r_s - \frac{1}{4\pi}\oiint \left[\phi(r_s)\nabla_s\frac{1}{r_{fs}} - \frac{1}{r_{fs}}\nabla_s\phi(r_s)\right]\cdot d\mathbf{S}_s. \tag{5-1.16}$$

Let the integration be over all space; then for large r_f, the surface integrand varies as r^{-3} while the surface element varies as r^2. Therefore, the surface integral vanishes as r_s goes to infinity, and ϕ becomes the Coulomb potential.

5-2   MULTIPOLE EXPANSION

Although the Coulomb potential in equation (5-1.16) satisfies Poisson's equation, the charge distribution is a completely arbitrary function of position. Exact evaluation of the integral may be impossible for a given distribution. Numerical evaluation of the Coulomb integral may still be possible, but the procedure is time consuming when high accuracy is needed, and the result is a table of values which may be difficult to interpret theoretically. For these reasons the expansion of the integral into a *multipole* series may be preferable, since it requires only a small number of integrations and the multipoles are of theoretical interest in themselves. Truncation of the series to only a few terms (multipoles) is usually sufficient. For the moment we express the multipoles in Cartesian coordinates; expression in spherical coordinates is possible after discussing orthogonal functions.

The Taylor expansion of a function of several variables requires only a slight generalization of the familiar process. Suppose a function can be expressed as the general sum of powers of the variables:

$$f(xyz) = \sum_{i=0}^{\infty} \sum_{j=0}^{\infty} \sum_{k=0}^{\infty} A_{ijk} x^i y^j z^k \qquad (i, j, k \text{ are exponents}). \qquad (5\text{-}2.1)$$

The coefficients, A_{ijk}, are found by differentiating the appropriate number of times and then setting x, y, z equal to zero. The general coefficient is:

$$A_{ijk} = \frac{1}{i! \, j! \, k!} \left[\frac{\partial^{i+j+k}}{\partial x^{(i)} \, \partial y^{(j)} \partial z^{(k)}} f(xyz) \right]_{x=y=z=0}, \qquad (5\text{-}2.2)$$

where (i), (j), and (k) are exponents. For simplicity expand the function around the origin, $x = y = z = 0$.

Now apply this expansion to the function of the *source variable* $\mathbf{r}_s$ where the field variable $\mathbf{r}_f$ is understood to be fixed:

$$f(x_s y_s z_s) = \frac{1}{r_{fs}} = \frac{1}{|\mathbf{r}_f - \mathbf{r}_s|}. \qquad (5\text{-}2.3)$$

To second order the expansion becomes:

$$\frac{1}{r_{fs}} = \frac{1}{r_f} + \frac{\sum_{j=1}^{3} r_s^j r_f^j}{r_f^3} + \frac{1}{2!} \frac{\sum_{j=1}^{3} \sum_{k=1}^{3} \left(r_s^j r_s^k \right) \left(3 r_f^j r_f^k - |r_f|^2 \delta^{jk} \right)}{r_f^5} + \cdots .$$
$$(5\text{-}2.4)$$

Substitute this into the Coulomb potential, giving a sum of integrals. But the integration is over the source variables, i.e., functions of the field variables may be treated as constant, and taken outside the integrals. Thus, to first order the potential becomes:

$$\phi(r_f) = \iiint \rho(r_s) \left[\frac{1}{r_f} + \frac{\sum_{j=1}^{3} r_s^j r_f^j}{r_f^3} + \cdots \right] d^3 r_s$$

$$= \frac{\iiint \rho(r_s)\, d^3r_s}{r_f} + \frac{\sum_{j=1}^{3} r_f^j \left(\iiint \rho(r_s) r_s^j\, d^3r_s \right)}{r_f^3} + \cdots$$

$$= \frac{q}{r_f} + \frac{\mathbf{p} \cdot \mathbf{r}_f}{r_f^3} + \cdots, \tag{5-2.5}$$

where q and P^j are, respectively, the *monopole moment*:

$$q = \iiint \rho(r_s)\, d^3r_s, \tag{5-2.6}$$

which is the total charge inside the volume, the *dipole moment*:

$$p^j = \iiint r_s^j \rho(r_s)\, d^3r_s. \tag{5-2.7}$$

The next term in the expansion is the quadrupole term:

$$\phi_{\text{quad}} = \frac{1}{2!} \iiint \rho(r_s) \left[\frac{\sum_{j=1}^{3} \sum_{k=1}^{3} \left(3r_f^j r_f^k - \delta^{jk} |r_f|^2 \right) r_s^j r_s^k}{r_f^5} \right] d^3r_s$$

$$= \frac{1}{2!} \frac{\sum_{j=1}^{3} \sum_{k=1}^{3} \left(3r_f^j r_f^k - \delta^{jk} |r_f|^2 \right) Q^{jk}}{r_f^5}, \tag{5-2.8}$$

expressed in terms of a *quadrupole moment*:

$$Q^{jk} = \iiint r_s^j r_s^k \rho(r_s)\, d^3r_s$$

$$= \begin{bmatrix} \iiint x_s^{\,2} \rho(r_s)\, d^3r_s & \iiint x_s y_s \rho(r_s)\, d^3r_s & \iiint x_s z_s \rho(r_s)\, d^3r_s \\ \iiint y_s x_s \rho(r_s)\, d^3r_s & \iiint y_s^{\,2} \rho(r_s)\, d^3r_s & \iiint y_s z_s \rho(r_s)\, d^3r_s \\ \iiint z_s x_s \rho(r_s)\, d^3r_s & \iiint z_s y_s \rho(r_s)\, d^3r_s & \iiint z_s^{\,2} \rho(r_s)\, d^3r_s \end{bmatrix}. \tag{5-2.9}$$

However, due to the identity:

$$\sum_{j=1}^{3} \sum_{k=1}^{3} \delta^{jk} \frac{\partial^2}{\partial x_s^j \partial x_s^k} \left(\frac{1}{r_{fs}} \right) = \sum_{j=1}^{3} \sum_{k=1}^{3} \delta^{jk} \frac{\partial^2}{\partial x_f^j \partial x_f^k} \left(\frac{1}{r_{fs}} \right) = \nabla^2 \frac{1}{r_{fs}} = 0, \tag{5-2.10}$$

we can replace $\mathbf{r}_f$ with $\mathbf{r}_s$ in the δ^{jk} term in equation (5-2.8). Then, since the other term in the integrand is obviously symmetric in $\mathbf{r}_f$ and $\mathbf{r}_s$, the multipole expansion may be written:

$$\phi(r_f) = \frac{q}{r_f} + \frac{\mathbf{p} \cdot \mathbf{r}_f}{r_f^3} + \frac{1}{2!} \frac{\sum_{j=1}^{3} \sum_{k=1}^{3} \tilde{Q}^{jk} r_f^j r_f^k}{r_f^5} + \cdots \tag{5-2.11}$$

in terms of the modified quadrupole moment:

$$\tilde{Q}^{jk} = \iiint \rho(r_s) \left[3r_s^j r_s^k - |r_s|^2 \delta^{jk} \right] d^3 r_s. \qquad (5\text{-}2.12)$$

The modified quadrupole moment is obviously symmetric, but in addition, the sum of the diagonal terms (the trace) vanishes, leaving only five independent components. Furthermore, by rotating the coordinate axes, it is possible to make the matrix diagonal. Then, applying the zero trace, only two independent components remain in the quadrupole moment.

Higher multipole moments are expressed by higher-order tensors. The problem of evaluating the Coulomb integral at each field point has been reduced to evaluating the multipole integrals. A multitude of numerical integrals can be reduced to just ten, or fewer when there are symmetries. When exact integration is impossible, the integrals may be evaluated numerically. For example, Monte Carlo integration is treated in the problems.

The preceding multipole expansion was in Cartesian coordinates. However, it is frequently convenient to express it in spherical (or polar) coordinates. This could be done by expanding $1/r_{fs}$ in spherical coordinates right from the beginning. Alternatively, we may transform our previous expressions to spherical coordinates using:

$$x = r \sin \theta \cos \phi,$$
$$y = r \sin \theta \sin \phi, \qquad (5\text{-}2.13)$$
$$z = r \cos \theta.$$

EXAMPLE

Let the charge density be symmetric around the z-axis. Then in terms of polar coordinates, r and θ, each multipole reduces to a single term, and the expansion (5-2.5) becomes:

$$\phi(r_f) = \frac{q}{r_f} + \frac{p \cos \theta}{r_f^2} + \frac{1}{2!} \frac{Q(3 \cos^2 \theta - 1)}{r_f^3} + \cdots$$

$$= \frac{q P_0(\cos \theta)}{r_f} + \frac{p P_1(\cos \theta)}{r_f^2} + \frac{1}{2!} \frac{Q P_2(\cos \theta)}{r_f^3} + \cdots, \qquad (5\text{-}2.14)$$

where the first three *Legendre polynomials* are:

$$P_0(\mu) = 1,$$
$$P_1(\mu) = \mu,$$
$$P_2(\mu) = \frac{1}{2}(3\mu^2 - 1). \qquad (5\text{-}2.15)$$

If the charge density lacks symmetry, all three spherical coordinates are necessary, and the expansion requires the associated Legendre polynomials. A complete discussion of the multipole expansion in spherical coordinates is lengthy and complicated, and we will not pursue it further. There is ample treatment in the literature, particularly in Jackson, Morse and Feshbach, and Rose.

In practice the series must be truncated after a few terms. However, in general, the multipole expansion is an infinite series, and the conditions for convergence must be considered. This method is particularly suitable when the sources are confined to a small region and the field is wanted at large distances. That is, when:

$$\mathbf{r}_f \gg \mathbf{r}_s. \tag{5-2.16}$$

EXAMPLE

A general charge distribution contains all multipoles; however, certain charge distributions may reduce to a finite number of terms. Thus a spherically symmetric distribution of charge contains only the monopole term when expanded around the origin. However, the same distribution translated from the origin contains both monopole and dipole terms. Take, for example, charge distributions: $\rho = q\delta(r)$ and $\rho = q\delta(r + \Delta r)$. A pure dipole can be constructed from a pure monopole by placing an identical monopole distribution but of opposite sign at some displacement from it. If a charge distribution has a potential:

$$\phi = \frac{\iiint \rho \, d^3 x}{r}, \tag{5-2.17}$$

then introducing an identical charge distribution with opposite sign at a displacement $\Delta \mathbf{x}$ from it produces the potential:

$$\phi = \iiint \frac{\rho \, d^3 r}{|\mathbf{r} + \Delta \mathbf{r}|} - \iiint \frac{\rho \, d^3 r}{r}$$

$$\approx \iiint \left[\frac{1}{r} + \frac{\mathbf{r} \cdot \Delta \mathbf{r}}{r^3} \right] \rho \, d^3 r - \frac{\iiint \rho \, d^3 r}{r}$$

$$= \frac{\mathbf{p} \cdot \mathbf{r}}{r^3}. \tag{5-2.18}$$

Thus the monopole term vanishes and by the symmetries of the integral all the higher terms vanish as well. Similarly, a pure quadrupole results from the introduction of an identical but opposite sign dipole displaced from the original. Figure 5-1 shows some elementary multipoles.

The vector potential A^k can also be expressed as a multipole expansion. Assuming unit permeability, the particular solution of Poisson's

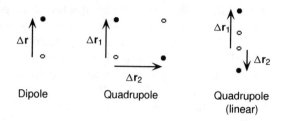

FIGURE 5-1. ELEMENTARY MULTIPOLES. The dark circles and light circles represent positive and negative charges, respectively. The two quadrupole displacements are independent of each other.

equation for the vector potential is:

$$A^k(r_f) = \frac{1}{c} \iiint \frac{J^k(r_s)}{|\mathbf{r}_f - \mathbf{r}_s|} \, d^3r_s. \qquad (5\text{-}2.19)$$

The expansion becomes:

$$A^k(r_f) = \frac{\iiint J^k(r_s)\, d^3r_s}{c|\mathbf{r}_f|} + \frac{\mathbf{r}_f \cdot \iiint \mathbf{r}_s J^k(r_s)\, d^3r_s}{c|\mathbf{r}_r|^3} + \cdots. \qquad (5\text{-}2.20)$$

They can be expressed in familiar terms.

The first, zeroth-order, term can be written in either of two ways depending on the current distribution. If the current density represents a moving charged particle, i.e., where:

$$\mathbf{J}(r_s) = q\delta^3(\mathbf{r}_s - \mathbf{r}_{\text{path}}(t))\mathbf{v}(\mathbf{r}_s), \qquad (5\text{-}2.21)$$

then the first-order term of the potential may be evaluated:

$$\mathbf{A}(r_f) = \frac{1}{c} \frac{q\mathbf{v}(\mathbf{r}_{\text{path}})}{|\mathbf{r}_f - \mathbf{r}_{\text{path}}|}. \qquad (5\text{-}2.22)$$

A relativistic form of this potential will be introduced in Section 8-5.

If the current distribution is a steady-state closed loop, then the integration is over the toroidal volume of the loop. The integral becomes:

$$\iiint \mathbf{J}(r_s)\, d^3r_s = \iiint \mathbf{J}(r_s)\, d\mathbf{S} \cdot d\mathbf{r}$$

$$= I \oint d\mathbf{r}$$

$$= 0, \qquad (5\text{-}2.23)$$

where we have used the fact that the direction of the current density, **J**, at any point, is parallel to the line element $d\mathbf{r}$. The *closed* line integral of $d\mathbf{r}$ vanishes identically. In this case the zeroth-order term vanishes.

The integral in the second (first-order) term in (5-2.20) is evaluated as follows. Separate the integrand into symmetric and antisymmetric terms:

$$\iiint r_s^j J^k(r_s)\, d^3 r_s = \iiint \frac{1}{2}\left[r_s^j J^k(r_s) - r_s^k J^j(r_s)\right] d^3 r_s$$
$$+ \iiint \frac{1}{2}\left[r_s^j J^k(r_s) + r_s^k J^j(r_s)\right] d^3 r_s.$$

$$(5\text{-}2.24)$$

The antisymmetric integral gives:

$$\sum_{j=1}^{3} r_f^j \iiint \frac{1}{2}\left[r_s^j J^k(r_s) - r_s^j J^k(r_s)\right] d^3 r_s$$

$$\rightarrow \iiint \frac{1}{2}\left[\mathbf{r}_s \times \mathbf{J}(r_s)\right] d^3 r_s \times \mathbf{r}_f = c\mathbf{m} \times \mathbf{r}_f. \quad (5\text{-}2.25)$$

The *magnetic dipole moment*, **m**, was defined in Section 3-2. The symmetric integral is evaluated using the divergence theorem. Applying the equation of continuity gives:

$$\sum_k \partial_k(r^i r^j J^k) = \sum_k (\partial_k r^i) r^j J^k + \sum_k r^i (\partial_k r^j) J^k + \sum_k r^i r^j (\partial_k J^k)$$

$$= r^j J^i + r^i J^j - r^i r^j\, \partial_t \rho. \quad (5\text{-}2.26)$$

But the integral:

$$\iiint \sum_{k=1}^{3} \partial_k(r^i r^j J^k)\, d^3 r = \oiint \sum_{k=1}^{3} r^i r^j J^k\, dS_k = 0 \quad (5\text{-}2.27)$$

vanishes because currents vanish on a sufficiently distant surface. Now combine equations (5-2.26) and (5-2.27) to evaluate the symmetric integral in equation (5-2.24). This gives:

$$\iiint r_s^j J^k(r_s)\, d^3 r_s = \iiint \left[r_s^j r_s^k\, \partial_t \rho^k(r_s)\right] d^3 r_s$$

$$= \partial_t Q^{jk}. \quad (5\text{-}2.28)$$

Strictly speaking, time derivatives vanish from steady-state expressions, but if the motion of the electric source charges is slow enough (adiabatic), the term may be useful. It also serves as a reminder of the close connection between electric and magnetic fields.

After making all the substitutions, the expansion of the vector potential becomes:

$$\mathbf{A}(r_f) = \frac{\mathbf{m} \times \mathbf{r}_f}{|r_f|^3} + \frac{1}{2c}\frac{\mathbf{r}_f \cdot \partial_t \mathbf{Q}}{|r_f|^3} + \cdots. \quad (5\text{-}2.29)$$

In the second term the dot product between the matrix $\mathbf{Q}$ and the vector $\mathbf{r}_f$ is a common short notation for the summation over the j-indices of Q^{ij} and r^j. The first-order term for $\mathbf{A}$ has r^{-2} dependence, like the zeroth-order term in ϕ. Corresponding relations hold for higher-order terms in each expansion.

5-3   Laplace's Equation: Separation of Variables

There are many methods of solution of Laplace's equation; however, we showed in Section 3-5 that any solution satisfying a particular set of boundary conditions is unique. This gives us confidence in using whatever method is convenient.

Probably the most powerful and convenient method of solution is the expansion in orthogonal series, usually referred to as separation of variables. For simplicity, we will consider the two-dimensional case. Then Laplace's equation becomes:

$$\frac{\partial^2 \phi}{\partial x^2} + \frac{\partial^2 \phi}{\partial y^2} = 0. \tag{5-3.1}$$

Because this is a homogeneous equation we may suppose the general solution is a sum of solutions. As a guess, try the product solution:

$$\phi(x, y) = X(x)Y(y). \tag{5-3.2}$$

Substituting and rearranging gives:

$$\frac{\partial_x^2 X}{X} + \frac{\partial_y^2 Y}{Y} = 0. \tag{5-3.3}$$

Since x and y are independent variables, the first expression is an arbitrary function of x, and the second is an arbitrary function of y. The only way this equation is possible is for each term to be constant. Thus equation (5-3.3) becomes two ordinary differential equations:

$$\frac{d^2 X}{dx^2} + k^2 X = 0, \tag{5-3.4}$$

$$\frac{d^2 Y}{dy^2} - k^2 Y = 0, \tag{5-3.5}$$

where the constant k is undetermined; for convenience it is written as a square and may take any value from zero to infinity. The choice of which equation takes the negative sign is arbitrary, but one of them must have it to satisfy equation (5-3.3). Equations of this form are called *eigenvalue equations*, and the allowed values of the constant k are called *eigenvalues*. The solutions to the eigenvalue equations are:

$$X = A_0 + B_0 x, \qquad k = 0,$$

$$X = \begin{cases} A_k \cos kx + B_k \sin kx, \\ \\ A_k e^{ikx} + B_k e^{-ikx}, \end{cases} \qquad k \neq 0,$$

$$Y = C_0 + D_0 y, \qquad k = 0,$$

$$Y = \begin{cases} C_k \cosh ky + C_k \sinh ky, \\ \\ C_k e^{ky} + D_k e^{-ky}, \end{cases} \qquad k \neq 0,$$

$$(5\text{-}3.6)$$

The sum of the products of these functions over all k is a solution of Laplace's equation. This can be seen by substitution. Thus, the general solution is:

$$\phi = [A_0 + B_0 x][C_0 + D_0 y]$$

$$+ \sum_{k \neq 0}^{\infty} [A_k \cos kx + B_k \sin kx][C_k \cosh ky + D_k \sinh ky]. \quad (5\text{-}3.7)$$

Periodic functions in the variable x may be represented by letting $k = 2\pi n/L$ where L is the period. Alternatively, for nonperiodic functions, the summation may be replaced by an integral over k. The question now becomes whether this series (or integral) can represent an arbitrary function on the boundary. Since there is an infinite number of undetermined parameters (the A's, B's, C's, and D's) in the expansion it seems reasonable. We will begin with a simple case.

EXAMPLE

Consider a parallel plate capacitor with potential Φ on the upper plate and zero potential on the lower. We suppose the separation is d, and the size of the plates is large enough to ignore edge effects. Since the potential vanishes when $y = 0$ and the boundary potential is an even function of x, we must have:

$$A_k = 0, \qquad k \neq 0, \tag{5-3.8}$$

$$B_k = 0. \tag{5-3.9}$$

Applying the last boundary condition gives:

$$\phi = \frac{\Phi}{d} y, \tag{5-3.10}$$

which is the expected expression.

Example

Consider similar parallel plate boundaries where the boundary potentials are:

$$\phi(x, d) = f(x), \tag{5-3.11}$$

where $f(x)$ is an even function of x and:

$$\phi(x, 0) = 0. \tag{5-3.12}$$

Application of the second boundary condition leaves the single sum:

$$\phi = D_0 y + \sum_{k \neq 0}^{\infty} D_k \cos kx \sinh ky. \tag{5-3.13}$$

Now apply the first boundary condition. The potential on the upper boundary becomes:

$$f(x) = D_0 d + \sum_{k \neq 0}^{\infty} \left[D_k \sinh kd \right] \cos kx$$

$$= E_0 + \sum_{k \neq 0}^{\infty} E_k \cos kx. \tag{5-3.14}$$

We thus represent an arbitrary (even) function as a series of cosine functions, i.e., by a *Fourier series*. The potential is illustrated by Figure 5-2.

The preceding may be generalized to include functions of odd or no symmetry in the variable, x. Since, sine and cosine are periodic functions, the function represented is periodic also. Then, assume that we may write for a given function $f(x)$ with period L the series

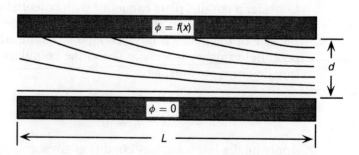

Figure 5-2. Solution to Laplace's equation with parallel boundaries.   The potentials on the two boundaries are independent functions represented by a Fourier series. The mathematical function repeats outside the region $-L < 2x < L$. Infinite boundaries in the x-direction would require a Fourier integral.

representation:

$$f(x) = E_0 + \sum_{n=1}^{\infty} E_n \cos \frac{2\pi n}{L} x + \sum_{n=1}^{\infty} F_n \sin \frac{2\pi n}{L} x$$

$$= \sum_{n=0}^{\infty} E_n \cos \frac{2\pi n}{L} x + \sum_{n=0}^{\infty} F_n \sin \frac{2\pi n}{L} x. \tag{5-3.15}$$

We must find a way to determine the coefficients E_n and F_n. First, note that the sine and cosine functions have the integrals:

$$\int_{-L/2}^{L/2} \left[\sin \frac{2\pi n}{L} x \right] dx = \int_{-L/2}^{L/2} \left[\cos \frac{2\pi n}{L} x \right] dx$$

$$= \int_{-L/2}^{L/2} \left[\sin \frac{2\pi n}{L} x \right] \left[\cos \frac{2\pi m}{L} x \right] dx$$

$$= 0, \tag{5-3.16}$$

$$\int_{-L/2}^{L/2} \left[\sin \frac{2\pi m}{L} x \right] \left[\sin \frac{2\pi n}{L} \right] x \, dx = \int_{-L/2}^{L/2} \left[\cos \frac{2\pi m}{L} x \right] \left[\cos \frac{2\pi n}{L} x \right] dx$$

$$= \delta_{mn} L/2. \tag{5-3.17}$$

Functions with these properties are said to be *orthogonal functions*, in analogy with vectors for which the orthogonality condition is:

$$\mathbf{V}_m \cdot \mathbf{V}_n = \sum_{j=1}^{N} V_m^j V_n^j = |V_n|^2 \delta_{mn}, \tag{5-3.18}$$

where $|V_n|$ is the magnitude, a constant.

The procedure for finding the coefficients is as follows: First, because the function is periodic, we may choose the limits of integration to be 0 and L instead of $L/2$ and $-L/2$. Now multiply equation (5-3.15) by a sine function and integrate over these limits:

$$\int_0^L f(x) \sin \frac{2\pi m}{L} x \, dx = \sum_{n=1}^{\infty} \int_0^L \left(F_n \sin \frac{2\pi m}{L} \sin \frac{2\pi n}{L} x \right) dx = \frac{L}{2} F_m, \tag{5-3.19}$$

which determines the constants F_m. Similarly, the constants E_m are found by using the cosine function, and the constant E_0 is found by simply integrating the function over the interval L. Summarizing the results:

$$E_0 = \frac{1}{L} \int_0^L f(x) \, dx, \tag{5-3.20}$$

$$E_m = \frac{2}{L} \int_0^L f(x) \cos \frac{2\pi m}{L} x \, dx, \tag{5-3.21}$$

$$F_m = \frac{2}{L} \int_0^L f(x) \sin \frac{2\pi m}{L} x \, dx. \qquad (5\text{-}3.22)$$

These are the *Fourier coefficients* for the function $f(x)$. If the function is even, all the F_m vanish, and if the function is odd, all the E_m vanish. Using these integrals, equation (5-3.14) is now the solution to Laplace's equation everywhere, including the boundaries. The conditions for convergence, the set of functions that can be represented by Fourier series, and related mathematical topics are beyond the scope of our discussion. For further discussion of these topics the reader should consult the references.

EXAMPLE

Find the Fourier coefficients for the function:

$$f(x) = x, \qquad -1 \le x \le 1. \qquad (5\text{-}3.23)$$

Then:

$$F_m = \int_{-1}^{1} x \sin m\pi x \, dx = \frac{-2\cos(m\pi)}{(m\pi)}, \qquad (5\text{-}3.24)$$

$$E_m = 0. \qquad (5\text{-}3.25)$$

Since the function $f(x) = x$ can be expressed as a Fourier series, it is not absolutely necessary to include it explicitly in equations (5-3.7) for $k = 0$, although that may be convenient for some specific calculations.

EXAMPLE

Figure 5-3 shows the Fourier representation, truncated after eight terms, of a square wave function of period $L = 1$ in the interval $-2 < x < 2$.

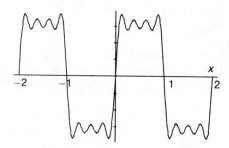

FIGURE 5-3. FOURIER EXPANSION OF THE SQUARE WAVE FUNCTION.   Two periods are calculated to 8 terms and plotted from $x = -2$ to $x = 2$.

A Fourier series can represent practically any piecewise continuous periodic function. This is sufficient to handle most boundary conditions involving a finite region when it does not matter if the boundary condition repeats outside the region of interest. However, representing a truly nonperiodic function (such as a function with infinite domain) requires a generalization, i.e., a Fourier transform. Let us write the Fourier series in terms of complex exponentials:

$$f(x) = \sum_{k=1}^{\infty} E_k e^{ikx} + \sum_{k=1}^{\infty} F_k e^{-ikx}, \tag{5-3.26}$$

where the constants E_k and F_k may be complex. Since we now wish to sum over the continuous values of k, the summation becomes an integral. Furthermore, a single integral handles negative and positive values of k. We write the *Fourier transform*:

$$f(x) = \int_{-\infty}^{\infty} F(k) e^{ikx} \, dk, \tag{5-3.27}$$

where constant coefficients E_k have become a function $E(k)$. The problem now is how to find this function. The coefficients of the Fourier series were found by the orthogonal properties of the sine and cosine. By analogy, we multiply by e^{-ikx} and integrate with respect to dx:

$$\int_{-\infty}^{\infty} f(x) e^{-ik'x} \, dx = \int_{-\infty}^{\infty} F(k) \int_{-\infty}^{\infty} e^{i(k-k')x} \, dx \, dk. \tag{5-3.28}$$

The inner integral is a δ-function of k. That is:

$$\int_{-\infty}^{\infty} e^{i(k-k')x} \, dx = 2\pi\delta(k - k'). \tag{5-3.29}$$

Therefore the *inverse Fourier transform* gives the coefficient:

$$F(k) = \frac{1}{2\pi} \int_{-\infty}^{\infty} f(x) e^{-ikx} \, dx. \tag{5-3.30}$$

This single equation is equivalent to the three equations (5-3.20), (5-3.21), (5-3.22).

The method of separation of variables with Fourier expansion presented above works efficiently with rectangular boundaries. If the boundary can be separated into rectangular regions, the coefficients can be matched over the boundaries between the regions. However, if the boundary is curvilinear, the problem is more difficult. One might naively attempt to apply equation (5-3.7) to a curved boundary. For example, one might express the boundary curves as parametric equations:

$$x = f(s), \tag{5-3.31}$$

$$y = g(s). \tag{5-3.32}$$

That is, the potential on the boundary is expressed as a function of s:

$$\phi(s) = [A_0 + B_0 f(s)][C_0 + D_0 g(s)]$$

$$+ \sum_{k \neq 0}^{\infty} [A_k \cos kf(s) + B_k \sin kf(s)]$$

$$\times [C_k \cosh kg(s) + D_k \sinh kg(s)]. \qquad (5\text{-}3.33)$$

Finding values for the coefficients in this case is more difficult. Although the expression is capable of representing an arbitrary function on the boundary, the new functions of s will not be orthogonal, in general.[1]

However, use of the *Gram–Schmidt procedure* gives an orthogonal basis. We illustrate this procedure for a set of three vectors in three-dimensional space.

EXAMPLE

Suppose the vectors $\mathbf{V}^1$, $\mathbf{V}^2$, and $\mathbf{V}^3$ are linearly independent vectors but are not orthogonal to each other, or of unit length. Define a new set of three vectors. First:

$$\mathbf{W}^1 = \mathbf{V}^1. \qquad (5\text{-}3.34)$$

Second, define a vector $\mathbf{W}^2$ by subtracting from $\mathbf{V}^2$ the part of $\mathbf{V}^2$ that is parallel to $\mathbf{V}^1$:

$$\mathbf{W}^2 = \mathbf{V}^2 - \left[\mathbf{V}^2 \cdot \frac{\mathbf{V}^1}{|\mathbf{V}^1|^2} \right] \mathbf{V}^1. \qquad (5\text{-}3.35)$$

Then $\mathbf{W}^1$ is orthogonal to $\mathbf{W}^2$:

$$\mathbf{W}^2 \cdot \mathbf{W}^1 = \mathbf{V}^2 \cdot \mathbf{V}^1 - \left[\mathbf{V}^2 \cdot \frac{\mathbf{V}^1}{|\mathbf{V}^1|^2} \right] \mathbf{V}^1 \cdot \mathbf{V}^1 = 0. \qquad (5\text{-}3.36)$$

Third, define a new vector, $\mathbf{W}^3$, by subtracting from $\mathbf{V}^3$ the parts of $\mathbf{V}^3$ that are parallel to $\mathbf{W}^1$ and $\mathbf{W}^2$:

$$\mathbf{W}^3 = \mathbf{V}^3 - \left[\mathbf{V}^3 \cdot \frac{\mathbf{W}^1}{|\mathbf{W}^1|^2} \right] \mathbf{W}^1 - \left[\mathbf{W}^3 \cdot \frac{\mathbf{W}^2}{|\mathbf{W}^2|^2} \right] \mathbf{W}^2. \qquad (5\text{-}3.37)$$

---

[1] In mathematical language, the sine and cosine functions of x are sufficient to span the function space on the boundary. The new functions of s also span the function space on the boundary, but are not necessarily orthogonal to each other.

Then $\mathbf{W}^3$ is orthogonal to $\mathbf{W}^1$ and $\mathbf{W}^2$. For example:

$$\mathbf{W}^3 \cdot \mathbf{W}^2 = \mathbf{V}^3 \cdot \mathbf{W}^2 - \left[\mathbf{V}^3 \cdot \frac{\mathbf{W}^1}{|\mathbf{W}^1|^2} \right] \mathbf{W}^1 \cdot \mathbf{W}^2 - \left[\mathbf{V}^3 \cdot \frac{\mathbf{W}^2}{|\mathbf{W}^2|^2} \right] \mathbf{W}^2 \cdot \mathbf{W}^2$$

$$= \mathbf{V}^3 \cdot \mathbf{W}^2 - 0 - \mathbf{V}^3 \cdot \mathbf{W}^2 = 0.$$

(5-3.38)

We can normalize $\mathbf{W}^j$ by dividing it by its magnitude $|W^n|$:

$$\mathbf{w}^n = \frac{\mathbf{W}^n}{|\mathbf{W}^n|}.$$

(5-3.39)

The orthogonality condition is analogous to the dot product in N dimensions:

$$\mathbf{w}^m \cdot \mathbf{w}^n = \sum_{j=1}^{N} w_j^m w_j^n = \delta^{mn},$$

(5-3.40)

where we sum over the components (j) of the vectors $\mathbf{w}^m$ and $\mathbf{w}^n$. Now the vectors $\mathbf{w}^n$ can be used to express any vector $\mathbf{u}$ in the N-dimensional space as a linear combination of the normalized basis vectors:

$$\mathbf{u} = \sum_{n=1}^{N} a_n \mathbf{w}^n.$$

(5-3.41)

The $\mathbf{w}^n$ are the *unit basis vectors*. The coefficients are found by using the orthogonality of the basis vectors:

$$\mathbf{w}^m \cdot \mathbf{u} = \sum_{n=1}^{N} a_n \mathbf{w}^m \cdot \mathbf{w}^n = \sum_{n=1}^{N} a_n \delta^{mn}$$

$$= a_m.$$

(5-3.42)

Comparing equations (5-3.16) through (5-3.22) with equations (5-3.34) through (5-3.42) shows an isomorphism between ordinary vectors and Fourier functions. The sinusoidal (Fourier) functions are generalized basis vectors and the function $u(x)$ represents an arbitrary vector in the Fourier space. Fourier space is an infinite-dimensional space, and it is necessary to include sine, cosine, and unit functions as basis vectors. In this scheme, the variable x is the index for the vector $u(x)$ and the basis vectors are $\sin(n\pi x/L)$, where the index n denumerates a specific basis vector. Integration over the continuous index, x, replaces summation over an integer index.

$$
\mathbf{w}^n \Longleftrightarrow
\begin{cases}
\sqrt{\dfrac{1}{L}}, & n = 0, \\[3mm]
\sqrt{\dfrac{2}{L}} \sin\left(\dfrac{2\pi n}{L}x\right), & \text{odd}, \\[3mm]
\sqrt{\dfrac{2}{L}} \cos\left(\dfrac{2\pi n}{L}x\right), & \text{even};
\end{cases}
\tag{5-3.43}
$$

$$\mathbf{w}^m \cdot \mathbf{w}^m = \delta^{mm}$$

$$
\Longleftrightarrow
\begin{cases}
\displaystyle\int_0^L \sqrt{\dfrac{2}{L}} \sin\left(\dfrac{2\pi m}{L}x\right)\sqrt{\dfrac{2}{L}} \sin\left(\dfrac{2\pi n}{L}x\right) dx \\[3mm]
\displaystyle\int_0^L \sqrt{\dfrac{2}{L}} \cos\left(\dfrac{2\pi m}{L}x\right)\sqrt{\dfrac{2}{L}} \cos\left(\dfrac{2\pi n}{L}x\right) dx
\end{cases} = \delta^{mn} \tag{5-3.44}
$$

$$
\mathbf{u} = \sum_{n=1}^{N} a_n \mathbf{w}^n \Longleftrightarrow u(x) = \sum_{n=1}^{\infty} a_n \sqrt{\dfrac{2}{L}} \sin\left(\dfrac{n\pi}{L}x\right).
\tag{5-3.45}
$$

Where the function is not periodic, the Fourier series becomes a Fourier integral.

If we have a set of N linearly independent functions in a N-dimensional function space, it is generally possible to find a set of mutually orthogonal functions by means of a Gram–Schmidt process generalized to function space. Thus, if we have a set of N linearly independent functions, $f^n(x)$, we may define orthogonal functions, $F^n(x)$:

$$F^1(x) = f^1(x),$$

$$F^2(x) = f^2(x) - \frac{\int_a^b f^2(x)f^1(x)\,dx}{\int_a^b f^1(x)f^1(x)\,dx} f^1(x),$$

$$F^3(x) = f^3(x) - \frac{\int_a^b f^3(x)f^1(x)\,dx}{\int_a^b f^1(x)f^1(x)\,dx} f^1(x)$$

$$\qquad\qquad - \frac{\int_a^b f^3(x)f^2(x)\,dx}{\int_a^b f^2(x)f^2(x)\,dx} f^2(x), \tag{5-3.46}$$

N components,

analogous to equations (5-3.34) through (5-3.37). They can be normalized as above.

EXAMPLE

The Taylor series can also represent a large number of functions. This suggests that for the functions $f''(x)$ we may take: $1, x, x^2, x^3, \ldots, x^n, \ldots$. However, they are not orthogonal. We may orthogonalize them over the interval, $-1 < x < +1$, as above. Thus:

$$F^0(x) = 1,$$

$$F^1(x) = x - \frac{\int_{-1}^{1} x(1)\, dx}{\int_{-1}^{1}(1)(1)\, dx}(1) = x,$$

$$F^2(x) = x^2 - \frac{\int_{-1}^{1} x^2(1)\, dx}{\int_{-1}^{1}(1)(1)\, dx}(1) - \frac{\int_{-1}^{1} x^2(x)\, dx}{\int_{-1}^{1}(x)(x)\, dx} x = x^2 - \frac{1}{3},$$

$$F^3(x) = x^3 - \frac{3}{5}x, \tag{5-3.47}$$

etc.

These functions are the (unnormalized) Legendre polynomials. Then, one can express an arbitrary function in the interval, $-1 < x < +1$, as a linear combination of the Legendre polynomials. The Legendre expansion coefficients of a given function can be found by use of the orthogonality of the polynomials, analogous to the Fourier expansion. Much of the labor of the integrations can be reduced by using symbolic computer programs like *Maple* or *Mathematica*. See Appendix 5-2.

EXAMPLE

Find the solution to Laplace's equation in a region with parabolic boundaries:

$$y = \pm x^2. \tag{5-3.48}$$

Let the boundary potential be even in x and odd in y; then the solution is expressed by:

$$\phi = A_0 y + \sum_{k=1}^{\infty} A_k \cos kx \sinh ky. \tag{5-3.49}$$

The boundary curves are given by the parametric equations $x = s$ and $y = \pm s^2$. We will express the potential on the boundary by:

$$\phi_{\text{boundary}}(x, y) = \phi(x, x^2) = b(x),$$

where $b(x)$ is given. Because it is an odd function in y, it is only necessary to specify it on the upper boundary. Then, on the boundary, equation

5-3.49 becomes:

$$b(x) = A_0 x^2 + \sum_{k=1}^{\infty} A_k \cos kx \sinh kx^2$$

$$= \sum_{k=0}^{\infty} A_k f^k(x). \tag{5-3.50}$$

That is, the set of functions $f^k(x)$ is the nonorthogonal basis of the single variable x:

$$f^k(x) = \left[x^2, \cos(\pi x) \sinh\left(\pi x^2\right), \cos(2\pi x) \sinh\left(2\pi x^2\right), \ldots \right]. \tag{5-3.51}$$

Now orthogonalize this set of functions on the interval, $-1 < x < +1$. For practical purposes it is necessary to truncate the series after N terms. Then the orthonormalized basis $F^k(x)$ obtained by numerical integration is:

$F^k(x)$

$$= \begin{bmatrix} 1.581, x^2 \\ 2.996x^2 + 0.479 \cos(\pi x) \sinh(\pi x^2), \\ 3.530x^2 + 1.206 \cos(\pi x) \sinh(\pi x^2) + 0.051 \cos(2\pi x) \sinh(2\pi x^2), \\ N \text{ components} \end{bmatrix}.$$

$$\tag{5-3.52}$$

The Gram–Schmidt calculations are shown in Appendix 5-2. Now express the boundary potential in the orthogonalized basis $F^k(x)$:

$$b(x) = \sum_{k=0}^{N} B_k F^k(x). \tag{5-3.53}$$

The expansion coefficients B_k are determined using the orthogonal properties of the $F^m(x)$ and integrate; this gives:

$$\int_{-1}^{1} b(x) F^m(x) \, dx = \sum_{k=0}^{N} B_k \int_{-1}^{1} F^k(x) F^m(x) \, dx = B_m. \tag{5-3.54}$$

Finally, we find the coefficients A_k in equation (5-3.49) in terms of the coefficients B_m in equation (5-3.53). The boundary potential can be expressed as a series in either basis:

$$b(x) = \sum_{k=0}^{N} B_k F^k(x) = \sum_{k=0}^{N} A_k f^k(x). \tag{5-3.55}$$

Then, multiplying by $F^m(x)$ and integrating gives the coefficient:

$$B_m = \sum_{k=0}^{N} B_k \int_{-1}^{1} F^k(x) F^m(x)\, dx$$

$$= \sum_{k=0}^{N} A_k \int_{-1}^{1} f^k(x) F^m(x)\, dx$$

$$= \sum_{k=0}^{N} A_k M^{km}, \qquad (5\text{-}3.56)$$

where the M^{im} are constants. This set of linear equations can now be solved for A_k.

Finding the coefficients A_k completes the solution of Laplace's equation by separation of variables in a region with curvilinear boundaries. The necessity for such involved calculations reflects the fact that the sine and cosine functions in the Fourier series are not well suited to curvilinear boundaries. Although such brute-force methods are useful when all else fails, acceptable accuracy may require lengthy calculation.

In many cases, problems with curvilinear boundaries can be solved by using matching curvilinear coordinate systems. For example, spherical boundaries are well expressed in the spherical coordinate system. In that case, the appropriate form of the Laplacian must be used, giving a set of orthogonal functions. The most useful cases of curvilinear coordinates are orthogonal coordinates, where Laplace's equation is solved by separation of variables into sets of orthogonal eigenfunctions. In this case the potential on the curved surfaces can be expressed as a sum (integral) of the orthogonal functions as with the Fourier functions. The general expression of the Laplacian operator in curvilinear coordinates was given in Chapter 1. See Table 1-2. Further representative examples are shown in Table 5-1.

TABLE 5-1. EXAMPLES OF THE LAPLACIAN IN CURVILINEAR COORDINATES

Cylindrical	$\dfrac{1}{r}\left[\dfrac{\partial}{\partial r}\left(r\dfrac{\partial \Phi}{\partial r}\right) + \dfrac{\partial^2 \Phi}{\partial \theta^2} + r\dfrac{\partial^2 \Phi}{\partial z^2}\right]$
Spherical	$\dfrac{1}{r^2 \sin\theta}\left[\dfrac{\partial}{\partial r}\left(r^2 \sin\theta \dfrac{\partial \Phi}{\partial r}\right) + \dfrac{\partial}{\partial \theta}\left(\sin\theta \dfrac{\partial \Phi}{\partial \theta}\right) + \dfrac{1}{\sin\theta}\dfrac{\partial^2 \Phi}{\partial \psi^2}\right]$
Paraboloidal	$\dfrac{1}{(p^2 + q^2)pq}\left[\dfrac{\partial}{\partial p}\left(pq\dfrac{\partial \Phi}{\partial p}\right) + \dfrac{\partial}{\partial q}\left(pq\dfrac{\partial \Phi}{\partial q}\right) + \dfrac{\partial}{\partial \theta}\left(\dfrac{p^2 + q^2}{pq}\dfrac{\partial \Phi}{\partial \theta}\right)\right]$

Example

In polar coordinates, Laplace's equation becomes:

$$r\frac{\partial}{\partial r}\left[r\frac{\partial \phi}{\partial r}\right] + \frac{\partial^2 \phi}{\partial \theta^2} = 0, \tag{5-3.57}$$

which is separated into functions of r and θ:

$$\frac{r\dfrac{d}{dr}\left[r\dfrac{dR(r)}{dr}\right]}{R(r)} + \frac{\dfrac{d^2\Theta(\theta)}{d\theta^2}}{\Theta(\theta)} = 0 \tag{5-3.58}$$

and has solutions:

$$\phi = [A_0 + B_0\theta][C_0 + D_0 \ln r] + \sum_{k\neq 0}^{\infty}[A_k \cos k\theta + B_k \sin k\theta][C_k r^k + D_k r^{-k}].$$

$$\tag{5-3.59}$$

As before, the coefficients can be found by fitting the boundary values and using orthogonality.

This coordinate system is convenient for applications having a singularity at the origin.

5-4 The Cauchy–Riemann Equations and Conformal Mapping

As we have seen, the most difficult part in solving the Laplace equation is fitting the solution to the boundary—particularly to a curved boundary. Choosing a curvilinear coordinate system, in which the boundary surface is specified by a constant coordinate, is a satisfactory way to treat the boundary problem. Unfortunately, it is generally difficult to find such a set of surfaces. However, there is a powerful tool for doing this in two dimensions, or three dimensions with symmetry on the third axis.

A complex number, $z = x + iy$, may be represented as a point in the x–y plane, i.e., as a vector with components x and y, and unit basis vectors 1 and i along the x- and y-axes, respectively. Defining a function of the complex number:

$$w = f(z)$$

$$= u(x,y) + iv(x,y), \tag{5-4.1}$$

implies a mapping between points on the z plane into points on the w plane. Depending on the specific function, the mapping may be single or multiple valued. A unique inverse requires a single-valued mapping.

Example

The mapping:

$$w = z^2 \qquad (5\text{-}4.2)$$

shown in Figure 5-4 has real and imaginary components:

$$u = (x^2 - y^2), \qquad (5\text{-}4.3)$$

$$v = 2xy, \qquad (5\text{-}4.4)$$

in the w-plane which form two sets of orthogonal hyperbolae.

We can show that the functions $u = u(x,y)$ or $v = v(x,y)$ each satisfy Laplace's equation. Consider the derivative:

$$\frac{dw}{dz} = \frac{du}{dz} + i\frac{dv}{dz}. \qquad (5\text{-}4.5)$$

Since the variable z has two degrees of freedom, the path approaching a limit point in the x–y plane is not unique; therefore the derivative of the complex function $w(z)$ is not unique unless the limit is independent of the path. Specifically, we require the derivative to be the same whether approaching the limit along the x or the iy axis. That is:

$$\frac{\partial w}{\partial x} = \frac{\partial w}{\partial iy}. \qquad (5\text{-}4.6)$$

This put conditions on the components u and v:

$$\frac{\partial u}{\partial x} + i\frac{\partial v}{\partial x} = \frac{\partial u}{\partial iy} + i\frac{\partial v}{\partial iy}. \qquad (5\text{-}4.7)$$

Equating the real and imaginary components gives the *Cauchy–Riemann* equations:

$$\frac{\partial u}{\partial x} = \frac{\partial v}{\partial y}, \qquad (5\text{-}4.8)$$

$$\frac{\partial u}{\partial y} = -\frac{\partial v}{\partial x}. \qquad (5\text{-}4.9)$$

Taking the second derivatives gives:

$$\frac{\partial^2 u}{\partial x^2} + \frac{\partial^2 u}{\partial y^2} = 0, \qquad (5\text{-}4.10)$$

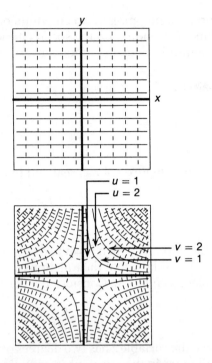

FIGURE 5-4. **Conformal mapping of the function** $w = z^2$.   The first diagram is the z-plane, and the second diagram is the w-plane. Note that the u and v curves are mutually orthogonal. This is a general property of such mappings, which makes them useful to represent equipotentials and lines of flux.

$$\frac{\partial^2 v}{\partial x^2} + \frac{\partial^2 v}{\partial y^2} = 0. \tag{5-4.11}$$

That is, both real and imaginary parts of the mapped function satisfy Laplace's equation.

Now consider the derivatives of a curve of constant u and constant v. The slopes of a u-curve and a v-curve are, respectively:

$$\left[\frac{\partial y}{\partial x}\right]_v = -\frac{\partial u}{\partial x} \bigg/ \frac{\partial u}{\partial y}, \tag{5-4.12}$$

$$\left[\frac{\partial y}{\partial x}\right]_u = \frac{\partial v}{\partial x} \bigg/ \frac{\partial v}{\partial y} = -\frac{\partial u}{\partial y} \bigg/ \frac{\partial u}{\partial x} = -1 \bigg/ \left[\frac{\partial x}{\partial y}\right]_v, \tag{5-4.13}$$

where the Cauchy–Riemann equations were used in the second equation. Equation (5-4.13) is the condition that two curves are orthogonal; therefore the two mapped families of curves are everywhere orthogonal.

Because the electrostatic field is orthogonal to the equipotential surfaces, the curves of constant u and constant v can be used to represent equipotential surfaces and the electric lines of flux, respectively.

EXAMPLE

Use the mapping $w = Az^2$ to find the equipotential surfaces and electric field lines inside a quadrupole lens with hyperbolic surfaces with a maximum aperture of 0.2 cm and charged to potential ±500 volts. The potential is $\phi = 2Axy$. Evaluating the constant A on the boundary gives:

$$\phi = 50,000xy \tag{5-4.14}$$

and the electric lines of flux are given by the curves:

$$25,000(y^2 - x^2) = u, \tag{5-4.15}$$

where the value of the constant u determines each line of flux.

A complex number may be written in polar form:

$$z = re^{i\theta}, \tag{5-4.16}$$

where θ is the *argument* (arg z), and r is the *absolute value* $r = (x^2+y^2)^{1/2}$. When the argument is an odd multiple of π, the complex number is real and negative; in particular:

$$e^{i\pi} = -1. \tag{5-4.17}$$

Now consider the mapping function:

$$w = z^\delta, \tag{5-4.18}$$

where δ is a constant. Such mapping functions have a useful property: points in the upper half of the z plane map into an angular region defined by the value of the exponent δ. This is shown in Figures 5-4 and 5-5. Points on the x-axis in the z plane map into two straight lines in the w plane, with the discontinuity of slope at the origin. Thus for points on the positive x-axis, $\theta = 0$, while for points on the negative x-axis, $\theta = \pi$, because of the change in sign in equation (5-4.17). Then the mapping of points on the x-axis obeys:

$$w = \begin{cases} r^\delta, & x > 0, \\ r^\delta e^{i\pi\delta}, & x < 0, \end{cases} \tag{5-4.19}$$

Other points in the upper half plane are mapped into the inside of the angular region shown in Figure 5-5. The mapping also involves changes of scale, i.e., a curve in the x–y plane having a given length is mapped into another curve having a different length in the w plane.

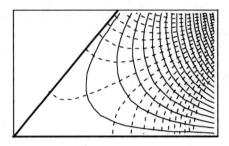

FIGURE 5-5. MAPPING OF THE FUNCTION $w = z^4$.   The solid lines (constant u) may be used to represent equipotentials, and the dashed lines (constant v) to represent lines of flux.

The relation between the exponent δ and the angle of bend can be understood from the following. The value $\delta = 1$ is the identity transformation in which $w = z$ and the x-axis is unchanged. When $\delta = 2$ the incremental angle is $\pi/2$ radians. In general:

$$\theta = \frac{\pi}{\delta}. \qquad (5\text{-}4.20)$$

Appendix 5-3 gives a *Mathematica* routine to calculate and plot various mappings. Figures 5-4, 5-5, and 5-7 were created with this routine and some post-processing. In general, the entire z plane is mapped into the w plane, but only that part of the mapping is shown in the diagrams which seems useful. Thus, for example, in t, Figure 5-5 only the interior of the boundary appears.

The term *conformal* refers to the fact that two curves intersecting in the x–y plane at some angle will intersect in the u–v plane at the same angle. The angle of intersection is preserved. To see this take the mapping $w = f(z)$ and consider two curves in the z plane intersecting at point z_0. Let Δz_1 and Δz_2 be short segments of a curve mapping into the corresponding segments Δw_1 and Δw_2 at the corresponding point, w_0. Then the segments are related by:

$$\Delta w_k = \frac{df}{dz} \Delta z_k, \qquad k = 1, 2. \qquad (5\text{-}4.21)$$

If we express the segments in polar form, e.g.:

$$\Delta z_k = |\Delta z_k| e^{i\theta}, \qquad k = 1, 2, \qquad (5\text{-}4.22)$$

then from equation (5-4.21) the arguments can be expressed:

$$\arg(\Delta w_k) = \arg\left(\left[\frac{df}{dz}\right]_{z=z_0}\right) + \arg(\Delta z_k), \qquad k = 1, 2. \qquad (5\text{-}4.23)$$

After subtracting, the angle of intersection between the two curves at w_0 becomes:

$$\Delta\theta_w = \arg(\Delta w_2) - \arg(\Delta w_1) = \arg(\Delta z_2) - \arg(\Delta z_1)$$
$$= \Delta\theta_z. \tag{5-4.24}$$

That is, the angle of intersection between two curves in the z plane at z_0 equals the angle of intersection of the corresponding curves in the w plane at w_0. Figure 5-6 illustrates the mapping of two curves and the preservation of the angle of intersection.

Conformal mapping is a powerful method of getting exact solutions to two-dimensional problems where the boundaries form lines of constant potential. However, its practical use depends on being able to find a mapping function and its inverse, in which the boundaries match the needed geometry. There are lists of transformations available from which it may be possible to find a mapping needed for a particular application. Since the parameters u and v are orthogonal coordinates, all the methods of curvilinear coordinates discussed in Section 1-4 can be used. Then solution by separation of variables and orthogonal functions may be possible in the curvilinear coordinates.

When the boundaries are straight-line segments, the *Schwartz transformation*:

$$\frac{dw}{dz} = (z - x_0)^\delta \tag{5-4.25}$$

may be particularly useful. There is a change of slope at the point x_0, as we saw above. The transformation is obtained by integration. The generalization:

$$\frac{dw}{dz} = (z - x_1)^{\delta_1}(z - x_2)^{\delta_2} \tag{5-4.26}$$

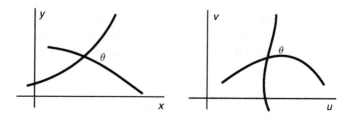

FIGURE 5-6. THE CONFORMAL PROPERTY. Conformal mapping preserves the angles of intersection between corresponding curves in the x–y and u–v planes.

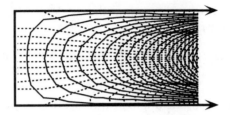

FIGURE 5-7. EQUIPOTENTIAL LINES FOR A SCHWARTZ TRANSFORMATION. The transformation is given by equation 5-4.26 with $\delta_1 = \delta_2 = 2$ and $x_2 = -x_1 = 1$. Due to hardware limitations, the field lines do not appear perfectly orthogonal to the equipotentials.

maps points on the x-axis onto points on the u-axis provided $x > x_1$. At this point there will be a change of slope. Similarly, at the point x_2, there will be a second change of slope. In this way, the upper half of the z plane is mapped into a region with a boundary composed of straight-line segments. An example is illustrated by Figure 5-7. This method can be very useful for finding solutions to the Laplace equation. However, as more sides are added to the polygon, exact integration may be difficult.

Once the Schwartz transformation is integrated, we can pick off the functions $u(x, y)$ and $v(x, y)$. Ideally, one can express the Laplacian in u and v and apply separation of variables with orthogonal functions. Of course, circumstances determine the practicality of the scheme. However, much research has gone into this method and there are lists of functions of which Kober is a good example.

5-5   Numerical Solutions by Finite-Element Analysis

When an exact solution to Laplace's equation is impossible, a numerical solution (implemented by computer) may be possible. Euler's method of finite differences is particularly simple. We approximate the derivatives for small intervals by the ratios of differences. Thus suppose we want to determine the potential only at discrete points (x_m, y_n). Then the partial derivatives can be approximated:

$$\frac{\partial \phi}{\partial x} \approx \frac{\phi_{m+1,n} - \phi_{m-1,n}}{\Delta x}, \tag{5-5.1}$$

$$\frac{\partial \phi}{\partial y} \approx \frac{\phi_{m,n+1} - \phi_{m,n-1}}{\Delta y}, \tag{5-5.2}$$

where the subscripts refer to points on a rectangular grid. Similarly, the second derivatives of the function are approximated using the second differences, i.e., the differences of the differences. They may be expressed symmetically around the point (x_m, y_n):

$$\frac{\partial^2 \phi}{\partial x^2} \approx \frac{\phi_{m+1,n} - 2\phi_{m,n} + \phi_{m-1,n}}{[\Delta x]^2}, \qquad (5\text{-}5.3)$$

$$\frac{\partial^2 \phi}{\partial y^2} \approx \frac{\phi_{m,n+1} - 2\phi_{m,n} + \phi_{m,n-1}}{[\Delta y]^2}. \qquad (5\text{-}5.4)$$

If we assume a square grid, i.e., $\Delta x = \Delta y$, then Laplace's equation is approximated by the recursion formula:

$$\phi_{m,n+1} + \phi_{m+1,n} - 4\phi_{m,n} + \phi_{m,n-1} + \phi_{m-1,n} = 0. \qquad (5\text{-}5.5)$$

Now form a rectangular grid and indicate the boundary curve (or surface in three dimensions). See Figure 5-8. If the boundary potential is given, then the $\phi_{m,n}$ are known there. Then equation (5-5.5) is used to determine $\phi_{m,n}$ on the interior grid points. The procedure for evaluation is: (1) choose some initial (seed) value at each point. (2) Use equation (5-5.5) to recalculate the values of each point (substituting the boundary values where needed). (3) Repeat the substitution with the newly calculated values as seeds. If the method is to be successful, the calculated values converge. The rate of convergence may depend on the order of substitution and the seed values. A spreadsheet is a convenient way to do these calculations. An example is given in Appendix 5-1.

EXAMPLE

Find the potential inside a rectangle having constant potentials on the sides of 10, 10, 0, 0 volts, respectively. Figure 5-9 shows the output calculated by a spreadsheet. Note the diagonal 5 volt equipotential and the symmetry of the values as is expected from the given boundary

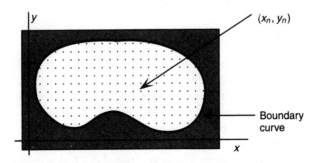

FIGURE 5-8. FINITE-ELEMENT GRID. The potential at points nearest to the boundary are given fixed values. Then the potential at points inside are calculated from the recursion formula (5-5.5).

10.00 10.00 10.00 10.00 10.00 10.00 10.00 10.00
10.00 09.54 09.09 08.60 07.98 06.99 05.00 00.00
10.00 09.09 08.20 07.32 06.33 05.00 03.01 00.00
10.00 08.60 07.32 06.16 05.00 03.67 02.02 00.00
10.00 07.98 06.33 05.00 03.84 02.68 01.40 00.00
10.00 06.99 05.00 03.67 02.68 01.80 00.91 00.00
10.00 05.00 03.01 02.02 01.40 00.91 00.46 00.00
10.00 00.00 00.00 00.00 00.00 00.00 00.00 00.00

Figure 5-9. Numerical solution of Laplace's equation.   The values of the potential on the edges of the rectangle were chosen to be 10, 10, 0, 0. Then initial values of the potential were chosen in the interior (0, to be specific). Then new values of the interior points were calculated using the recursion relation. The order of evaluation of the interior points was an inward spiral. The procedure was iterated until the values of the interior points were unchanged (to the accuracy shown).

values. The values on the outer edges represent the given boundary values.

There are many special methods for finding the potential in the literature. We mention one, the method of images, because it has considerable practical application. The surface of a conductor has constant potential, and the electric field is everywhere normal to the surface (in steady-state conditions). First, consider a grounded, infinite, conducting plane. Now, suppose that a charge q is brought to a distance d from it. The charges in the conductor must rearrange themselves so that the electric lines of flux are normal to it and its potential remains zero. However, on the vacuum side of the conductor, the electric field is just the same as if there were a charge $-q$ at a symmetric distance d inside the conductor, and the conductor is removed. The actual charge is reflected by the conducting "mirror." Any collection of charges can be treated this way. See Figures 5-10 and 5-11.

Example

Consider an electrostatic lens used in some types of electron microscopes. The radial component of the electric field acts to focus a beam of charged particles passing through the lens. There is a plane of symmetry between the two half lenses. If a conducting sheet (at the given potential) is placed in this plane, then the electric field in each half lens is unchanged. However, the sheet has a certain thickness, in which the electric field vanishes; this field-free region modifies the optical properties of the lens. This feature can be useful in the construction of beam optics in electron microscopes or particle physics experiments.

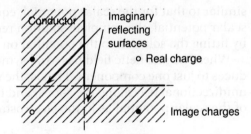

FIGURE 5-10. SOLUTION BY REFLECTIONS. The potential in the corner region is the same as for the set of four charges. Note the changes in sign of the image charges. There are four image charges because the reflected charges also have their reflected charges. This configuration gives constant potential at the boundary surface.

5-6 MAGNETIC FIELDS

Much of the above discussion about the electrostatic potential ϕ applies to the vector potential **A** as well, since they both must satisfy the Poisson and Laplace equations, albeit with different source terms. However, the potentials are not uniquely defined—there is the gauge condition to satisfy! In the steady-state case, one usually applies the Coulomb gauge. The vector potential must satisfy:

$$\nabla^2 \mathbf{A} = 0, \tag{5-6.1}$$

$$\nabla \cdot \mathbf{A} = 0. \tag{5-6.2}$$

The gauge condition thereby reduces the degrees of freedom from 3 to 2. The solution can be expressed by a (vector) orthogonal expansion

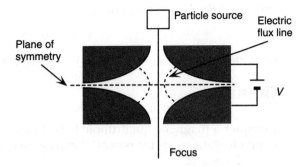

FIGURE 5-11. ELECTROSTATIC LENS. A charged particle beam passing through the lens is be accelerated by the longitudinal component of the electric field. The transverse (radial) component fields produces a converging effect on the particle beam. The shape of the lens surface determines the spherical aberration.

similar to that for the scalar potential, equation (5-3.13). Then, like the scalar potential, the complete solution requires finding the coefficients by fitting the solution to the potential on the boundary.

When the magnetic field has axial symmetry, the vector potential reduces to just one component. That is, the vector potential is everywhere unidirectional (say along the z-axis) and the problem is reduced to that of the scalar potential as in the electrostatic case. We write:

$$\nabla^2 A_z(x, y) = -\frac{4\pi}{c} J_z(x, y), \tag{5-6.3}$$

$$A_x = A_y = 0, \tag{5-6.4}$$

which is the Poisson equation. When the current vanishes, it becomes the Laplace equation, and the nonvanishing components of the magnetic field become:

$$B_x = \frac{\partial A_z}{\partial y}, \tag{5-6.5}$$

$$B_y = -\frac{\partial A_z}{\partial x}. \tag{5-6.6}$$

In any region in which the currents vanish, the Maxwell equations for the magnetic field reduce to the simple form:

$$\nabla \cdot \mathbf{B} = 0, \tag{5-6.7}$$

$$\nabla \times \mathbf{B} = 0. \tag{5-6.8}$$

Then the magnetic field can be expressed as gradients of a *magnetic scalar potential* ψ:

$$\mathbf{B} = -\nabla \psi \tag{5-6.9}$$

satisfying the Laplace equation:

$$\nabla^2 \psi = 0 \tag{5-6.10}$$

subject to appropriate (magnetic) boundary conditions.

EXAMPLE

Consider a magnetic quadrupole lens shown in Figure 5-12. Either the magnetostatic or vector potential can be used in this case; we will show both. They are:

$$A_z = \frac{G}{2}[x^2 - y^2], \tag{5-6.11}$$

$$\psi = Gxy, \tag{5-6.12}$$

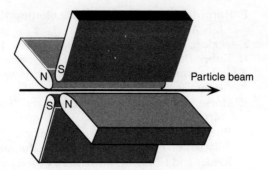

FIGURE 5-12. MAGNETIC QUADRUPOLE LENS. The direction of the particle beam is indicated by the arrow. The variation of the magnetic field in the interior region depends on the shape of the pole faces. A hyperbolic cross section will produce a field similar to that shown in Figure 5-4.

which both satisfy Laplace's equation. The magnetic field components are the same for either potential:

$$B_x = -Gy, \tag{5-6.13}$$

$$B_y = -Gx, \tag{5-6.14}$$

$$B_z = 0. \tag{5-6.15}$$

They satisfy the Maxwell and gauge equations:

$$\nabla \cdot \mathbf{B} = \partial_x B_x + \partial_y B_y + \partial_z B_z$$

$$= -G[\partial_x y + \partial_y x + \partial_z 0] = 0, \tag{5-6.16}$$

$$\nabla \cdot \mathbf{A} = \partial_x A_x + \partial_y A_y + \partial_z A_z$$

$$= G\left[\partial_x 0 + \partial_y 0 + \partial_z \frac{G}{2}(x^2 - y^2)\right] = 0. \tag{5-6.17}$$

The magnetic lines of flux and the equipotential surfaces are those shown in Figure 5-4. In Chapter 7 we will consider how the lens is used in particle beams.

5-7 SELECTED BIBLIOGRAPHY

E. Butkov, *Mathematical Physics*, Addison-Wesley, 1966. Treats orthogonal functions and theory of distributions.

A. Crewe et al., "A Scanning Electron Microscope," Scientific American, March 1971.

E. Horowitz, "A General Field-Line Plotting Algorithm," Computers in Physics, July/August 1990, p. 418.

J. Jackson, *Classical Electrodynamics*, Wiley, 1962.

H. Kober, *Dictionary of Conformal Representations*, Dover, 1957.

J. Mathews and R. Walker, *Mathematical Methods of Physics*, 2nd ed., Benjamin, 1970.

P. Morse and H. Feshbach, *Methods of Theoretical Physics*, Vols. 1 and 2, McGraw-Hill, 1953.

W. Press et al., *Numerical Recipes*, Cambridge University Press, 1986. Detailed discussion of computer algorithms for numerical analysis.

W. Rieder and H. Busby, *Introductory Engineering Modeling*, Wiley, 1986.

M. Rose, *Multipole Fields*, Wiley, 1955.

W. Smythe, *Static and Dynamic Electricity* , 3rd ed., McGraw-Hill, 1968.

M. Spiegel, *Fourier Analysis*, Schaum, 1974.

J. Stratton, *Electromagnetic Theory*, McGraw-Hill, 1941.

C. Wylie, *Advanced Engineering Mathematics*, McGraw-Hill, 1951.

5-8   PROBLEMS

1. Find the finite element recursion formula for Laplace's equation in polar coordinates. (See equation (5-5.5).)

2. Find the electric octupole term in expansion of the scalar potential, ϕ.

3. The Monte Carlo method is an efficient method of numerical integration. To find the area of an irregular closed curve, enclose the curve in a rectangle larger than the curve; then choose N random points in the rectangle. The ratio of points in the curve to those in the rectangle equals the ratio of the areas (to within an error that depends on N). The importance of using random points in the integration is that they will be less likely to miss a singularity, as a regular array of points might. A computer is recommended.

     A. Find the area of a circle of radius 2 by the Monte Carlo method.
     B. Calculate the dipole moment for the charge density: $\rho = x - x^2 y$ inside a circle of radius 2 and 0 outside by the Monte Carlo method.
     C. Find the quadrupole moment for this charge density by the Monte Carlo method.
     D. Find the errors between the exact values and the Monte Carlo values in the above.

4. Expand the function x^2 as a Fourier series between $x = \pm 1$ to 4 terms.

5. For the functions $(x, x^{1/2}, x^{1/3} \ldots)$ use the Gram–Schmidt method to find a set of orthonormal functions in the interval: $0 < x < 1$.

6. A function of the discrete variable, x_n, can be expressed as a sum of step functions $(u(x) = 0, x < x_0, u(x) = 1, x > x_0)$ with steps at $x = 0, 1, 2, 3 \ldots$.

That is:

$$f(x_n) = \sum_{k=1}^{N+1} A_k u^k(x) = \sum_{k=1}^{N+1} A_k u\left(x - (k-1)\frac{L}{N}\right), \qquad 0 \le x_n \le L,$$

where $u(x) = u^0(x)$ is the unit step function with the step at $x = 0$.

A. By using the Gram–Schmidt procedure, express $u^k(x)$ as a series of orthonormal functions $F^k(x)$.

B. Find the normalization constant.

C. Describe the function $u^k(x_n)$ as the number of terms N approaches infinity?

7. The unit step function might be a useful basis in which to express a function as a transform:

$$f(x) = \int_{-L}^{L} g(k)u(x-k)\,dk.$$

Can you convert the unorthogonal $u(x-k)$ basis into an orthogonal basis $U(x,k)$ and expand $f(x)$ in it? Are there conditions on the value of L? Is there an inverse transform?

8. Find an explicit expression for the potential inside the rectangular region bounded by the values $(-L < x < L, -d < y < d)$ where the potential on the upper $(y = d)$ surface is given by $V_0(1 - x^2)$, and on the lower $(y = -d)$ surface by $V_0(1 - x^4)$.

9. Find the expression for the solution to Laplace's equation in Cartesian coordinates in three dimensions. It will require a double sum; are there any relations between the two indices of summation? See equation (5-3.7).

10. Using the method of finite element analysis, find the solution of Laplace's equation for boundary $y = \pm x^2$ where the potential on the upper and lower boundaries are the constant values $\pm V_0$, respectively.

11. Use the conformal mapping $w = e^z$ to find solutions to Laplace's equation in the interior of a wedge-shaped boundary of angle $45°$.

12. Use the Schwartz transformation to find solutions to Laplace's equation in the interior of a wedge–shaped boundary of angle $45°$.

13. Discuss the use of finite element analysis to find solutions of Poisson's equation.

14. Solution by reflection. A flat conductor lies in the x–y plane, $z = 0$. Above it, lying in the x–y plane, $z = d$, is a charged rod, with charge per unit length λ. Assuming the conducting plane and the rod are large compared to the length d, find the potential produced by this configuration.

15. Find the magnetic field inside an octupole lens. (This generalization of the quadrupole lens discussed in the text, is used to correct aberrations in charged particle beams.) Hint: The equation $x^2 - y^2 = \psi = \text{constant}$, defines four hyperbolic pole faces; what would the magnetic scalar potential be for eight pole faces?

16. Suppose a dielectric sphere of radius R and dielectric constant ε is placed in a uniform external electric field $\mathbf{E}_0 = (0, 0, E_0)$. For simplicity, let the origin of coordinates be at the center of the sphere.

A. Show that the potential outside the sphere is given by the external field plus that of an induced dipole moment, i.e., the boundary conditions are satisfied.

B. Express the dipole moment in terms of E_0, R, and ε.

C. Express the polarization surface charge density as a function of angle.

APPENDIX 5-1   RELAXATION SOLUTION OF LAPLACE'S EQUATION

The relaxation solution to Laplace's equation, can be solved in simple cases by implementing the relaxation method on a spreadsheet program. Although the spreadsheet software is relatively slow to execute, entering the data and formulas is extremely easy. It also has the advantage that the results are immediately visualized since the data is displayed on a rectangular grid. Finally, if the spreadsheet has the graphics capability, the equipotentials can be immediately plotted. The following example was done with the WINGZ spreadsheet on a Macintosh IIci computer.

Step 1. Switch the spreadsheet to MANUAL CALCULATION.

Step 2. Fill the boundary cells with the boundary values of the potential. For example, a grid of 5×5 cells might have the values on the boundary cells:

$$1.0000 \ \ 1.0000 \ \ 1.0000 \ \ 1.0000 \ \ 0.0000$$
$$1.0000 \qquad\qquad\qquad\qquad\qquad 0.0000$$
$$1.0000 \qquad\qquad\qquad\qquad\qquad 0.0000$$
$$1.0000 \qquad\qquad\qquad\qquad\qquad 0.0000$$
$$1.0000 \ \ 0.0000 \ \ 0.0000 \ \ 0.0000 \ \ 0.0000.$$

The five-place accuracy displayed here is more than adequate for this coarse grid.

Step 3. Fill the interior cells with the relaxation formula for calculating the values at each point. For example, the cell C3 should contain the expression:

$$C3 = \frac{(C2 + C4 + B3 + D3)}{4}.$$

From this point on, the spreadsheet absolutely must be on manual calculation. If it is on automatic calculation, the program will print error messages indicating circular definition:

$$1.0000 \quad 1.0000 \quad 1.0000 \quad 1.0000 \quad 0.0000$$
$$1.0000 \quad 0.5000 \quad 0.5000 \quad 0.5000 \quad 0.0000$$
$$1.0000 \quad 0.5000 \quad 0.5000 \quad 0.5000 \quad 0.0000$$
$$1.0000 \quad 0.5000 \quad 0.5000 \quad 0.5000 \quad 0.0000$$
$$1.0000 \quad 0.0000 \quad 0.0000 \quad 0.0000 \quad 0.0000.$$

Step 4. Manually recalculate the spreadsheet as many times as is necessary so that the values do not visibly change any more. In the example the values are:

$$1.0000 \quad 1.0000 \quad 1.0000 \quad 1.0000 \quad 0.0000$$
$$1.0000 \quad 0.8571 \quad 0.7143 \quad 0.5000 \quad 0.0000$$
$$1.0000 \quad 0.7143 \quad 0.5000 \quad 0.2857 \quad 0.0000$$
$$1.0000 \quad 0.5000 \quad 0.2857 \quad 0.1429 \quad 0.0000$$
$$1.0000 \quad 0.0000 \quad 0.0000 \quad 0.0000 \quad 0.0000.$$

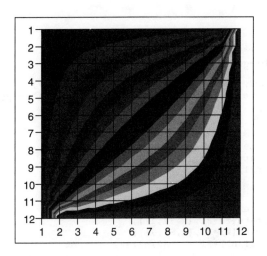

Note that the values are symmetrical about the diagonal, as expected from the symmetry of the boundary values.

Step 5. The potential distribution can now be plotted by the spreadsheet as a contour graph as shown. The contour fill patterns can be changed; these patterns were chosen to give good contrast. The equipotential lines are the boundaries between the contours.

APPENDIX 5-2   GRAM–SCHMIDT ORTHOGONALIZATION

Initialize

```
Remove["Global'@"]
```

Load orthogonalization package for use later.

```
<<LinearAlgebra'Orthogonalization'
```

Gram–Schmidt Orthogonalization

The function x^n, where n runs from 0 to infinity, can be used to express a function as an infinite series (Taylor series), but they are not orthogonal to each other. However, linear combinations of the x^n that are orthogonal can be found by the Gram–Schmidt process.

Define Original Function Set

```
f1 = 1;
f2 = x;
f3 = x^2;
f4 = x^3;
```

Calculate Orthogonal Functions

```
(* The n1, n2, etc. are normalization factors *)
(* The cjk are factors for subtracting parallel components *)
F1 = f1

1

n1 = Integrate[F1*F1,{x,-1,1}];
c21 = Integrate[f2*F1,{x,-1,1}];
F2 = f2 - (c21/n1)*F1

x

n2 = Integrate[F2*F2,{x,-1,1}];
c31 = Integrate[f3*F1,{x,-1,1}];
c32 = Integrate[f3*F2,{x,-1,1}];
F3 = f3 - (c31/n1)*F1 - (c32/n2)*F2

    1     2
 -(-) + x
    3

n3 = Integrate[F3*F3,{x,-1,1}];
c41 = Integrate[f4*F1,{x,-1,1}];
c42 = Integrate[f4*F2,{x,-1,1}];
c43 = Integrate[f4*F3,{x,-1,1}];
```

```
F4 = f4 - (c41/n1)*F1 - (c42/n2)*F2 - (c43/n3)*F3
```

```
-3 x     3
---- + x
 5
```

```
n4 = Integrate[F4*F4,{x,-1,1}];
```

Display Orthonormalized Functions
Display Orthonormalized Functions $FN[[j]]$, $j = 1$ to 4:

```
Fun = {F1,F2,F3,F4};
norm = {n1,n2,n3,n4};
FF = {F1/Sqrt[n1],F2/Sqrt[n2],F3/Sqrt[n3],
     F4/Sqrt[n4]};
```

Write the normalized function in standard form

```
FN = Together[FF]
```

```
                               5         2
                        Sqrt[-] (-1 + 3 x )
      1            3         2
{--------, Sqrt[-] x, -------------------,
  Sqrt[2]         2                2
```

```
      7              3
 Sqrt[-] (-3 x + 5 x )
      2
 --------------------}
          2
```

Check Orthogonormality and Compare
Check Orthogonormality of $FN[[j]]$, $j = 1$ to 4

```
ortho=Table[Integrate[FN[[j]]*FN[[k]],
      {x,-1,1}],{j,1,4},{k,1,4}];
mf = MatrixForm[ortho]
```

```
1    0    0    0
0    1    0    0
0    0    1    0
0    0    0    1
```

Compare this with *Mathematica*'s math packages:

```
GramSchmidt[{1,x,x^2,x^3,x^4},
InnerProduct ->
(Integrate[#1 #2,{x,-1,1}]&)] //Together
```

```
                               5         2
                        Sqrt[-] (-1 + 3 x )
      1            3         2
{--------, Sqrt[-] x, -------------------,
  Sqrt[2]         2                2
```

$$\text{Sqrt}[\frac{7}{2}] \; \frac{(-3 \; x + 5 \; x^3)}{2}, \; \frac{3 \; (3 - 30 \; x^2 + 35 \; x^4)}{8 \; \text{Sqrt}[2]}\}$$

Orthogonalizing on a Curve

Create a List of Functions

The following functions were used in the text to express a function on a curved line, $y = x^2$:

```
f=
{x^2, Cos[Pi*x]*Sinh[Pi*x^2], Cos[2*Pi*x]*Sinh[2*Pi*
x^2]}
\verbatim
          2                      2                          2
{x , Cos[Pi x] Sinh[Pi x ], Cos[2 Pi x] Sinh[2 Pi x ]}
```

Create Orthogonal Functions

```
GramSchmidt[f, InnerProduct -> (NIntegrate[#1 #2,{x,-1,1}]&)];
FN = Together[%];
MatrixForm[%]
```

$$1.58114 \; x^2$$

$$2.99628 \; x^2 + 0.479153 \; \text{Cos}[\text{Pi } x] \; \text{Sinh}[\text{Pi } x^2]$$

$$3.52962 \; x^2 + 1.20595 \; \text{Cos}[\text{Pi } x] \; \text{Sinh}[\text{Pi } x^2] +$$

$$0.0511054 \; \text{Cos}[2 \; \text{Pi } x] \; \text{Sinh}[2 \; \text{Pi } x^2]$$

Test for Orthogonality by Integrating

```
ortho=Table[NIntegrate[FN[[j]]*FN[[k]],{x,-1,1},
AccuracyGoal -> 4], {j,1,3},{k,1,3}];
mf = MatrixForm[Chop[ortho, 10^-8]]
```

```
1.   0    0
0    1.   0
0    0    1.
```

APPENDIX 5-3   SCHWARTZ TRANSFORMATIONS

The Schwartz transformation can be useful for solving the Laplace equation in regions having rectilinear boundaries. The following finds the real and imaginary parts of mapping functions and plots them over different ranges.

In the first part of the notebook, we consider the mapping $w = z^\delta$, where δ is a constant. This mapping can be used for boundaries with a single in angle in the rectilinear boundaries.

In the second half of the notebook, we consider a transformation with two bends. This transformation is useful for the solution of the Laplace equation inside a rectangular slot.

Initialize

```
(* Call complex functions package *)
Needs["Algebra'ReIm'"]
```

Define the Complex Variable and Mapping Function

```
(* Complex variable in Cartesian form *)
z = x+I*y/.{Re[x]->x,Im[x]->0,Re[y]->,Im[y]->0};
(* Mapping function *)
delta = 5/4
w = z^delta
(* Pick off real and imaginary parts of w *)
u = Re[w];
v = Im[w];
```

```
          5/4
(x + I y)
```

Graphs of the Mapped Function

The contours correspond to equipotentials and lines of flux. In these plots both the upper and the lower half planes are shown.

```
ContourPlot[v,{x,-2,2},{y,-2,2},Contours->20,
ContourShading->False];
ContourPlot[u,{x,-2,2},{y,-2,2},Contours->20,
ContourShading->False];
```

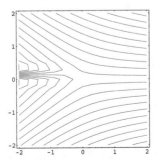

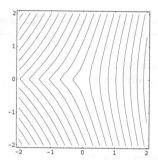

Restrict the Plot Range to the Upper Half Plane
This restricted range shows the discontinuity in the angle on the horizontal axis (x-axis).

```
vplot = ContourPlot[v,{x,-2,2},{y,0,2},Contours->20,
         AspectRatio->3/5,ContourShading->False];
uplot = ContourPlot[u,{x,-2,2},{y,0,2},Contours->20,
         AspectRatio->3/5,ContourShading->False];
Show[uplot,vplot]
```

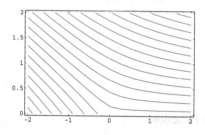

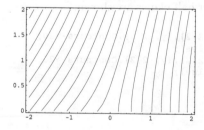

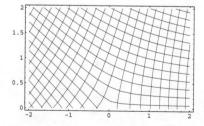

Transformation with Two Bends

Define the Mapping Function

```
(* Mapping function for rectangular slot *)
ff = ((s-1)^2)*((s+1)^2);
funct = Integrate(ff,s];
www = funct/.s->z
(* Find the real and imaginary parts of w.  *)
(* Mathematica has a function for this.     *)
(* However, it is faster for a polynomial   *)
(* to expand it to find the conjugate then  *)
(* subtract to find the real and imginary.  *)
wwXpnd = Expand[www];
wwConj = wwXpnd/.y->(-y);
(* Pick off real and imaginary parts.  *)
uuu = Expand[(wwXpnd+wwConj)/2]
vvv = Expand[(wwXpnd-wwConj)/(2*I)]
```

$$x - \frac{2(x+Iy)^3}{3} + \frac{(x+Iy)^5}{5} + Iy$$

$$x - \frac{2x^3}{3} + \frac{x^5}{5} + 2xy^2 - 2x^3y^2 + xy^4$$

$$y - 2x^2y + x^4y + \frac{2y^3}{3} - 2x^2y^3 + \frac{y^5}{5}$$

Plot the Real and Imaginary Parts
The contours to equipotentials and lines of flux.

```
ContourPlot[vvv,{x,-2,2},{y,-2,2},
    Contours->40,ContourShading->False];
ContourPlot[uuu,{x,-2,2},{y,-2,2},
    Contours->40,ContourShading->False];
```

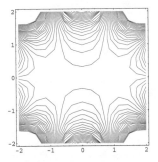

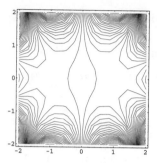

Change the Plot Range to Show the Rectangular Slot

By choosing the range of the transformation, the potentials and field lines in the interior of a rectangular slot can be mapped.

```
uuuplot = ContourPlot[uuu, {x, -0.75, 0.75}, {y, 0, 1},
          Contours->40, ContourShading->False];
vvvplot = ContourPlot[vvv, {x, -0.75, 0.75}, {y, 0, 1},
          Contours->40, ContourShading->False];
Show[uuuplot, vvvplot]
```

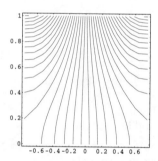

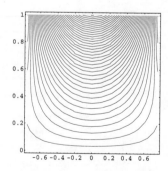

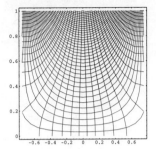

APPENDIX 5-4   FOURIER SERIES

The Fourier series is a powerful technique for representing an arbitrary function. It is particularly useful for representing the solutions to differential equations. The amplitudes (coefficients) of the Fourier series of a function of time represent the frequency distribution of the function. The method depends on the orthogonality property of the sine and cosine functions; the coefficients are found by using the orthogonality property.

Intialization

```
Remove["Global'@"]
```

Orthogonality of Sine and Cosine Functions

Show that the sine and cosine functions are orthogonal.

Series to Third Order with Displaced Limits of Integration

A periodic function remains periodic when its argument is displaced by an arbitrary constant. Choose some value for the displacement (expressed here in terms of the period).

```
disp = T/4
```

$$\frac{T}{4}$$

Only those integrals containing the same-order cosine function have nonzero values when integrated over one period. In this sense the cosine functions of different order are orthogonal in analogy with vectors. Note that the zeroth-order cosine function actually is constant and has a different integral from the higher-order cosine functions.

```
intcos := Integrate[Cos[(2*Pi*n/T)*t]*Cos[(2*Pi*m/T)*t],
          {t,disp,disp+T}]

Table[intcos,   {n,0,3},{m,0,3}];
TableForm[%]
```

```
T     0    0    0
```

$$
\begin{array}{cccc}
0 & \dfrac{T}{2} & 0 & 0
\end{array}
$$

$$
\begin{array}{cccc}
0 & 0 & \dfrac{T}{2} & 0
\end{array}
$$

$$
\begin{array}{cccc}
0 & 0 & 0 & \dfrac{T}{2}
\end{array}
$$

The sine functions of different order are also orthogonal. The zeroth-order sine function vanishes and need not be used, but we include it here for illustration.

```
intsin := Integrate[Sin[(2*Pi*n/T)*t]*Sin[(2*Pi*m/T)*t],
            {t,disp,disp+T}]

Table[intsin,    {n,0,3},{m,0,3}];
TableForm[%]
```

$$
\begin{array}{cccc}
0 & 0 & 0 & 0
\end{array}
$$

$$
\begin{array}{cccc}
0 & \dfrac{T}{2} & 0 & 0
\end{array}
$$

$$
\begin{array}{cccc}
0 & 0 & \dfrac{T}{2} & 0
\end{array}
$$

$$
\begin{array}{cccc}
0 & 0 & 0 & \dfrac{T}{2}
\end{array}
$$

The sine functions and cosine functions are orthogonal for all orders.

```
intcsn := Integrate[Cos[(2*Pi*n/T)*t]*Sin[(2*Pi*m/T)*t],
            {t,disp,disp+T}]

Table[intsin,    {n,0,3},{m,0,3}];
TableForm[%]
```

$$
\begin{array}{cccc}
0 & 0 & 0 & 0 \\
0 & 0 & 0 & 0 \\
0 & 0 & 0 & 0 \\
0 & 0 & 0 & 0
\end{array}
$$

Curve Fitting with Sine and Cosine Functions

An arbitrary function that is at least piecewise continuous can be represented accurately by a Fourier series (truncated to sufficiently high order). The Fourier coefficients are found by using the orthogonality of the cosine and sine functions. We will use the normalized functions.

Example: Square Wave with Sine and Cosine Functions

When the function to be expanded by the Fourier series is piecewise continuous, it is necessary to split the integration into several parts as we do here.

The coefficients (amplitudes) for the sine and cosine are found by separate calculations and put into two lists (vectors).

```
termno = 6

6
```

Sine Coefficients

```
normsin[n_] := 1/Sqrt[Integrate[Sin[(2*Pi*n/T)*t]^2,{t,0,T}]]
g[n_,t_] := normsin[n]*Sin[(2*Pi*n/T)*t]
gtable = Table[g[a,t],{a,1,termno}]
```

```
            2 Pi t                 4 Pi t                 6 Pi t
  Sqrt[2] Sin[------]     Sqrt[2] Sin[------]     Sqrt[2] Sin[------]
              T                      T                      T
{-------------------, -------------------, -------------------,
      Sqrt[T]              Sqrt[T]              Sqrt[T]
```

```
           8 Pi t               10 Pi t
 Sqrt[2] Sin[------]    Sqrt[2] Sin[-------]
             T                      T
-------------------, -------------------,
      Sqrt[T]              Sqrt[T]
```

```
           12 Pi t
 Sqrt[2] Sin[-------]
             T
-------------------}
      Sqrt[T]
```

```
A[m_] := Integrate[1*g[m,t],{t,0,T/2}]+
         Integrate[0*g[m,t],{t,T/2,T}]
atable = Table[A[a],{a,1,termno}]
```

```
 Sqrt[2] Sqrt[T]      Sqrt[2] Sqrt[T]      Sqrt[2] Sqrt[T]
{---------------, 0, ---------------, 0, ---------------, 0}
      Pi                 3 Pi                 5 Pi
```

Cosine Coefficients

```
normsin[n_] :=
1/Sqrt[Integrate[Cos[(2*Pi*n/T)*t]^2,{t,0,T}]]
f[n_,t_] := normcos[n]*Cos[(2*Pi*n/T)*t]
ftable = Table[f[a,t],{a,1,termno}]
```

```
                       2 Pi t               4 Pi t
             Sqrt[2] Cos[------]    Sqrt[2] Cos[------]
    1                    T                      T
{-------, -------------------, -------------------,
 Sqrt[T]        Sqrt[T]              Sqrt[T]
```

```
           6 Pi t               8 Pi t
 Sqrt[2] Cos[------]    Sqrt[2] Cos[------]
             T                      T
-------------------, -------------------,
      Sqrt[T]              Sqrt[T]
```

```
                 10 Pi t                     12 Pi t
    Sqrt[2] Cos[-------]        Sqrt[2] Cos[-------]
                T                             T
    -------------------- ,     -------------------}
         Sqrt[T]                     Sqrt[T]

    B[m_] := Integrate[1*f[m,t],{t,0,T/2}]+
                Integrate[0*f[m,t],{t,T/2,T}]
    btable = Table[B[a],{a,0,termno}]

     Sqrt[T]
    {-------, 0, 0, 0, 0, 0, 0}
        2
```

Make Polynomial and Plot

The expanded function is found by taking the dot product of the amplitude list with the orthogonal function list (for the sine and cosine functions separately). At this point we evaluate numerically the period T and the value of Pi, in order to plot the function.

```
funct = N[(atable.gtable + btable.ftable)/.T->0.5]
```

```
0.5 + 0.63662 Sin[12.5664 t] + 0.212217 Sin[37.6991 t] +
0.127324 Sin[62.8319 t]
```

The sixth-order expansion gives a very recognizable representation of the square wave. The undershoot and overshoot around the values 0 and 1 are typical of Fourier series; they become smaller as higher-order terms are included.

```
Plot[funct,{t,0,1.5}]
```

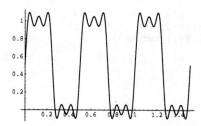

Plotting the Frequency Distribution

We need only to plot the coefficient of the sine functions, since only non-vanishing cosine coefficient corresponds to the constant term (frequency zero).

```
(* Show the values for the amplitudes and plot them *)
(* Note: the lowest order amplitude is n = 1 *)
(* Like the function plot, the period is set to 0.05 *)
(* The plot point size is increased for visibility *)


N[atable/.T>0.5]
ListPlot[N[atable/.T->0.5], AxesLabel->{n,Amplitude},
  Prolog -> AbsolutePointSize[4]]

{0.31831, 0, 0.106103, 0, 0.063662, 0}
```

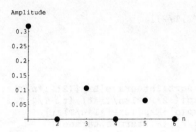

Complex Exponential Functions

Define the complex exponential functions and their corresponding conjugate functions. Show that they have orthogonal properties.

Series to Third Order
Integrate the product of an exponential function with the conjugate of another. Only those integrals containing the same-order exponential function produce nonzero values when integrated over one period.

```
intconj := Integrate[Exp[(2*Pi*I*n/T)*t]*Exp[(-
2*Pi*I*m/T)*t],
              {t,-T/2,T/2}]
Table[intconj,  {n,0,3},{m,0,3}];
TableForm[%]
```

T	0	0	0
0	T	0	0
0	0	T	0
0	0	0	T

All exponential functions of the same sign are orthogonal except for the zeroth-order function. However, the zeroth-order exponential function is a special case that is included with the conjugate functions.

```
intexp :=
Integrate[Exp[(2*Pi*I*n/T)*t]*Exp[(-2*Pi*I*m/T)*t],
            {t,-T/2,T/2}]
Table[intconj,  {n,0,3},{m,0,3}];
TableForm[%]
```

T	0	0	0
0	0	0	0
0	0	0	0
0	0	0	0

Example: Saw Tooth with Exponential Functions
Expand the square wave in terms of the complex exponential functions. This calculation is slightly shorter than in the sine and cosine expansion because all of the orthogonal functions can be expressed by a single notation. Because we include both positive and negative exponentials, the number of terms in the exponentials expansion is 2*termno + 1.

```
termno = 4
```

4

```
Integrate[Exp[(2*Pi*I*n/T)*t]*Exp[(-2*Pi*I*m/T)*t],
```

```
                {t,-T/2,T/2}]

   T
```

Coefficients

```
normsexp[n_] := 1/Sqrt[Integrate[Exp[(2*Pi*n/T)*t]*
                   Exp[(-2*Pi*I*n/T)*t],{t,-T/2,T/2}]]
e[n_,t_] := normexp[n]*Exp[(2*Pi*I*n/T)*t]
etable = Table[e[a,t],{a,-termno,termno}]
```

```
   (-8 I Pi t)/T      (-6 I Pi t)/T      (-4 I Pi t)/T
  E                  E                  E
{----------------, ----------------, ----------------,
    Sqrt[T]            Sqrt[T]            Sqrt[T]


   (-2 I Pi t)/T                (2 I Pi t)/T    (4 I Pi t)/T
  E                     1      E               E
----------------, -------, ----------------, ----------------,
    Sqrt[T]        Sqrt[T]     Sqrt[T]            Sqrt[T]


   (6 I Pi t)/T    (8 I Pi t)/T
  E               E
----------------, ----------------}
    Sqrt[T]           Sqrt[T]
```

```
A[m_] := Integrate[t*e[m,t],{t,-T/2,T/2}]
aetable = Table[Simplify[A[a]],{a,-termno,termno}]
```

```
  I   3/2  -I   3/2   I   3/2  -I   3/2     I   3/2  -I   3/2   I   3/2
  - T      -- T       - T      -- T         - T      -- T       - T
  8         6         4         2           2         4         6
{------, -------, ------, -------, 0, ------, -------, ------,
   Pi       Pi      Pi      Pi          Pi       Pi      Pi


 -I   3/2
 -- T
  8
-------}
   Pi
```

Make Polynomial and Plot

The expanded function is found by taking the dat product of the amplitude list with the orthogonal function list (for the sine and cosine functions separately).

```
funct = aetable.etable
```

```
 -I   (-2 I Pi t)/T      I   (2 I Pi t)/T      I   (-4 I Pi t)/T
 -- E                    - E                   - E
 2                  T    2                 T   4                  T
------------------ + ------------------ + ------------------ +
         Pi                   Pi                   Pi


  I   (4 I Pi t)/T      I   (-6 I Pi t)/T      I   (6 I Pi t)/T
  - E                   - E                    - E
  4                T    6                 T    6                T
---------------- + ------------------ + ---------------- +
        Pi                   Pi                  Pi
```

```
I  (-8 I Pi t)/T        I  (-2 I Pi t)/T
- E                T   - E                T
8                       8
------------------ +  ------------------
      Pi                     Pi
```

Now evaluate *T* and *Pi* numerically, in order to plot the function. The conversion of the exponential functions to sines and cosines gives only the sine terms as expected, and the factor of *I* makes the function pure real.

```
funct2 = N[0.5*Pi*T*(Sin[2*Pi*t/T] -
                     Sin[4*Pi*t/T]/2+
                     Sin[6*Pi*t/T]/3-
                     Sin[8*Pi*t/T]/4)/.T->0.5]

0.785398 (Sin[12.5664 t] - 0.5 Sin[25.1327 t] +
    0.333333 Sin[37.6991 t] - 0.25 Sin[50.2655 t])

Plot[funct2,{t,-0.5,0.5}]
```

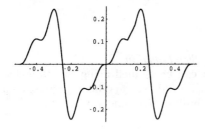

Using the *Mathematica* Package

Load the Fourier Transform Package

```
Needs["Calculus'FourierTransform'"]
```

Example: Saw Tooth

```
orderno = 8

8

FourierTrigSeries[x,  {x,  -1,  1},  orderno]
asdf = N[%]
Plot[asdf,{x,-2,2}]

2 Sin[Pi x]    Sin[2 Pi x]    2 Sin[3 Pi x]    Sin[4 Pi x]
----------- - ----------- + ------------- - ----------- +
    Pi             Pi            3 Pi             2 Pi
```

```
 2 Sin[5 Pi x]     Sin[6 Pi x]    2 Sin[7 Pi x]    Sin[8 Pi x]
 -------------  -  -----------  +  -------------  -  -----------
     5 Pi            3 Pi             7 Pi             4 Pi
```

```
 0.63662 Sin[3.14159 x] - 0.31831 Sin[6.28319 x] +
 0.212207 Sin[9.42478 x] - 0.159155 Sin[12.5664 x] +
 0.127324 Sin[15.708 x] - 0.106103 Sin[18.8496 x] +
 0.0909457 Sin[21.9911 x] - 0.0795775 Sin[25.1327 x]
```

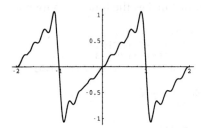

Example: Step Function

```
orderno = 8
```

8

Define and Plot the Unit Step Function
The unit step function can be defined using the *Mathematica* conditionals. This method can be generalized to form other piecewise continuous functions. *Mathematics* also has a built-in step function.

```
u[x_] := 0/;x<0
u[x_] := 1/;x>=0
```

```
Plot[u[x],{x,-1,1}]
```

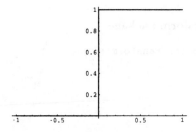

Find the Fourier Series and Plot
Note that although the step function is nonperiodic, its Fourier series representation is periodic because of the properties of Fourier series. The series therefore becomes a square wave.

```
FourierTrigSeries[u[x], {x, -1, 1}, orderno]
asdf = Chop[N[%]]
Plot[asdf,{x,-2,2}]
(* Several lines of error messages associated with *)
```

(* step function discontinuities have been deleted *)

```
Integrate[u[x], {x, -1, 1}]
--------------------------- +
            2
```

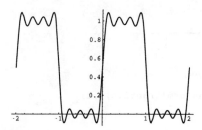

```
0.5 + 0.63662 Sin[3.14159 x] + 0.212207 Sin[9.42478 x] +
0.127324 Sin[15.708 x] + 0.0909457 Sin[21.9911 x]
```

Nonperiodic Functions: The Fourier Transform

The Fourier series is a series of complex exponential functions of discrete frequencies. A nonperiodic function can be represented if the series is replaced by an integral over the frequency from −Infinity to +Infinity. This integral is called the Fourier transform. The coefficients of the transform (corresponding to the coefficients of the Fourier series) are given by the inverse Fourier transform.

Define a Function to be Transformed and Plot

```
fun = (1+t^2)^(-1/2)
Plot[fun,{t,0,5},  PlotRange->{0,1}]
```

```
       1
------------
           2
Sqrt[1 + t ]
```

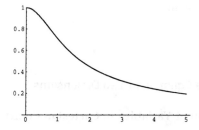

Make the Transform and Plot

```
transfun = FourierTransform[fun,  t,  w]
Plot[transfun,{w,0,5},AxesLabel->{w,Transform}]
```

```
             2
2 BesselK[0,  Sqrt[w ]]
```

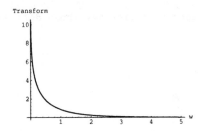

Find the Inverse Transform

```
InverseFourierTransform[transfun,w,t]
        1
     - - - - - - - - - - - -
              2
     Sqrt[1 + t ]
```

APPENDIX 5-5 LAPLACE EQUATION

The method of separation of variables gives a solution of the Laplace equation with a set of undetermined constants. We consider the case of Cartesian coordinates, but the method also applies to any separable coordinate system. The method is particularly powerful in that the coefficients can be determined by expressing the boundary potentials as a Fourier series.

Initialize

Remove symbols:

```
Remove["Global'@"]
```

Number of terms in the solution:

```
orderno = 6

6
```

Solution of the Laplace Equation in Two Dimensions

Solution by Separation of Variables in Cartesian Coordinates

```
phi = Sum[(Sx[n]*Sin[k*n*x]+Cx[n]*Cos[k*n*x])*
      (Sy[n]*Sinh[k*n*y]+Cy[n]*Cosh[k*n*y]),
      {n,0,orderno}];
Short[phi,4]

Cx[0] Cy[0] + (Cos[k x] Cx[1] + Sin[k x] Sx[1])
  (Cosh [k y] Cy[1] + Sinh[k y] Sy[1]) + <<4>> +
  (Cos[6 k x] Cx[6] + Sin[6 k x] Sx[6])
  (Cosh[6 k y] Cy[6] + Sinh[6 k y] Sy[6])
```

Proof of the Solution

Show that this is a solution of the Laplace equation by using the equality test.

```
Equal[D[D[phi,s]+D[D[phi,y],]]==0]
```

```
True
```

Expanded Form

For some applications the expanded form is faster.

```
expphi = Expand[phi];
Short[expphi,4]
```

```
Cx[0] Cy[0] + Cos[k x] Cosh[k y] Cx[1] Cy[1] +
  Cos[2 k x] Cosh[2 k y] Cx[2] Cy[2] + Cos[3 k x] <<3>> +
  <<19>> + Cos[6 k x] Cx[6] Sinh[6 k y] Sy[6] +
  Sin[6 k x] Sinh[6 k y] Sx[6] Sy[6]
```

Proof that the Expanded Form is a Solution

The application of the Laplacian operator directly to the series shows that it vanishes.

```
D[D[expphi,x],x]+D[D[expphi,y],y]
```

```
0
```

Determining the Coefficients by Fourier Expansion

To illustrate the method of finding the coefficients by Fourier analysis, assume that the potential is symmetric in x and antisymmetric in y. Then the solution to the Laplace equation takes a simple form, giving relatively quick *Mathematica* calculations.

Define the Potential with the Given Symmetries

Generic Potential Term

This term gives a solution to the Laplace equation, which can be shown by substitution. It has the symmetries mentioned above.

```
potTerm = A[n]*Sinh[2*Pi*n*y/L]*Cos[2*Pi*n*x/L]
```

$$A[n] \; Cos\left[\frac{2\,n\,Pi\,x}{L}\right] \; Sinh\left[\frac{2\,n\,Pi\,y}{L}\right]$$

The Potential Expressed by the Separation of Variables Formula

Because the Laplace equation is a linear equation, any linear combination of the preceding terms for arbitrary values of n is also a solution. We assume that n takes on only positive integer values, or zero. The integer values lead to Fourier series expression of the boundary conditions, and the solution is periodic. If n were chosen to be a continuous variable, Fourier transform methods would apply, and the boundary value need not be periodic.

```
pot = Sum[potTerm, {n,0,orderno}];
  Short[pot,4]
```

$$A[1] \; Cos\left[\frac{2\,Pi\,x}{L}\right] \; Sinh\left[\frac{2\,Pi\,y}{L}\right] + A[2] \; Cos\left\{\frac{4\,Pi\,x}{L}\right\} \; Sinh\left\{\frac{4\,Pi\,y}{L}\right\} +$$

```
            6 Pi x          6 Pi y
A[3] Cos[------] Sinh[------] + <<1>> +
             L               L
```

```
            10 Pi x         10 Pi y
A[1] Cos[-------] Sinh[-------] +
              L               L
```

```
            12 Pi x         12 Pi y
A[6] Cos[-------] Sinh[-------]
              L               L
```

Application of the Boundary Potentials

The coefficients can be found by the following procedure: (1) Multiply the series expression for the potential by the cosine. (2) Then integrate this over the period of the equation L. (3) Because of the orthogonality of the cosines, all of the terms vanish except for one term. (4) Therefore, the integration of the entire series can be replaced by just one term. Because the symbolic integration is slow, we will do this. (5) Solve this for the corresponding coefficient.

Choose the Potential on the Upper Boundary ($y = +L/2$)
The boundary function must be symmetric in x to agree with the symmetry of the cosine series chosen. We choose a semicircular function.

```
g = L^2-x^2
```

```
 2   2
L  - x
```

Evaluate the Generic Potential Term on the Upper Boundary

```
PotTermUp = potTerm/. y -> L/2
```

```
            2 n Pi x
A[n] Cos[--------] Sinh[n Pi]
              L
```

Multiply by the Cosine Function, and Integrate the Generic Term

```
potInt = Integrate[PotTermUp+Cos[2*Pi*n*x/L],
            {x,-L/2,L/2}]
```

```
L A[n] (2 n Pi + Sin[2 n Pi]) Sinh[n Pi]
----------------------------------------
                4 n Pi
```

Multiply by the Cosine Function, and Integrate the Boundary Potential

```
functInt = Integrate[g*Cos[2*Pi*n*x/L]
            {x,-L/2.L/2}]
```

```
   3                                                 2   2
-(L  (2 n Pi Cos[n Pi] - 2 Sin[n Pi] - 3 n  Pi  Sin[n Pi]))
----------------------------------------------------------- +
                            3   3
                         8 n  Pi
```

$$
\frac{(L \ (2 \ n \ Pi \ Cos[n \ Pi] + 2 \ Sin[n \ Pi] + 3 \ n^2 \ Pi^2 \ Sin[n \ Pi]))^3}{8 \ n^3 \ Pi^3}
$$

Equate the Two Integrals for the Potential on the Boundary

```
potEqn = potInt == functInt
```

$$
\frac{L \ A[n] \ (2 \ n \ Pi + Sin[2 \ n \ Pi]) \ Sinh[n \ Pi]}{4 \ n \ Pi} ==
$$

$$
\frac{-(L^3 \ (2 \ n \ Pi \ Cos[n \ Pi] - 2 \ Sin[n \ Pi] - 3 \ n^2 \ Pi^2 \ Sin[n \ Pi]))}{8 \ n^3 \ Pi^3} +
$$

$$
\frac{(L^3 \ (2 \ n \ Pi \ Cos[n \ Pi] + 2 \ Sin[n \ Pi] + 3 \ n^2 \ Pi^2 \ Sin[n \ Pi]))}{8 \ n^3 \ Pi^3}
$$

Solve the Equation for the Generic Coefficient

```
coef = Flatten[Solve[potEqn,A[n]]]
```

$$
\{A[n] \ \to \ -((L^2 \ Csch[n \ Pi]
$$
$$
(2 \ n \ Pi \ Cos[n \ Pi] - 2 \ Sin[n \ Pi] - 3 \ n^2 \ Pi^2 \ Sin[n \ Pi])) \ /
$$
$$
(n^2 \ Pi^2 \ (2 \ n \ Pi + Sin[2 \ n \ Pi]))) \}
$$

```
coefRule = Flatten[Table[coef,{n,0,orderno}]];
   Short[coefRule,4]
```

$$
\{A[0] \ \to \ \text{Indetermine}, \ A[1] \ \to \ \frac{L^2 \ Csch[Pi]}{Pi^2},
$$

$$
A[2] \ \to \ \frac{-(L^2 \ Csch[2 \ Pi])}{4 \ Pi^2}, \ A[3], \ \to \ \frac{L^2 \ Csch[3 \ Pi]}{9 \ Pi^2},
$$

$$
A[4] \ \to \ \frac{-(L^2 \ Csch[4 \ Pi])}{16 \ Pi^2}, \ A[5], \ \to \ \frac{L^2 \ Csch[5 \ Pi]}{25 \ Pi^2},
$$

```
              2
          -(L  Csch[6 Pi])
  A[6]  ->  ----------------}
                  2
              36 Pi
```

Apply These Coefficients to the Series Expression for the Potential

```
generalPot = pot/.coefRule;
   Short[generalPot,4]

 2     2 Pi x                2 Pi y
L  Cos[------] Csch[Pi] Sinh[------]
         L                     L
-------------------------------------  -
                 2
               Pi


 2     4 Pi x                4 Pi y
L  Cos[------] Csch[2 Pi] Sinh[------]
         L                     L
-------------------------------------- +
                 2
               4 Pi


 2     6 Pi x                6 Pi y
L  Cos[------] Csch[3 Pi] Sinh[------]                2
         L                     L            L  <<3>>
---------------------------------------- + <<1>> + -------- -
                 2                                     2
               9 Pi                                 25 Pi


 2    12 Pi x                12 Pi y
L  Cos[-------] Csch[6 Pi] Sinh[-------]
         L                      L
--------------------------------------
                 2
              36 Pi
```

Plot the Potential in the Unit Box
In order to plot the potential, it is necessary to choose numerical values for the period *L*.

```
boxSize = Pi

Pi

numericalPot = N[generalPot/.L->boxSize];
   Short[numericalPot,4]

0.0865895 Cos[2. x] Sinh[2. y] -
 0.000933725 Cos[4. x] Sinh[4. y] +
 0.0000179332 Cos[6. x] Sinh[6. y] + <<1>> +
           -8
 1.20561 10    <<1>> Sinh[10. y] -
           -10
 3.61801 10     Cos[12. x] Sinh[12. y]
```

```
plotPot = Plot3D[numericalPot,{x,-boxSize/2,boxSize/2},
                              {y,-boxSize/2,boxSize/2}]
```

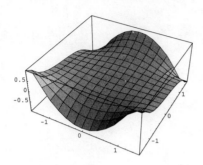

Show the Potentials on the Boundaries of the Unit Box

```
Plot3D[numericalPot,{x,-boxSize/2,boxSize/2},
                    {y,0,boxSize/2},
                    ViewPoint->{-0.000,10.000,0.000}]
Plot3D[numericalPot,{x,-boxSize/2,boxSize/2},
                    {y,0,boxSize/2},
                    ViewPoint->{-10.000,2.000,0.000}]
```

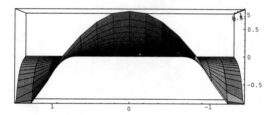

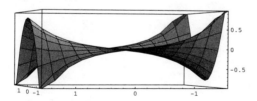

Radiation and Optics in Dielectric Media

The intense atom glows a moment, then is
quenched in a most cold repose.

> Percy Shelly
> *Adonais [1821]*

In Nature's infinite book of secrecy a little I can
read.

> William Shakespeare
> *Antony and Cleopatra*

Maxwell's discovery of the equations of electromagnetism is one of the greatest triumphs of the scientific method, and one of the great unifications. Not only were electricity and magnetism shown to be two aspects of a single phenomenon, but physical optics was included as well. In this chapter we discuss the radiative solutions of Maxwell's equations in charge-free regions. We show how these solutions have bearing on the refraction, reflection, and diffraction of light, and also show how energy and momentum are carried by the electromagnetic field.

6-1 WAVE EQUATION IN UNIFORM MEDIA

In Chapter 3 we derived Maxwell's field equations for a medium with permittivity ε and permeability μ. However, in order to find solutions, it is desirable to manipulate the Maxwell equations so as to uncouple them, i.e., as equations in **E** only and **B** only. For example, we found the uncoupled set of equations (4-4.36) and (4-4.37) for **E** and **B** in vacuum.

Let us find the uncoupled field equations in a uniform medium with Maxwell's equations (3-3.9) through (3-3.15) as the starting point. The curl of equation (3-3.12) becomes

$$\nabla \times \nabla \times \mathbf{E} = -\frac{1}{c}\frac{\partial \nabla \times \mathbf{B}}{\partial t} = -\frac{1}{c}\frac{\partial \nabla \times \mu\mathbf{H}}{\partial t}$$

$$= -\frac{1}{c}\frac{\partial}{\partial t}\left(\frac{\mu}{c}\frac{\partial \mathbf{D}}{\partial t} + \frac{4\pi\mu}{c}\mathbf{J}\right)$$

$$= -\frac{\varepsilon\mu}{c^2}\frac{\partial^2 \mathbf{E}}{\partial t^2} - \frac{4\pi\mu}{c^2}\frac{\partial \mathbf{J}}{\partial t}. \tag{6-1.1}$$

Using equations (3-3.9) and (1-3.32), the term on the left can be written:

$$\nabla \times \nabla \times \mathbf{E} = \nabla(\nabla \cdot \mathbf{E}) - \nabla^2\mathbf{E}$$

$$= \frac{4\pi}{\varepsilon}\nabla\rho - \nabla^2\mathbf{E}. \tag{6-1.2}$$

Combining these two equations gives the uncoupled equation for the electric field:

$$\frac{\varepsilon\mu}{c^2}\frac{\partial^2 \mathbf{E}}{\partial t^2} - \nabla^2\mathbf{E} = -\frac{4\pi}{\varepsilon}\nabla\rho - \frac{4\pi\mu}{c^2}\frac{\partial \mathbf{J}}{\partial t}. \tag{6-1.3}$$

An analogous calculation gives the uncoupled equation for the magnetic field:

$$\frac{\varepsilon\mu}{c^2}\frac{\partial^2 \mathbf{B}}{\partial t^2} - \nabla^2\mathbf{B} = \frac{4\pi\mu}{c}\nabla \times \mathbf{J}. \tag{6-1.4}$$

Equations (6-1.3) and (6-1.4) reduce to equations (4-4.36) and (4-4.37), in vacuum, i.e., when:

$$\varepsilon \to 1,$$
$$\mu \to 1. \tag{6-1.5}$$

In future calculations it will be useful to define the quantity:

$$c_m = \frac{c}{\sqrt{\varepsilon\mu}}, \tag{6-1.6}$$

which occurs in these equations. It will turn out to be the wave velocity in the medium. Since we particularly want to discuss the relationship between the electric and magnetic fields, we will use the nonrelativistic representation of Maxwell's equations, and discuss the fields $\mathbf{E}$ and $\mathbf{B}$ rather than the potentials ϕ and $\mathbf{A}$.

In this chapter, we consider fields in source-free regions. That is, ρ and $\mathbf{J}$ vanish,[1] the field equations for $\mathbf{E}$ and $\mathbf{B}$ are homogeneous, and

---

[1] Except for the charge and current densities induced by the presence of the fields. These densities are handled in the field equations by ε and μ.

have identical form. We will consider the inhomogeneous equations in Chapter 8. Homogeneous solutions of the equations can be found by means of separation of variables, similar to finding solutions of the Laplace equation. Assume the solution can be separated into temporal and spatial functions in the form of the product:

$$E = T(t)S(x, y, z), \tag{6-1.7}$$

where E is any component of the electric field. Due to the linearity of the field equation, the general solution is a linear combination of such terms. Substitution into equation (6-1.3) gives the separated equation:

$$\frac{1}{c_m^2} \frac{\partial_t^2 T}{T} - \frac{\nabla^2 S}{S} = 0. \tag{6-1.8}$$

Since the spatial variables and time are independent of each other, the functions T and S must satisfy equations:

$$\partial_t^2 T + \omega^2 T = 0 \tag{6-1.9}$$

$$\nabla^2 S + k^2 S = 0 \tag{6-1.10}$$

$$\omega = c_m k, \tag{6-1.11}$$

where ω is the angular frequency $\omega = 2\pi f$. Equation (6-1.10), called the *Helmholtz equation*, contains the spatial variables and can be further reduced by separation of variables in Cartesian or other coordinate systems. Note that equations (6-1.9) and (6-1.10) both contain a plus sign, in contrast to the Laplace equation.

If the Helmholtz equation is further separated in Cartesian coordinates, the general solution is a Fourier series or transform in three dimensions:

$$E(\mathbf{r}, t) = \begin{cases} \displaystyle\sum_{k_1=1}^{\infty} \sum_{k_2=1}^{\infty} \sum_{k_3=1}^{\infty} C_{k_1 k_2 k_3} \exp\left[i(\mathbf{k} \cdot \mathbf{r} - \omega t)\right], \\[2ex] \displaystyle\iiint_{-\infty}^{\infty} C(k_1, k_2, k_3) \exp\left[i(\mathbf{k} \cdot \mathbf{r} - \omega t)\right] d^3 k. \end{cases} \tag{6-1.12}$$

The amplitude, $C(k_1, k_2, k_3)$, is an arbitrary function of the *propagation vector*, $\mathbf{k}$. A constant phase angle was not written explicitly, since it can be absorbed into the amplitude. Since the four parameters are related by the Lorentz invariant condition:

$$\omega^2 - c_m^2 k_1^2 - c_m^2 k_2^2 - c_m^2 k_3^2 = 0, \tag{6-1.13}$$

it is only necessary to integrate over the three spatial parameters.

The simplest solution is the single term,[2] represented by:

$$\mathbf{E} = \mathbf{E}_0 \exp[i(\mathbf{k} \cdot \mathbf{r} - \omega t)], \qquad (6\text{-}1.14)$$

where there is no sum over $\mathbf{k}$ or ω. There is no loss of generality in defining the x-direction along $\mathbf{k}$; then in terms of the wavelength λ and period T, equation (6-1.14) is:

$$\mathbf{E} = \mathbf{E}_0 \exp\left[i\left(\frac{2\pi}{\lambda}x - \frac{2\pi}{T}t\right)\right] = \mathbf{E}_0 \exp\left[i\frac{2\pi}{\lambda}(x - c_m t)\right]. \quad (6\text{-}1.15)$$

Now define a *wave front* to be a surface which is orthogonal to the propagation vector, $\mathbf{k}$, at all points on the surface at a given point in time. Then we define a sequence of wave fronts for increasing values of t (usually at equal time increments, $\delta t = T$). The solution represented by equation (6-1.15) is called a *plane wave* because the wave fronts are planes; it is also *monochromatic* because the frequency, ω, is constant. The general solution, given by equation (6-1.12), may be interpreted as a linear combination of plane waves, forming wave fronts of arbitrary shape.

EXAMPLE

Evaluate the integral in equation (6-1.12) for the propagation constant, k in one dimension. This gives the solution:

$$\mathbf{E} = \int E_0(k) \exp[ik(x - c_m t)]\, dk$$

$$= \mathbf{f}(x - c_m t), \qquad (6\text{-}1.16)$$

where the function $\mathbf{f}$ is differentiable but otherwise arbitrary. This function can also be shown to be a solution of the wave equation by direct substitution. This function represents an arbitrary shape moving with constant velocity, c_m, as shown in Figure 6-1.

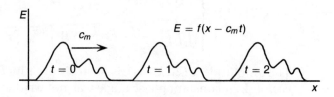

FIGURE 6-1. THE WAVE FUNCTION.   The curve is translated to the right without changing its shape, with velocity c_m.

---

[2] The other solution, $\mathbf{E} = \mathbf{E}_0 \exp[i(\mathbf{k} \cdot \mathbf{r} + wt)]$, containing a plus sign, represents a wave traveling in the opposite direction.

Maxwell's equations imply the existence of electromagnetic waves. The wave velocity in the medium, c_m, depends on the electric and magnetic properties of the medium—in vacuum it takes the value: 2.9979×10^{10} cm/s. Because it is also the speed of light in vacuum, this suggests that light itself is an electromagnetic wave! Since visible light has frequencies roughly between 0.5×10^{15} and 0.7×10^{15} Hz, there must exist other similar radiations at frequencies not visible to the eye. These are, of course, radio, microwaves, infrared, ultraviolet, x-rays, and gamma rays. Hertz's detection of electromagnetic radiation at radio frequencies (10^6 Hz) in 1886 verified Maxwell's theory.

Some general properties of electromagnetic waves can be deduced from these solutions of Maxwell's equations. First, in a sufficiently small region, any wave front may be treated as a plane wave. Then, substituting the plane wave expressions:

$$\mathbf{E} = \mathbf{E}_0 \exp[i(\mathbf{k} \cdot \mathbf{r} - \omega t)], \tag{6-1.17}$$

$$\mathbf{B} = \mathbf{B}_0 \exp[i(\mathbf{k} \cdot \mathbf{r} - \omega t)], \tag{6-1.18}$$

into Maxwell's equations (3-3.11) and (3-3.12) (but set $\mathbf{J} = 0$ in this dielectric medium):

$$\mathbf{n} \times \sqrt{\varepsilon\mu}\mathbf{E} = \mathbf{B}, \tag{6-1.19}$$

$$\mathbf{n} \times \mathbf{B} = -\sqrt{\varepsilon\mu}\mathbf{E}. \tag{6-1.20}$$

The vector $\mathbf{n}$ is a unit vector in the direction of $\mathbf{k}$. These equations imply that the vectors $\mathbf{E}$, $\mathbf{B}$, and $\mathbf{k}$ are mutually orthogonal. Note: this orthogonality refers to wave solutions; it does not apply to static fields, nor to resistive media, discussed in Section 6.3. Equations (6-1.19) and (6-1.20) also imply that the ratio of the magnitudes of $\mathbf{E}$ and $\mathbf{B}$ are constant in the medium, which becomes unity in vacuum (in Gaussian units). Let us define the following two ratios:

$$\frac{|\mathbf{B}|}{|\mathbf{E}|} = \sqrt{\varepsilon\mu} = n, \tag{6-1.21}$$

$$\frac{|\mathbf{E}|}{|\mathbf{H}|} = \sqrt{\frac{\mu}{\varepsilon}} = Z. \tag{6-1.22}$$

Their physical significance will be discussed in detail in Section 6-3. The electric and magnetic fields in a dielectric medium are illustrated in Figure 6-2.

Equations (6-1.19) and (6-1.20) show that electromagnetic radiation in uniform media is *transverse*, i.e., the direction of propagation is perpendicular to the direction of the fields. Transverse waves have a property that longitudinal waves do not—they can be polarized. This can be observed at visible frequencies by passing a beam of light through a filter composed of certain crystals. Incident light of random

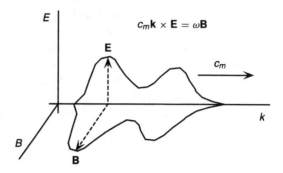

Figure 6-2. Electric and magnetic fields in a dielectric medium. Maxwell's equations require that **E** and **B** and **k** be mutually orthogonal in a dielectric medium, consistent with the equation shown in the diagram. Thus, if **E** lies in the x–y plane, and **k** along the x-axis, then **B** lies in the x–z plane. Since the direction of **E** (and **B**) is orthogonal to the direction of travel **k**, electromagnetic waves are transverse. The constant orientation of the electric vector shown here corresponds to plane polarization. Other polarizations are discussed in the text.

orientation is selectively absorbed, leaving a polarized beam exiting the filter. Observing the beam through a second filter shows maximum transmission only when the second filter has the same orientation as the first. A more general polarization results from superimposing a second beam, polarized at right angles but out of phase with the first. If the waves are sinusoidal, the electric vector will follow an elliptical helix in the direction of the propagation vector.

$$\mathbf{E} = \hat{\mathbf{i}} E_x \exp[ik(z - c_m t)] + \hat{\mathbf{j}} E_y \exp[ik(z - c_m t + \phi)]. \qquad (6\text{-}1.23)$$

The relative phase difference between the two components is ϕ. This equation represents *elliptical polarization*. In special cases elliptical polarization reduces to circular or plane polarization. That is, when the phase difference vanishes, the polarization becomes plane polarization (in the x–y plane); when the two amplitudes are equal, the polarization becomes circular polarization, with **E** precessing around the z-axis. This is shown in Figure 6-3.

6-2 SPHERICAL WAVES

Although plane wave solutions in Cartesian coordinates are useful for investigating certain general properties of electromagnetic radiation, in many ways spherical coordinates better describe radiation from a pointlike source. These spherical waves are useful for calculating scattering cross sections. The Helmholtz equation, (6-1.10), which

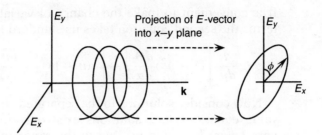

E_y

Projection of E-vector
into x–y plane

E_y

ϕ

k

E_x

E_x

FIGURE 6-3. ELLIPTICAL POLARIZATION. The electric field vector rotates in an elliptical helix which reduces to circular or plane polarization for special cases. The phase angle rotates the axes of the ellipse away from the coordinate axes. The magnetic vector is orthogonal to **E** and **k**, but is not shown.

determines the spatial contribution to the wave equation, holds for general coordinate systems, including spherical. In spherical coordinates it becomes:

$$\frac{1}{r^2}\frac{\partial}{\partial r}\left[r^2\frac{\partial S}{\partial r}\right] + \frac{1}{r^2\sin\theta}\frac{\partial}{\partial\theta}\left[\sin\theta\frac{\partial S}{\partial\theta}\right] + \frac{1}{r^2\sin^2\theta}\frac{\partial^2 S}{\partial\phi^2} + k^2 S = 0.$$

$$(6\text{-}2.1)$$

We solve this equation by separation of variables into radial and angular functions:

$$S(r, \theta, \phi) = R(r)Y(\theta, \phi). \qquad (6\text{-}2.2)$$

In these terms the separated equations become:

$$\frac{\partial}{\partial r}\left[r^2\frac{\partial R}{\partial r}\right] + [k^2 r^2 - n(n+1)]R = 0, \qquad (6\text{-}2.3)$$

$$\frac{\partial}{\partial\theta}\left[\sin\theta\frac{\partial Y}{\partial\theta}\right] + \frac{1}{\sin\theta}\frac{\partial^2 Y}{\partial\phi^2} + n(n+1)\sin\theta Y = 0, \qquad (6\text{-}2.4)$$

where we anticipated the form of the separation constant $n(n+1)$. The solutions of equation (6-2.4) are called *spherical harmonics*. Equation (6-2.4) can be further separated by writing $Y(\theta, \phi) = \Theta(\theta)\Phi(\phi)$ and substituting. This gives:

$$\frac{d^2\Phi}{d\phi^2} + m^2\Phi = 0, \qquad (6\text{-}2.5)$$

$$\frac{1}{\sin\theta}\frac{d}{d\theta}\left[\sin\theta\frac{d\Theta}{d\theta}\right] + \frac{1}{\sin^2\theta}m^2\Theta + n(n+1)\Theta = 0. \qquad (6\text{-}2.6)$$

It is convenient to make the change of variables $\mu = \cos\theta$. In these terms the second equation takes its standard form:

$$\frac{d}{d\mu}\left[(1 - \mu^2)\frac{d\Theta}{du}\right] + \left[n(n + 1) - \frac{m^2}{1 - \mu^2}\right]\Theta = 0. \qquad (6\text{-}2.7)$$

Now consider solutions to the separated equations. They are eigenvalue equations, i.e., solutions exist only for certain values of the parameters k, n, and m. Due to the range of values on the angles ϕ and θ there are conditions on the eigenvalues. The limits of ϕ are 0 and 2π; then in order for Φ to have single-valued solutions, m must be integer. The solutions to equation (6-2.5) can be written in complex exponential form:

$$\Phi = \{\Phi_0 \exp(im\phi)\} \qquad m = 0, \pm1, \pm2, \pm3, \ldots, \qquad (6\text{-}2.8)$$

or as sine and cosine. They are orthogonal functions as we saw in Chapter 5.

Solutions to equation (6-2.7) are finite only when n takes integer values. The solutions are the *associated Legendre functions*. We will give only a brief discussion of their properties. More complete treatments are given in the mathematical references in the Selected Bibliography. A generating algorithm for these functions is:

$$P_n^m(\mu) = \frac{[1 - \mu^2]^{m/2}}{2^n n!} \frac{d^{n+m}}{d\mu^{n+m}}[\mu^2 - 1]^n, \qquad n = 0, 1, 2, \ldots. \qquad (6\text{-}2.9)$$

The orthogonality integral:

$$\int_{-1}^{1} P_n^m(\mu)P_{n'}^{m'}(\mu)\, d\mu = \frac{2\delta^{nn'}\delta^{mm'}}{2n + 1}\frac{(n + m)!}{(n - m)!} \qquad (6\text{-}2.10)$$

gives the normalization. The eigenvalues are related by the condition $|m| \leq n$. Table 6-1 lists the first few spherical harmonics, $Y(\theta, \phi)$, showing their normalization constants. Examples are shown in Figure 6.4.

TABLE 6-1.   Normalized Spherical Harmonics

n	m	$Y(\theta, \phi)$
0	0	$\sqrt{1/4\pi}$
1	0	$\sqrt{3/4\pi}\cos\theta$
	1	$\sqrt{3/8\pi}\sin\theta e^{\pm i\phi}$
2	0	$\sqrt{3/16\pi}[3\cos^2\theta - 1]$
	1	$\sqrt{15/8\pi}[\sin\theta\cos\theta e^{\pm i\phi}]$
	2	$\sqrt{15/32\pi}[\sin^2\theta e^{\pm 2i\phi}]$

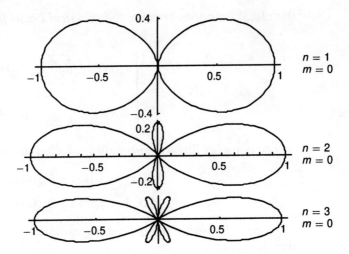

FIGURE 6-4. EXAMPLES OF THE SPHERICAL HARMONICS.   To express them in three dimensions rotate them around the vertical axis.

The radial equation:

$$\frac{d}{dr}\left[r^2 \frac{dR_n}{dr}\right] - [n(n+1) - k^2r^2]R_n = 0 \qquad (6\text{-}2.11)$$

can be simplified and put into standard form by a couple of transformations of variables. First, rewrite the equation in terms of the dimensionless variable:

$$u = kr. \qquad (6\text{-}2.12)$$

Second, transform the dependent variable:

$$U_n(u) = \frac{R_n(u)}{\sqrt{u}}. \qquad (6\text{-}2.13)$$

These transformations give the *Bessel equation*:

$$u^2 \frac{d^2 U_n}{du^2} + u \frac{dU_n}{du} - \left[\left(n + \frac{1}{2}\right)^2 - u^2\right] U_n = 0. \qquad (6\text{-}2.14)$$

The solutions are the *Bessel and Neumann functions* of half integer order. They are tabulated in various mathematical texts and handbooks and are available in computer software. Reversing the transformations, the radial functions become the *spherical Bessel functions*:

$$
R_n(kr) = \begin{cases}
j_n(kr) = \sqrt{\dfrac{\pi}{2kr}}\, J_{n+1/2}(kr), \\[2ex]
n_n(kr) = \sqrt{\dfrac{\pi}{2kr}}\, N_{n+1/2}(kr).
\end{cases}
\tag{6-2.15}
$$

The complex linear combinations, called the *Hankel functions*:

$$
\begin{aligned}
h_n(kr) &= j_n(kr) + i n_n(kr), \\
h_n^*(kr) &= j_n(kr) - i n_n(kr),
\end{aligned}
\tag{6-2.16}
$$

are especially useful for describing radiation, since at large distances they become:

$$
\begin{aligned}
h_n(kr) &\to \frac{\exp(ikr)}{kr}, \\[2ex]
h_n^*(kr) &\to \frac{\exp(-ikr)}{kr}.
\end{aligned}
\tag{6-2.17}
$$

They are therefore suitable for describing spherical waves. Table 6-2 gives some of the lower order functions. The half integer Bessel functions are orthogonal, but the details are more complicated than those of the angular functions. Therefore, they will not be discussed here in detail. At large distances the orthogonality of the Hankel functions is quite simple, being based on the orthogonality of the exponential functions. Figure 6.5 shows the lowest-order Bessel and Neumann functions.

The general solution to the wave equation for a scalar ψ in spherical coordinates can be written:

$$
\psi(r, \theta, \phi, t) = \sum_{m=0}^{\infty} \sum_{n=0}^{\infty} \sum_{k \neq 0}^{\infty} C_{mnk} h_n(kr) P_n^m(\cos\theta) e^{im\phi} e^{-i\omega t}.
\tag{6-2.18}
$$

Due to the relations between the quantities m, n, k, and ω, only three summations are required. The coefficients C_{mnk} can be found by means of the orthogonality of the functions (similar to Fourier series). Equation (6-2.18) gives the general solution to the scalar wave equation and, a fortiori, to the Helmholtz equation. It represents arbitrarily shaped wave fronts. Figure 6-6 illustrates the wave for just values $n = 1$, $m = 0$ at a fixed time.

Now consider the homogeneous solutions to equations (6-1.3) and (6-1.4) for $\mathbf{E}$ and $\mathbf{B}$ in spherical coordinates. Representing these vector functions is, unfortunately, much more complicated than in Cartesian coordinates. In Cartesian coordinates, the Laplacian acts on the

TABLE 6-2. SPHERICAL BESSEL, NEUMANN, AND HANKEL FUNCTIONS

Index n	Bessel $j_n(u)$	Neumann $n_n(u)$	Hankel $h_n(u)$
0	$\dfrac{\sin u}{u}$	$-\dfrac{\cos u}{u}$	$-\dfrac{i\,\exp(iu)}{u}$
1	$\dfrac{\sin u}{u^2} - \dfrac{\cos u}{u}$	$-\dfrac{\sin u}{u} - \dfrac{\cos u}{u^2}$	$-\left[\dfrac{1}{u} + \dfrac{i}{u^2}\right]\exp(iu)$
2	$\left[\dfrac{3}{u^3} - \dfrac{1}{u}\right]\sin u - \dfrac{3\cos u}{u^2}$	$-\left[\dfrac{3}{u^3} - \dfrac{1}{u}\right]\cos u - \dfrac{3\sin u}{u^2}$	$\left[\dfrac{1}{u} - \dfrac{3}{u^2} - \dfrac{3i}{u^3}\right]\exp(iu)$

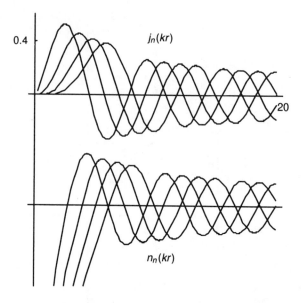

Figure 6-5. Spherical Bessel and Neumann functions.   The functions for $n = 0, 1$, and 2 are plotted.

components of a vector as independent functions. That is:

$$\nabla^2 \begin{pmatrix} V_x \\ V_y \\ V_z \end{pmatrix} = \begin{pmatrix} \nabla^2 V_x \\ \nabla^2 V_y \\ \nabla^2 V_z \end{pmatrix}, \tag{6-2.19}$$

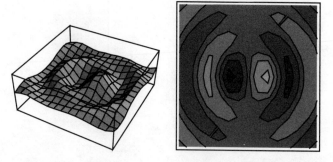

Figure 6-6. Radiation from a point source.   These diagrams are two representations of the same scalar wave, $\psi(r, \theta) = n_1(kr) P_1^0(\theta)$. The contour plot is a little rough due to hardware limitations. The contours represent constant values of ψ (like equipotentials). The wave travels radially outward.

with the consequence that the equations for the components are uncoupled equations for the vector wave equation in Cartesian coordinates. The components can be treated as three independent scalars. By contrast, the vector components lie along the directions $\mathbf{e}_r$, $\mathbf{e}_\theta$, and $\mathbf{e}_\phi$, which are functions of the coordinate variables r, θ, ϕ. Furthermore, the Laplacian itself contains the variables. Consequently, the equations for the components are coupled equations. These coupled equations for the vector components can be solved by brute-force (but cumbersome) methods. However, a simpler method of finding solutions is to express the vector wave solution in terms of the scalar wave solution, which we discussed above.

Let the electric field be expressed in terms of the scalar function, ψ:

$$\mathbf{E} = \mathbf{r} \times \nabla \psi. \tag{6-2.20}$$

As one can verify in Cartesian coordinates, if ψ satisfies the scalar Helmholtz equation, then this expression for $\mathbf{E}$ satisfies the vector Helmholtz equation:

$$[\nabla^2 + k^2](\mathbf{r} \times \nabla \psi) = 0. \tag{6-2.21}$$

The Helmholtz operator is scalar, and the electric field defined by equation (6-2.20) is an axial vector; thus equation (6-2.21) acts like a vector under continuous coordinate transformations. A general property of tensors is that a tensor that vanishes in one coordinate system, vanishes in all coordinate systems. The proof is a direct result of the transformation equations (4-1.8). Consequently, since equation (6-2.21) holds in Cartesian coordinates, it holds in spherical coordinates.

The electric field in expression (6-2.20), is orthogonal to the radius vector, $\mathbf{r}$. It therefore represents radiation in which the electric field is transverse to the direction of propagation, and we will call it *transverse electric mode* (TE). However, the magnetic field is not transverse to the direction of propagation. From Maxwell's equation (3-3.11), the magnetic field must take the form:[3]

$$\frac{-i\omega}{c} \mathbf{B} = -\nabla \times \mathbf{E} = -\nabla \times (\mathbf{r} \times \nabla \psi), \tag{6-2.22}$$

which is easily shown to contain a component parallel to $\mathbf{r}$. In place of the electric field in equation (6-2.20), we could have chosen the magnetic field to be:

$$\mathbf{B} = \mathbf{r} \times \nabla \psi \tag{6-2.23}$$

corresponding to a set of *transverse magnetic mode* (TM) solutions. The TE and TM solutions are special cases, but are of particular interest in certain cases, e.g., waveguides, where conducting boundaries constrain

---

[3] Remember that ω is constant, i.e., ψ is monochromatic.

the fields to these modes. A general solution can be expressed by the linear combination:

$$\mathbf{E} = C_1 \mathbf{r} \times \nabla\psi_1 + C_2 \nabla \times (\mathbf{r} \times \nabla\psi_2), \tag{6-2.24}$$

where ψ_1 and ψ_2 are independent solutions of Helmholtz's equation.

EXAMPLE

Take the first-order (dipole) term in equation 6-2.18 with $n = 1, m = 0$:

$$\psi = n_1(kr)P_1^0(\cos\theta) = \frac{1}{kr}\left[1 + \frac{i}{kr}\right]e^{ikr}\cos\theta$$

$$\rightarrow \frac{e^{ikr}}{kr}\cos\theta, \qquad kr \gg 1. \tag{6-2.25}$$

Then, using equations (1-4.47) and (1-4.28) for the gradient and cross product, the TE electric field becomes:

$$\mathbf{E} = r \times \nabla\psi$$

$$= -\left(\left[\frac{1}{kr} + \frac{i}{(kr)^2}\right]e^{ikr}\sin\theta\right)\mathbf{e}_\phi$$

$$\rightarrow -\frac{e^{ikr}\sin\theta}{kr}\mathbf{e}_\phi, \tag{6-2.26}$$

where the azimuthal unit vector is written $\mathbf{e}_\phi$. The corresponding magnetic field is obtained from equation (6-2.22). Then, using equation (1-4.58) for the curl, we have:

$$\mathbf{B} = \frac{c}{i\omega}\nabla \times \mathbf{E}$$

$$= \left(\frac{2c}{i\omega}\left[\frac{1}{kr^2} + \frac{i}{k^2r^3}\right]e^{ikr}\cos\theta\right)\mathbf{e}_r$$

$$+ \left(\frac{c}{i\omega}\left[\frac{1}{kr^2} + \frac{i}{k^2r^3} - \frac{i}{r}\right]e^{ikr}\sin\theta\right)\mathbf{e}_\theta$$

$$\rightarrow -\frac{e^{ikr}\sin\theta}{kr}\mathbf{e}_\theta. \tag{6-2.27}$$

The magnetic field has a radial (longitudinal) component, parallel to the propagation vector. However, in the limiting region where $kr \gg 1$, called the *radiation zone*, the largest terms in $\mathbf{E}$ and $\mathbf{B}$ are transverse, orthogonal to each other, and inversely proportional to r. In contrast, the steady-state electric field is inversely proportional to the *square* of *r*. This feature will reappear in Chapter 8, when we discuss radiation from accelerating charges.

6-3 Radiation in Conductive and Dispersive Media

The wave function discussed above is a function of the argument $kr - \omega t$. This quantity is frequently called the *phase of the wave function*. In general, we can add a constant; then the phase is written:

$$\phi = \mathbf{k} \cdot \mathbf{r} - \omega t + \phi_0$$

$$= \frac{2\pi}{\lambda} [\mathbf{n} \cdot \mathbf{r} - c_m t] + \phi_0 \qquad (6\text{-}3.1)$$

in terms of the wavelength:

$$\lambda = c_m T. \qquad (6\text{-}3.2)$$

The relative phase ϕ_0 is often absorbed into the amplitude. In general, the wavelength depends on the electric and magnetic properties of the media. Expressing the propagation constants as a 4-vector:

$$k_j = \frac{2\pi}{\lambda^j} \qquad (6\text{-}3.3)$$

the phase is invariant, i.e., constant under changes of the reference frame. Note that this expression allows a different wavelength in each direction, which may be the case when the magnetic or electric susceptibilities are tensors. Figure 6-7 illustrates the wave function in two dimensions; lines of constant phase are indicated.

For ideal dielectrics, as Figure 6-7 shows, wave motion is a rigid translation of the wave function at the wave velocity of the medium. The same velocity applies to waves of all wavelengths and is called the *phase velocity*:

$$c_{\text{phase}} = \frac{\omega}{k}. \qquad (6\text{-}3.4)$$

A wave of nonsinusoidal shape can be expressed as a linear combination of wavelengths—a Fourier series. Such a wave still translates rigidly with the phase velocity. If, however, the velocity depends on the magnetic and dielectric constants of the medium, then these "constants" may be frequency dependent:

$$\varepsilon = \varepsilon(\omega),$$

$$\mu = \mu(\omega). \qquad (6\text{-}3.5)$$

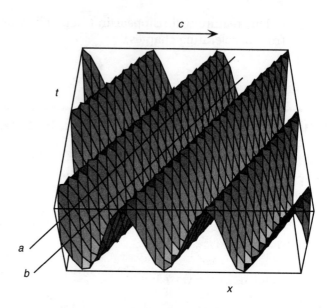

FIGURE 6-7. WAVE FUNCTION IN TWO DIMENSIONS. The wave moves in the x-direction with velocity c. For any point on the wave front, the phase ($\phi = kx - ct + \phi_0$) is constant. The motion of the wave is shown by taking successive sections through the function for increasing values of time. Then the sinusoidal function appears to translate rigidly. Points on a particular phase line remain on that line as the translation occurs.

In this case, each frequency has its own velocity and we define the *group velocity* which is a function of frequency:

$$c_{\text{group}} = \frac{1}{dk/d\omega} = f(\omega). \qquad (6\text{-}3.6)$$

Each Fourier component moves with its velocity; consequently, the wave function does not translate rigidly, but changes its shape as it translates. Then the group velocity represents the rate of spread. Such a medium is said to be *dispersive*. The index of refraction is frequency dependent as well, resulting in some interesting physical effects including chromatic aberration and rainbows. A mechanical analogy with water waves occurs in surf, where the velocity decreases due to the decreasing depth of the water—the wave becomes steeper and steeper, and finally the tip of the wave curls over and the wave breaks. Figure 6-8 illustrates the phase and group velocities.

The dependence of the propagation vector on frequency is called a *dispersion relation*. These relations appear in complex systems, such as elementary particles and condensed matter physics. The frequency dependence of the electric susceptibility is determined by the electronic properties of the material. Use of an elementary model of the elec-

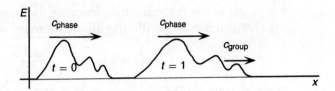

FIGURE 6-8. PHASE AND GROUP VELOCITY.   In a dispersive medium, the group velocity is a function of frequency. Consequently, each Fourier component moves with its own velocity, and the shape of the wave changes as its geometric center moves (with the phase velocity).

tronic binding potentials will enable us to find an approximation to the electric susceptability over a short frequency interval. Assume that an electron in the sample is bound in its place by a potential that can be approximated by a harmonic oscillator. See Figure 6-9.

Then if electromagnetic radiation passes through the sample, the electron will be subject to a periodic force with the frequency ω. The equation of motion of the electron will be:

$$m \frac{d^2x}{dt^2} + kx = -eE_0 \sin \omega t, \qquad (6\text{-}3.7)$$

where E_0 is the magnitude of the electric field of the incident radiation. The solution to this equation is:

$$x = x_0 \sin \omega t, \qquad (6\text{-}3.8)$$

where the amplitude of the displacement is:

$$x_0 = \frac{-(e/m)E_0}{\omega^2 - k/m} = \frac{-(e/m)E_0}{\omega^2 - \omega_R^2}, \qquad (6\text{-}3.9)$$

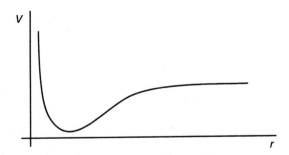

FIGURE 6-9. BINDING POTENTIAL.   One of the outer electrons in a molecule will be subject a potential of roughly this shape. If its energy is small, it will be bound with a potential roughly that of a harmonic oscillator.

where ω_R is the resonant frequency of the harmonic oscillator. Then the electron has an oscillating dipole moment with magnitude:

$$d_0 = -ex_0 = \frac{(e^2/m)E_0}{\omega_R^2 - \omega^2}. \qquad (6\text{-}3.10)$$

From the dipole moment we can calculate the polarization and thus the permittivity:

$$\varepsilon = \varepsilon_h + \frac{C}{\omega_R^2 - \omega^2}, \qquad (6\text{-}3.11)$$

where ε_h is the high-frequency value. See Figure 6-10. The preceding calculation was strictly classical, and assumed only one resonance. Other bound electrons having their own characteristic frequencies would produce corresponding contributions to the permittivity. A more exact calculation requires considerable knowledge of condensed matter physics and will not be pursued here.

We have so far mainly ignored electrical conductivity in dielectrics. Real materials have an enormous range of conductivity, amounting to 23 orders of magnitude between sulfur and silver. The propagation of electromagnetic waves through conductive media is of considerable practical as well as theoretical interest. As we saw in Chapter 3, the current density in a sample of material is proportional to the local electric field. In an isotropic sample, equation (3-2.1) becomes:

$$\mathbf{J} = \sigma\mathbf{E}. \qquad (6\text{-}3.12)$$

Then, substituting this into the field equations (6-1.3) and (6-1.4) gives:

$$\frac{\varepsilon\mu}{c^2}\partial_t^2\mathbf{E} - \nabla^2\mathbf{E} + \frac{4\pi\sigma\mu}{c^2}\partial_t\mathbf{E} = 0, \qquad (6\text{-}3.13)$$

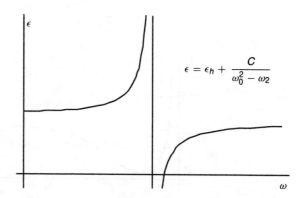

$$\epsilon = \epsilon_h + \frac{C}{\omega_0^2 - \omega_2}$$

FIGURE 6-10. TYPICAL DIELECTRIC CONSTANT NEAR A RESONANCE.   If there are other resonances in the sample, they will induce corresponding discontinuities in the permittivity at their characteristic frequencies.

$$\frac{\varepsilon\mu}{c^2}\partial_t^2\mathbf{B} - \nabla^2\mathbf{B} - \frac{4\pi\sigma}{c}\nabla \times \mathbf{E} = 0. \tag{6-3.14}$$

These equations can be solved by separation of variables as before.[4] First equation (6-3.13) is solved and then the electric field can be used in equation (6-3.14). Thus:

$$\frac{(\varepsilon\mu/c^2)\partial_t^2 T + (4\pi\sigma\mu/c^2)\partial_t T}{T} - \frac{\nabla^2 S}{S} = 0. \tag{6-3.15}$$

The spatial equation is the Helmholtz equation as it is for nonconducting dielectrics, but the equation for the time-dependent factor is:

$$\partial_t^2 T + \frac{1}{\tau}\partial_t T + c_m^2 k^2 T = 0, \tag{6-3.16}$$

expressed in terms of the relaxation time (see equation (5-1.4)):

$$\tau = \frac{\varepsilon}{4\pi\sigma}. \tag{6-3.17}$$

Let the solution of the time equation take the form:

$$T = T_0 \exp(i\omega t). \tag{6-3.18}$$

Substitution into (6-3.16) gives a *dispersion relation* with complex terms:

$$\omega^2 + \frac{i}{\tau}\omega - \sigma_m^2 k^2 = 0. \tag{6-3.19}$$

For plane waves, we choose k to be parallel to the x-axis with no loss of generality. We solve for the complex k by writing it explicitly as real and imaginary components:

$$k = \mathrm{Re}\,k + i\,\mathrm{Im}\,k. \tag{6-3.20}$$

Then substituting into equation (6-3.19), gives:

$$\mathrm{Re}\,k = \frac{\omega}{c_m}\sqrt{\frac{1}{2}\left[\sqrt{1+\left(\frac{1}{\omega\tau}\right)^2}+1\right]},$$

$$\tag{6-3.21}$$

$$\mathrm{Im}\,k = \frac{\omega}{c_m}\sqrt{\frac{1}{2}\left[\sqrt{1+\left(\frac{1}{\omega\tau}\right)^2}-1\right]}.$$

---

[4] We assume that the conductivity is constant. In general, it is frequency dependent, but may be taken constant over a sufficiently small frequency range. We also assume it is scalar.

These complicated expressions simplify in two limiting cases: (1) a good insulator, where the relaxation time is much larger than the period $\sigma/\varepsilon \ll \omega$; and (2) a good conductor, where the relaxation time is much smaller than the period $\sigma/\varepsilon \gg \omega$. That is:

$$k = \begin{cases} \dfrac{\omega}{c_m} + i\dfrac{2\pi\sigma}{c}Z & \textit{Insulator}: \omega\tau \gg 1, \\[4mm] \dfrac{\sqrt{2\pi\sigma\mu\omega}}{c}(1+i) & \textit{Conductor}: \omega\tau \ll 1, \end{cases} \qquad (6\text{-}3.22)$$

defined in terms of the intrinsic impedance (see equation (6-1.22)):

$$Z = \sqrt{\frac{\mu}{\varepsilon}}, \qquad (6\text{-}3.23)$$

and the *attenuation length* (skin depth):

$$\delta = \frac{c}{\sqrt{2\pi\sigma\mu\omega}}. \qquad (6\text{-}3.24)$$

By this criterion, the definition of conductor or insulator depends on the relative frequency and relaxation time. Since σ and ε are themselves functions of frequency, one must assume the dispersion relation (6-3.19) to be an approximation over limited frequency ranges. Typically, the skin depth and impedance are also strongly dependent on temperature.

Consider the electric field in a conductor. Written in terms of the attenuation length, the electric field, for a given frequency, is expressed:

$$\mathbf{E} = [\mathbf{E}_0 \exp(-x/\delta)] \exp[i((\mathrm{Re}\,k)x \pm \omega t)]. \qquad (6\text{-}3.25)$$

This decaying exponential function is shown in Figure 6-11. Since the propagation constant, $\mathrm{Re}\,k$, is frequency dependent, the system is dispersive.

We have seen that in vacuum, the ratio of the electric and magnetic fields E/B is unity, whereas in a dielectric medium it depends on ε and μ. To calculate this ratio substitute the plane wave electric and magnetic fields:

$$\begin{aligned} E_y &= E_0 \exp[i(kx \pm \omega t)], \\ B_z &= B_0 \exp[i(kx \pm \omega t - \varphi)], \end{aligned} \qquad (6\text{-}3.26)$$

into Maxwell's equation (3-3.11), which becomes:

$$\frac{\partial E_y}{\partial x} = -\frac{1}{c}\frac{\partial B_z}{\partial t}. \qquad (6\text{-}3.27)$$

This gives the intrinsic impedance for the medium:

$$Z = \frac{E_y}{H_z} = \frac{\mu\omega}{kc}e^{i\varphi} = \frac{\mu}{c}\frac{c}{\sqrt{\varepsilon\mu}}e^{i\varphi}$$

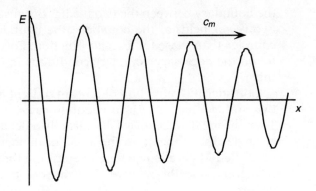

Figure 6-11. Attenuation of the electric field in a conductive dielectric. The wavelength is smaller than the attenuation length.

$$= \sqrt{\frac{\mu}{\varepsilon}} e^{i\varphi}, \tag{6-3.28}$$

which is complex when B and E are out of phase.

Example

In vacuum, the intrinsic impedance has the value:

$$Z = \begin{cases} 1, & \text{Heaviside,} \\ 1, & \text{Gaussian,} \\ 376.7, & \text{SI.} \end{cases} \tag{6-3.29}$$

When the medium is vacuum, there is no phase difference between the electric and magnetic fields and the impedance is pure resistance.

6-4 Refraction and Reflection at a Dielectric Boundary

A beam of light[5] incident on the boundary between two dielectric media is split into two beams, one reflected from the surface into the original medium, and the other refracted (transmitted) into the second medium. Since the strength and direction of a light beam is represented by its fields, the angles of reflection and refraction, and their intensities relative to the incident beam, can be calculated by matching the fields at

---

[5] Including EM radiation at all frequencies except x-rays and γ-rays because their wavelengths are too small for the media to be uniform.

the boundary between the two media. Because it is necessary to match two *vector* fields at the boundary, the boundary conditions are more complex than those for a scalar function. Fortunately, we have already found the necessary boundary conditions in Chapter 3; we now apply them.

Consider the refraction of a beam of light at the boundary between two uniform media with susceptibilities ε_1, μ_1 and ε_2, μ_2, respectively. Then the incident beam will be partially reflected and partially refracted at angles that depend on the angle of incidence and the susceptibilities. Then Maxwell's equations determine the angles of reflection and refraction, and the relative intensities of the reflected and refracted beams.

Assume the medium is piecewise uniform and the incident light is monochromatic; then each of the fields may be represented by a plane wave function. For example:

$$\mathbf{E} = \mathbf{E}_0 \exp[i(\mathbf{k} \cdot \mathbf{r} - \omega t)]. \tag{6-4.1}$$

We will use subscripts to label each field, i.e., I for incident, R for reflected, and T for transmitted (refracted). The total electric field in the incident medium is the sum of the incident and reflected fields. The total electric field in the second medium is only the refracted field. The corresponding quantities are shown in Figure 6-12.

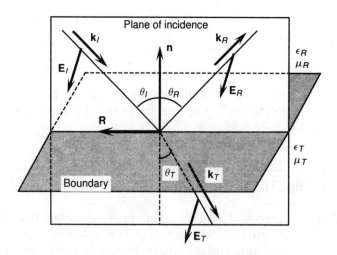

FIGURE 6-12. **RELATIONS BETWEEN QUANTITIES IN REFLECTION AT A BOUNDARY.** The subscript, R, labels the reflected beam and the subscript, T, labels the refracted beam. The unit vector, **n**, is normal to the plane defining the boundary surface between the two media, and lies in the plane of incidence. The vector, **R**, lies in the intersection of the two planes. The electric field is not necessarily orthogonal to the plane of incidence. Since the boundary conditions apply locally, the boundary surface is represented by a plane.

We can apply the boundary conditions obtained in Section 3-4. Since these boundary conditions apply for all points on the boundary and at all times, this constrains the arguments of the three fields. At each point in the intersection of the boundary plane and the plane of incidence in Figure 6-12, we choose $\mathbf{r} = \mathbf{R}$. Then in this intersection the incident, reflected, and transmitted fields, equation (6-4.1) becomes:

$$\mathbf{E} = \mathbf{E}_0 \exp[i(\mathbf{k} \cdot \mathbf{R} - \omega t)]. \tag{6-4.2}$$

At each point in the intersection the arguments are equal and constant, and we write:

$$(\mathbf{k}_I \cdot \mathbf{R} - \omega t) = (\mathbf{k}_R \cdot \mathbf{R} - \omega t) = (\mathbf{k}_T \cdot \mathbf{R} - \omega t) = \text{constant}. \tag{6-4.3}$$

Since the fields are monochromatic, we have the constraints on the propagation vectors:

$$\mathbf{k}_I \cdot \mathbf{R} = \mathbf{k}_R \cdot \mathbf{R} = \mathbf{k}_T \cdot \mathbf{R}. \tag{6-4.4}$$

Written in terms of the angles these become:

$$k_I \sin \theta_I = k_R \sin \theta_R, \tag{6-4.5}$$

$$k_I \sin \theta_I = k_T \sin \theta_T. \tag{6-4.6}$$

Because they are in the same medium, k_I and k_R are equal. Therefore the two angles must be equal. This is just the *law of reflection*:

The angle of reflection equals the angle of incidence.

Applying the dispersion equation, $c_m = \omega/k_m$, to the second equation gives *Snell's law*:

$$n_I \sin \theta_I = n_T \sin \theta_T, \tag{6-4.7}$$

where the *index of refraction* is defined for each medium:

$$n = \frac{c}{c_m} = \sqrt{\varepsilon \mu}. \tag{6-4.8}$$

It is conventional to define the angles to be measured from the normal to the boundary plane. This is particularly convenient when the boundary is a curved surface, since the normal has a unique direction. The best alternative is to measure the angle from one of the tangent lines, but nothing particularly useful is gained.

EXAMPLE

Evaluate the index of refraction in terms of the electromagnetic properties of the medium. For common dielectrics and almost any material at optical frequencies, the permeability is about 1.0; then the index of refraction can be approximated by the square root of the dielectric

TABLE 6-3.   REFRACTIVE INDICES

$T = 293°K$, $\lambda = 598.3$ nm. Substance	Refractive
NaCl	1.544
SiO$_2$ (Quartz)	1.544 (ordinary)
SiO$_2$ (Quartz)	1.553 (extraordinary)
Diamond	2.4173
H$_2$O	1.333

constant. The refractive indices of some common materials are given in Table 6-3.

Now let us apply the boundary conditions to find the reflected and refracted intensity ratios. Boundary equations (3-4.3), (3-4.5), (3-4.11), and (3-4.14) contain both the electric and magnetic fields. However, for a beam of light (or a plane wave), we can eliminate the magnetic field in favor of just the electric field. In our present problem, we have three beams to consider, which we must match on the boundary. That is, in the boundary surface, the sum of the incident and reflected fields must equal the transmitted field. With this in mind, the boundary conditions become:

$$(\varepsilon_I \mathbf{E}_I + \varepsilon_R \mathbf{E}_R - \varepsilon_T \mathbf{E}_T) \cdot \mathbf{n} = 0, \qquad (6\text{-}4.9)$$

$$(\mathbf{k}_I \times \mathbf{E}_I + \mathbf{k}_R \times \mathbf{E}_R - \mathbf{k}_T \times \mathbf{E}_T) \cdot \mathbf{n} = 0, \qquad (6\text{-}4.10)$$

$$(\mathbf{E}_I + \mathbf{E}_R - \mathbf{E}_T) \times \mathbf{n} = 0, \qquad (6\text{-}4.11)$$

$$\left(\frac{1}{\mu_I} \mathbf{k}_I \times \mathbf{E}_I + \frac{1}{\mu_R} \mathbf{k}_R \times \mathbf{E}_R - \frac{1}{\mu_T} \mathbf{k}_T \times \mathbf{E}_T \right) \times \mathbf{n} = 0, \qquad (6\text{-}4.12)$$

These equations apply only at points on the boundary surface. By choosing the origin of coordinates where the incident ray meets the boundary, the wave functions are represented by their amplitudes. Since the calculations are rather complex, it is useful to consider resolving the electric and magnetic fields into two components: one lying in the plane of incidence and the other normal to the plane of incidence.

$$\mathbf{E} = \mathbf{E}^N + \mathbf{E}^P. \qquad (6\text{-}4.13)$$

They are illustrated in Figure 6-13.

First consider the normal component, $\mathbf{E}^N$. Since this component of the field is orthogonal to the normal vector, $\mathbf{n}$, equation (6-4.9) is an identity. Only two of these equations, (6-4.11) and (6-4.12) are nontrivial for the normal component. They reduce to:

$$E_I^N + E_R^N = E_T^N, \qquad (6\text{-}4.14)$$

$$E_I^N \cos \theta_I - E_R^N \cos \theta_R = \frac{k_T \mu_R}{k_R \mu_T} E_T^N \cos \theta_T$$

$$= \frac{Z_R}{Z_T} E_T^N \cos \theta_T, \qquad (6\text{-}4.15)$$

in terms of the intrinsic impedances of the media. Solving these two equations for the transmission and reflection ratios gives:

$$T^N = \frac{E_T^N}{E_I^N} = \frac{2 \cos \theta_I}{\cos \theta_I + (Z_R/Z_T) \cos \theta_T}, \qquad (6\text{-}4.16)$$

$$R^N = \frac{E_R^N}{E_I^N} = \frac{\cos \theta_I - (Z_R/Z_T) \cos \theta_T}{\cos \theta_I + (Z_R/Z_T) \cos \theta_T}. \qquad (6\text{-}4.17)$$

Similarly, transmission and reflection formulas are obtained for the component lying in the plane of incidence. They are:

$$T^P = \frac{E_T^P}{E_I^P} = \frac{2 \cos \theta_I}{\cos \theta_I + (Z_T/Z_R) \cos \theta_T}, \qquad (6\text{-}4.18)$$

$$R^P = \frac{E_R^P}{E_I^P} = \frac{\cos \theta_I - (Z_T/Z_R) \cos \theta_T}{\cos \theta_I + (Z_T/Z_R) \cos \theta_T}, \qquad (6\text{-}4.19)$$

and are similar to (6-4.16) and (6-4.17) except for the inversion of the impedance ratio. These formulas are a form of *Fresnel's equations*. They appear to depend on both the angle of incidence and the angle of refraction. Since by Snell's law θ_T depends on θ_I, then $R^{P,N}$ and $T^{P,N}$ actually depend only on the electric and magnetic susceptibilities of the media and θ_I. In the common case of nonmagnetic dielectrics, the impedance

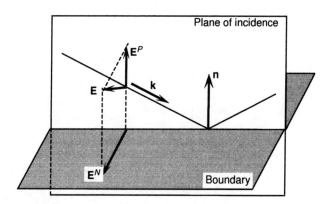

FIGURE 6-13. POLARIZATION COMPONENTS. The electric field is resolved into a component lying in the plane of incidence E^P and a component orthogonal to it E^N. Both are orthogonal to the propagation vector **k**. The corresponding magnetic field components are not shown.

ratios equal the reciprocal refractive index ratios:

$$\frac{Z_R}{Z_T} = \frac{Z_I}{Z_T} = \sqrt{\frac{\varepsilon_T \mu_I}{\varepsilon_I \mu_T}} \approx \sqrt{\frac{\varepsilon_T}{\varepsilon_I}} \approx \frac{n_T}{n_I}. \tag{6-4.20}$$

Then the Fresnel formulas can be given in terms of the refractive index ratios. Since the reflection medium is the same as the incident medium, we can write the impedances and indices of refraction $Z_R = Z_I$ and $n_R = n_I$.

We shall see in the next section that the energy flux of an electromagnetic field is proportional to the square of its field strength. For this reason the relative scattering intensity is proportional to the square of the reflection coefficient.

$$I_R = I_0 R^2. \tag{6-4.21}$$

Similar relations hold for both components of reflected and transmitted beams.

Inspection of Figure 6-14 shows that the reflected intensities $R^{N,P}$ vanish at an angle, called the *Brewster angle*. Light reflected at this angle will be polarized. This is a convenient way to produce polarized light and is the reason that Polaroid$^{\text{©}}$ sun glasses are effective in reducing the glare of reflected sunlight. The filter in the lenses is oriented to block out the remaining light.

From Snell's law we can calculate a maximum angle for reflection. That is, consider a light beam inside a medium with refractive index n_I, being reflected at a boundary with a medium of a smaller index of refraction, n_T. Solving for the angle of refraction gives:

$$\sin \theta_T = \frac{n_I}{n_T} \sin \theta_I \leq \frac{n_I}{n_T}. \tag{6-4.22}$$

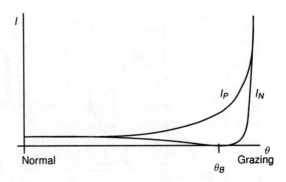

Figure 6-14. Reflection Intensity at Interface Between Dielectrics. The vertical axis represents intensity which is proportional to the square of the amplitude. The horizontal axis is the angle of incidence.

The limit determines the *critical angle*,

$$\theta_{\text{critical}} = \sin^{-1} \frac{n_T}{n_I}. \tag{6-4.23}$$

This formula seems to imply that at the critical angle, the refracted ray will travel parallel to the boundary surface. In fact, if the intensities in the Fresnel equations are considered, the intensity of the refracted ray vanishes. Therefore, all of the incident light is reflected. Since this commonly occurs for light going from some dielectric into air, this phenomenon is called *total internal reflection*. It is the physical basis for some useful technology, including fiber optics. It should also be popular with brides (and jewelers) since it accounts for the glitter of diamonds.

EXAMPLE

For normal incidence we may, without loss of generality, choose the plane of incidence to be normal to the electric vector. Then the Fresnel equations reduce to:

$$T^N = \frac{2}{1 + Z_R/Z_T}, \tag{6-4.24}$$

$$R^N = \frac{1 - Z_R/Z_T}{1 + Z_R/Z_T}. \tag{6-4.25}$$

When the two media have the same impedance (they need not be the same material) the reflection vanishes and the transmission is total. This can be seen by immersing a glass rod in a sugar solution. At just the right concentration, the glass rod is invisible. When applied to wave guides or transmission lines, this phenomenon is called *impedance matching*. For example, a signal will be transmitted without reflection across the connection between two coaxial cables of the same impedance. On the other hand, the reflection is a useful way to find the location of a flaw in a long piece of cable—send a pulse down the cable and measure the position of the reflection point on an oscilloscope. If one wants to transmit a signal down a piece of cable without reflection at the end, it is necessary to *terminate the cable* with an impedance equal to the characteristic impedance of the cable. For most cables the impedance is mainly resistive, so a resistor is sufficient. For high frequencies, efficient termination can be somewhat more complicated. Electrical engineering texts are useful references for this application. Equation (6-4.25) also shows a sign change (change of phase = 180°) observed in the reflected component when the radiation enters a higher-density medium.

EXAMPLE

For certain kinds of crystals, the velocities of light are slightly different for each polarization, depending on the direction of travel through the crystal. This phenomenon is known as *double refraction*. It can be seen, for example, in large calcite ($CaCO_3$) crystals. Because of the different velocities, the angles of refraction are different, and two images pass through the crystal, slightly displaced from each other, with different polarizations.

6-5   MOMENTUM AND ENERGY

We have not heretofore dealt with energy or momentum carried by electromagnetic fields. It is clear, however, that since electromagnetic fields are so intimately related to forces, charges moving in electromagnetic fields must have an associated energy and momentum. Consider how energy and momentum may be expressed in terms of the fields. First, the work done per unit time on a system of charged particles by external forces is:

$$\frac{dW}{dt} = -\sum_{k=1}^{N} q_k \mathbf{v}_k \cdot \mathbf{E}_k \rightarrow -\iiint \mathbf{J} \cdot \mathbf{E}\, d^3 r. \qquad (6\text{-}5.1)$$

The minus sign refers to the work done *on the charge*; when several forces act the expression may be separated into work done *on* the system and work done *by* the system.

The potential energy of a collection of point charges is given as the sum:

$$PE = \frac{1}{2} \sum_{k=1}^{n} q_k \phi_k. \qquad (6\text{-}5.2)$$

The factor, 1/2, is needed to compensate for including the energy of each pair of charges twice. To see the need for this factor, express the energy of just two charges, q_1 and q_2, as such a sum. This expression excludes the self-energy of a charge in its own field. Equation (6-5.2) may be generalized to the Lorentz invariant expression:[6]

$$PE = \iiint \frac{1}{2} \left(\rho\phi + \frac{1}{c} \mathbf{J} \cdot \mathbf{A} \right) d^3 r$$

$$= \iiint \frac{1}{2c} J^k A_k\, d^3 r. \qquad (6\text{-}5.3)$$

---

[6] The plus sign in the first line is the result of defining both $\mathbf{J}$ and $\mathbf{A}$ as contravariant vectors. The inner product in the second line is invariant.

It is possible to express the energy in terms of the electric and magnetic fields alone by eliminating the charge densities directly from the two Maxwell equations with sources. For steady-state fields (specifically $\partial \mathbf{D}/\partial t = \partial \mathbf{B}/\partial t = 0$), this is:

$$PE = \frac{1}{8\pi} \iiint (\phi \nabla \cdot \mathbf{D} + \nabla \times \mathbf{H} \cdot \mathbf{A}) \, d^3r. \qquad (6\text{-}5.4)$$

Now apply vector identities (1-3.27) and (1-3.29), and the divergence theorem to get:

$$PE = \frac{1}{8\pi} \iiint \nabla \cdot (\phi \mathbf{D}) - \nabla \phi \cdot \mathbf{D} + \nabla \times \mathbf{A} \cdot \mathbf{H} + \nabla \cdot (\mathbf{A} \times \mathbf{H}) \, d^3r$$

$$= \iiint \frac{1}{8\pi} (\mathbf{D} \cdot \mathbf{E} + \mathbf{B} \cdot \mathbf{H}) \, d^3r + \frac{1}{8\pi} \oiint (\phi \mathbf{D} + \mathbf{A} \times \mathbf{H}) \cdot d\mathbf{S}.$$

$$(6\text{-}5.5)$$

The surface term vanishes for a sufficiently distant boundary, where the fields vanish. This equation suggests that the *electromagnetic energy density* may be written:

$$U = \frac{1}{8\pi} (\mathbf{D} \cdot \mathbf{E} + \mathbf{B} \cdot \mathbf{H}). \qquad (6\text{-}5.6)$$

This calculation is suggestive, but not completely satisfactory, since it is limited to steady-state fields and an infinite boundary. A more systematic calculation is indicated.

Take the two Maxwell equations and multiply each equation by the corresponding field:

$$\mathbf{E} \cdot \nabla \times \mathbf{H} = \frac{1}{c} \mathbf{E} \cdot \partial_t \mathbf{D} + \frac{4\pi}{c} \mathbf{E} \cdot \mathbf{J}, \qquad (6\text{-}5.7)$$

$$\mathbf{H} \cdot \nabla \times \mathbf{E} = -\frac{1}{c} \mathbf{H} \cdot \partial_t \mathbf{B}. \qquad (6\text{-}5.8)$$

Adding the two equations gives:

$$[\mathbf{H} \cdot \nabla \times \mathbf{E} - \mathbf{E} \cdot \nabla \times \mathbf{H}] = -\frac{1}{c} [\mathbf{E} \cdot \partial_t \mathbf{D} + \mathbf{H} \cdot \partial_t \mathbf{B}] - \frac{4\pi}{c} \mathbf{E} \cdot \mathbf{J}. \quad (6\text{-}5.9)$$

Now define the *Poynting vector*:

$$\mathbf{N} = \frac{c}{4\pi} \mathbf{E} \times \mathbf{H}, \qquad (6\text{-}5.10)$$

which has units of energy per unit time per unit area. From its units at least, it can be considered a *flux*, like current density. By analogy with current density, the surface integral of the Poynting vector should be interpreted as energy transmitted through the surface per unit time:

$$\frac{d\mathcal{E}}{dt} = \iint \mathbf{N} \cdot d\mathbf{S}. \qquad (6\text{-}5.11)$$

For uniform media, the Poynting vector is parallel to the unit vector $\mathbf{n} = \mathbf{k}/k$:

$$\mathbf{N} = \frac{c}{4\pi} E^2 \mathbf{n}. \tag{6-5.12}$$

Applying vector identity (1-3.27), we get an equation called *Poynting's theorem* (in differential form):

$$\nabla \cdot \mathbf{N} + \frac{\partial U}{\partial t} = -\mathbf{E} \cdot \mathbf{J} = \frac{dW}{dt}. \tag{6-5.13}$$

Now consider the physical interpretation. We integrate over some volume and use the divergence theorem. We also separate the work into work done on the system (W_{in}) and work done by the system (W_{out}).

$$\iiint -\mathbf{E}_{in} \cdot \mathbf{J}\, d^3r = \oiint \mathbf{N} \cdot d\mathbf{S} + \frac{d}{dt} \iiint U\, d^3r$$
$$+ \iiint \mathbf{E}_{out} \cdot \mathbf{J}\, d^3r. \tag{6-5.14}$$

This equation is illustrated by Figure 6-15. The first term on the right may be interpreted as the energy transmitted per unit time through the boundary surface. The second term represents the rate of increase of potential energy of the electromagnetic field in the interior. The third term represents the rate at which the system does work on the exterior. This equation represents a form of the *first law of thermodynamics*, where the radiation represents heat transfer. Compare it to the integral form of charge conservation, i.e.:

$$\frac{d}{dt} \iiint \rho\, d^3r + \oiint \mathbf{J} \cdot d\mathbf{S} = 0. \tag{6-5.15}$$

Poynting's theorem deals with the energy balance for electromagnetic interactions. However, since energy is the zeroth component of a four-dimensional vector, there should be an invariant formulation. We

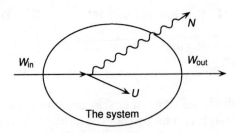

Figure 6-15. Energy balance. The work done on the system by external forces is shared three ways: increasing the internal energy; radiation through the surface; and work done by the system. The surface serves only to distinguish the system from the exterior; it is otherwise imaginary and arbitrary.

take the charge and current densities from Maxwell's equations, and find the inner product with the field. First, we make a useful calculation. Take the source-free equation (4-4.26) and find the inner product with the field.

$$F^{j\sigma} \partial_k F_{j\sigma} = - F^{j\sigma} \partial_j F_{\sigma k} - F^{j\sigma} \partial_\sigma F_{kj}$$

$$= - F^{j\sigma} \partial_j F_{\sigma k} - F^{\sigma j} \partial_j F_{k\sigma}$$

$$= - 2F^{j\sigma} \partial_j F_{\sigma k}. \tag{6-5.16}$$

In step two, the dummy indices, σ and j were interchanged. Then, in step three, both sets of indices were interchanged, preserving the sign. Now, express the term on the left as the derivative of the square, and relabel the k-index using the Kronecker delta. Then equation (6-5.16) becomes:

$$F^{j\sigma} \partial_j F_{\sigma k} = - \frac{1}{4} \partial_k \left[F^{j\sigma} F_{j\sigma} \right]$$

$$= - \partial_j \left[\frac{\delta_k^j}{4} F^{\alpha\beta} F_{\alpha\beta} \right], \tag{6-5.17}$$

where we have also renamed the dummy indices $j \to \alpha, \sigma \to \beta$ for convenience.

Now multiply the Maxwell equation (4-4.25) by the field, giving:

$$\left(\partial_j F^{j\sigma} \right) F_{\sigma k} = \frac{4\pi}{c} J^\sigma F_{\sigma k}. \tag{6-5.18}$$

Next, add and subtract a term:

$$\left[\left(\partial_j F^{j\sigma} \right) F_{\sigma k} + F^{j\sigma} \partial_j F_{\sigma k} \right] - F^{j\sigma} \partial_j F_{\sigma k} = \frac{4\pi}{c} J^\sigma F_{\sigma k}. \tag{6-5.19}$$

Last, apply equation (6-5.17) and get:

$$\partial_j \left[F^{j\sigma} F_{\sigma k} + \frac{\delta_k^j}{4} F^{\alpha\beta} F_{\alpha\beta} \right] = \frac{4\pi}{c} J^\sigma F_{\sigma k} = 4\pi f_k. \tag{6-5.20}$$

We define the *electromagnetic stress tensor*:

$$T_k^j = \frac{1}{4\pi} \left[F^{j\sigma} F_{\sigma k} + \frac{\delta_k^j}{4} F^{\alpha\beta} F_{\alpha\beta} \right]$$

$$
= \frac{1}{4\pi}
\begin{bmatrix}
B^2 - E^2 - \dfrac{B^2 + E^2}{2} & E_3B_2 - E_2B_3 & E_1B_3 - E_3B_1 & E_2B_1 - E_1B_2 \\[2ex]
E_3B_2 - E_2B_3 & B^2 + E^2 - \dfrac{B_1^2 + E_1^2}{2} & E_1E_2 - B_1B_2 & E_1E_3 - B_1B_3 \\[2ex]
E_1B_3 - E_3B_1 & E_2E_1 - B_2B_1 & B^2 + E^2 - \dfrac{B_2^2 + E_2^2}{2} & E_2E_3 - B_2B_3 \\[2ex]
E_2B_1 - E_1B_2 & E_3E_1 - B_3B_1 & E_3E_2 - B_3B_2 & B^2 + E^2 - \dfrac{B_3^2 + E_3^2}{2}
\end{bmatrix},
$$

$$(6\text{-}5.21)$$

and the *electromagnetic force density*:

$$
f_k = \frac{1}{c} F_{\sigma k} J^\sigma = \frac{1}{c} \rho U^\sigma F_{\sigma k}. \tag{6-5.22}
$$

For radiation, the magnetic and electric fields have equal magnitude $|B| = |E|$; then the term T^{00} can be written just as: $T^{00} = -(B^2 + E^2)/8\pi$. For static fields, the time derivatives vanish from the divergence, and equation (6-5.20) effectively reduces to the three spatial dimensions. Unlike the field tensor, F^{jk}, the stress tensor, T^{jk}, is symmetric; the off-diagonal spatial components take the form:

$$
T_k^j = \frac{1}{4\pi} \left(E_j E_k + B_j B_k \right), \qquad j \neq k \neq 0, \tag{6-5.23}
$$

and the zeroth components include the Poynting vector:

$$
T_k^0 \approx \frac{1}{4\pi} \mathbf{E} \times \mathbf{B} = \frac{1}{c} \mathbf{N}. \tag{6-5.24}
$$

The zeroth component of the force density is proportional to the work density:

$$
f_0 = \frac{1}{c} F_{\sigma 0} J^\sigma = \frac{1}{c} \rho \mathbf{E} \cdot \mathbf{v}. \tag{6-5.25}
$$

In terms of the stress tensor, the four-dimensional equivalent of Poynting's theorem becomes:

$$
\partial_j T_k^j = f_k. \tag{6-5.26}
$$

The spatial and time components are, respectively:

$$
-\partial_t \mathbf{N}_k - \partial_\sigma T_k^\sigma = f_k, \qquad \sigma, k \neq 0, \tag{6-5.27}
$$

$$
-\frac{1}{c} \nabla \cdot \mathbf{N} - \frac{1}{c} \frac{\partial U}{\partial t} = \frac{1}{c} \rho \mathbf{E} \cdot \mathbf{v}, \qquad k = 0. \tag{6-5.28}
$$

Thus the zeroth component is Poynting's theorem, and the spatial components refer to the momentum balance of the system. Take the integral over some volume and apply the divergence theorem. Then:

$$
\oiint T_k^\sigma \, dS_\sigma + \iiint f_k \, d^3r = 0. \tag{6-5.29}
$$

The sign of the surface integral is determined by the direction of the surface element dS_σ. By convention, it is directed outward from a closed surface. The surface integral represents the rate of momentum transfer through the surface. That is, *electromagnetic radiation carries momentum as well as energy*. This is expected because of the relativistic connection between momentum and energy. Then equation (6-5.29) is a statement of Newton's second law generalized to include the forces of radiation.

An alternative interpretation is that the surface integral represents the force transmitted through the surface. When it is combined with the second force density term, the equation implies that the total force acting on an isolated system is zero. The stress tensor has some applications in calculating electromagnetic forces by analogy with the elastic stress tensor. Consider a rod made of an isotropic material of tensile modulus Y, length L, and cross section A, subjected to a force F. Then it will stretch by an amount ΔL given by the empirical formula:

$$\frac{F}{A} = Y\frac{\Delta L}{L}. \qquad (6\text{-}5.30)$$

The quantities, F/A and $\Delta L/L$, are called the stress and strain, respectively. The strain is the fractional deformation. Thus, the law of elasticity is stated: stress is proportional to strain. For an isotropic material, the deformation has the same direction as the applied force. But in what direction is the direction of deformation if the material is not isotropic (e.g., in a low-symmetry crystal)? In this case, the strain may be composed of stretch, rotation, and shear, and we must replace the scalar strain with a *strain tensor*, L^{jk}. Now consider the stress; since both force and area are vectors, stress must, in general, be a two-index quantity, or *stress tensor*, T^{mn}. Then the elasticity equation may be written for nonisotropic materials by the tensor equation:

$$T^{jk} = Y^{jk}_{mn}L^{mn}, \qquad (6\text{-}5.31)$$

where the rank four, mixed tensor, Y^{jk}_{mn}, expresses the elastic moduli (Young's modulus and shear modulus) of the material.

The stress tensor, in its primitive form, represents the force acting per unit area on some region and has the same units as pressure. This suggests that, in general, one can calculate the force transmitted across some surface by integrating the stress tensor over the surface:

$$\Delta F^j = \iint T^{jk}\,dS_k. \qquad (6\text{-}5.32)$$

Since stress is a tensor, the transmitted force may include forces tangent to the surface, i.e., shear forces. With a closed surface, the integral can be used to calculate the total force acting on the system of objects inside the surface.

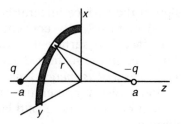

FIGURE 6-16. STRESS TENSOR CALCULATION.   The force is transmitted through the surface by the stress tensor. The closed surface chosen here is an infinite plane. Any surface completely surrounding one of the charges would do as well.

EXAMPLE

Find the force on a static charge q at the point $z = -a$ due to another charge $-q$, at the point $z = a$. For the closed surface we will take the x–y plane at $z = 0$. This is shown in Figure 6-16.

The electric field on the x–y surface is:

$$E_x = E_y = 0,$$

$$E_z = \frac{2qa}{\left[\sqrt{r^2 + a^2}\right]^3}.$$

$$(6\text{-}5.33)$$

Then the only nonvanishing component of the stress tensor is:

$$T^{zz} = \frac{1}{8\pi} E_z^2 = \frac{1}{8\pi} \frac{q^2 a^2}{[r^2 + a^2]^3}.$$

$$(6\text{-}5.34)$$

Now, integrate this over the boundary plane. We can use cylindrical coordinates. This gives:

$$F^z = \int\!\!\int_{0,0}^{\infty,2\pi} \frac{1}{8\pi} \frac{q^2 a^2 r \, dr \, d\theta}{[r^2 + a^2]^3} = \int_0^\infty \frac{1}{4} \frac{q^2 a^2 r \, dr}{[r^2 + a^2]^3}$$

$$= \int_{a^2}^\infty \frac{q^2 a^2}{4} \frac{1}{2} \frac{dp}{p^3},$$

$$(6\text{-}5.35)$$

where we have changed variables $p = r^2 + a^2$. This evaluates to:

$$F^z = \frac{q^2}{(2a)^2},$$

$$(6\text{-}5.36)$$

which agrees with Coulomb's law.

For ease of calculation, we chose the surface to be the x–y plane. However, any surface surrounding one of the charges would do. Note that this treatment does not specify how the force on the "interior" of the surface is applied to objects inside. Rather, the stress through the surface is calculated. If there are several charged bodies inside the surface, only the net force on the bodies is obtained—not the individual forces on each.

EXAMPLE

Apply energy density to find the energy of a sphere with charge uniformly distributed on its surface. The magnetic components vanish and the electric field has only a radial component:

$$
\text{PE} = \frac{1}{8\pi} \int_R^\infty E^2 4\pi r^2 \, dr = \frac{1}{2} \int_R^\infty \frac{q^2}{r^4} r^2 \, dr = \frac{q^2}{2} \left[\frac{-1}{r} \right]_R^\infty
$$

$$
= \frac{q^2}{2R}. \tag{6-5.37}
$$

This agrees with the energy needed to assemble the charge on the sphere by moving the charge to the surface of the sphere from an infinite distance:

$$
\text{PE} = \int_0^q \phi \, dq = \int_0^q \frac{q}{R} \, dq
$$

$$
= \frac{q^2}{2R}. \tag{6-5.38}
$$

We can use this as a model of the electron, in which its mass is assumed to be entirely due to its electrostatic energy. We do not know the exact charge distribution of a real electron. However, our model is representative of models having finite charge densities. Certainly this model is a natural one, in that this charge distribution gives the lowest energy (see the charge distribution on an isolated conducting sphere). In the simplest model of this sort we define an effective radius, called the *classical electron radius*:

$$
R_C = \frac{e^2}{mc^2}
$$

$$
= 2.82 \times 10^{-13} \text{ cm}, \tag{6-5.39}
$$

which is comparable to the radius of the proton. A finite charge distribution has not been experimentally observed for electrons; scattering experiments are consistent with a charge distribution that is point-like, not extended. In that case, equation (6-5.39) implies an infinite electromagnetic mass. This problem will be considered further in Chapter 9.

The mechanical model used in illustrating the stress tensor was believed by physicists in the nineteenth century to have more than just academic significance. That is, an aether was postulated to exist in all space and to carry the stress from one point to another. The aether would have to have a very large elastic coefficient (i.e., be very rigid) in order to account for the enormous velocity of light. At the same time, it had to be perfectly permeable by matter, since no viscous drag caused by it was observed on celestial bodies. The Michelson–Morley experiment and relativity removed the need for this mechanical model of electromagnetic fields by changing the concept of space–time. However, it is interesting that modern theories, especially the relativistic string theories,[7] have restored interest in the microscopic details of space–time. *Plus ca change, plus c'est la meme chose.*

6-6   Huygens' Principle and Diffraction

In 1690, Christiaan Huygens published a seminal work on the properties of light, the *Traite de la Lumiere*. In this work he treated light as a wave phenomenon (contradicting Isaac Newton's particle model). He introduced a very sophisticated concept, now known as *Huygens' principle*:

> *Each point on a wave-front may be treated as the source of a secondary, spherical, wave-front. Then the wave-front at any later time is the envelope of all the secondary wave-fronts.*

Huygens' principle is illustrated in Figure 6-17.

In this early work, Huygens showed great insight into the behavior of waves; however, the treatment was essentially qualitative. Fresnel's memoir of 1818, put Huygens' principle in mathematical form, by interpreting diffraction of light in terms of a wave equation and its boundary values. We will follow Kirchhoff's extension (in 1882) of Fresnel's scheme. The program is as follows: First, we will find the Green function for the wave equation. Second, we will apply the Green function method to find the solution to the wave equation in a source-free region. Third, we will obtain the Kirchhoff diffraction integral. Finally, we will apply the Kirchhoff integral to some examples. According to Born and Wolf, diffraction is among the most difficult problems in optics because of its complexity. Thus, at several points in the program, it will be necessary to use approximations because of the difficulty of the calculations. The interested reader will find more information in Marion and in Born and Wolf.

---

[7] An elementary account of string theory is given in Peat.

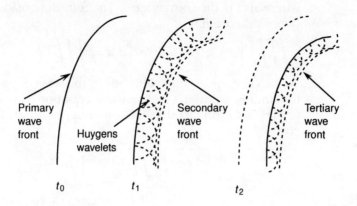

Primary
wave
front

Huygens
wavelets

Secondary
wave
front

Tertiary
wave
front

t_0 t_1 t_2

Figure 6-17. Huygens' principle. Only the converging wave is shown. In the limit of zero curvature (plane wave) there is no convergence. That is, a plane wave produces a plane wave.

In the following calculations it is desirable to use both three-dimensional and four-dimensional expressions. In order to indicate which of the two cases we are using, we will adopt the following conventions. The symbol x^k represents points in four-dimensional space–time, where k runs from 0 to 3, while the boldface symbol $\mathbf{r}$ represents points in three-dimensional space. If it is necessary to refer to the components of $\mathbf{r}$, then in the symbol r^k the index runs from 1 to 3. Of course, the spatial components of x^k are identical to r^k. Functions of space and time may be written either as $f(x^k)$ or as $f(\mathbf{r}, t)$. Occasionally, we will suppress the vector index k in x^k or r^k, in order to distinguish between field points and source points. The differential volume in 3-space is written d^3r and the differential hypervolume in 4-space is written d^4x. Since these expressions are familiar, there should be no confusion.

In Section 4-4 we derived wave equations for $\mathbf{E}$ and $\mathbf{B}$. For the electric field in vacuum the homogeneous wave equation is:

$$\partial^j \partial_j E^k = 0. \tag{6-6.1}$$

In Section 6-2 we introduced a scalar wave representation of the electric and magnetic fields. In discussing diffraction it is convenient to continue to use this representation, and find a Green function solution of the wave equation. That is, let $\psi(x)$ be a scalar field and let $G(x)$ be the Green function satisfying the equations:

$$\partial^j \partial_j \psi = s(x), \tag{6-6.2}$$

$$\partial^j \partial_j G = \delta^4(x), \tag{6-6.3}$$

where $s(x)$ is the source term. The four-dimensional generalization of Green's theorem is:

$$\iiiint \left[\psi \, \partial^j \partial_j G - G \, \partial^j \partial_j \psi \right] d^4 x = \oiiint \left[\psi \, \partial_j G - G \, \partial_j \psi \right] dH^j. \quad (6\text{-}6.4)$$

We can use this to find solutions to the wave equation. The arguments for finding solutions to the wave equation closely parallel those for finding solutions to the Poisson equation given in Section 3-5. They give:

$$\psi(x_f) = \iiiint G(x_{fs}) s(x_s) \, d^4 x_s$$

$$+ \oiiint \left[\psi \, \partial_j G(x_{fs}) - G(x_{fs}) \, \partial_j \psi \right] dH^j. \quad (6\text{-}6.5)$$

The first and second terms are the particular solution and the homogeneous solution, respectively. We are concerned here only with the homogeneous term. We will consider the particular solution, involving the radiation of moving charges, in Chapter 8.

In order to find a useful representation of the Green function we will find its Fourier expansion. It is frequently useful to consider monochromatic radiation; in that case, the problem is reduced to finding the Green function solution to the Helmholtz equation (the wave equation at fixed frequency). We can directly verify that the Green function for the Helmholtz equation is (now in three dimensions):

$$G_{\text{Helmholtz}}(\mathbf{r}) = \frac{1}{4\pi} \frac{\exp(i\mathbf{k} \cdot \mathbf{r})}{r}$$

$$\rightarrow \frac{1}{4\pi} \frac{1}{r}$$

$$= -G_{\text{Poisson}}(\mathbf{r}). \quad (6\text{-}6.6)$$

The limit of the Helmholtz–Green function is the negative of the Poisson–Green function when k vanishes. See equation (3-5.8). This sign difference is a direct result of the sign convention in the D'Alembertian operator:

$$\partial^j \partial_j = \frac{\partial^2}{c^2 \partial t^2} - \nabla^2. \quad (6\text{-}6.7)$$

That is, the spatial derivatives in the D'Alembertian are negative, but those in the Laplacian are positive.

We show that equation (6-6.6) is a solution of the Helmholtz equation by direct calculation. At all points excluding $r = 0$ we have:

$$\left[\nabla^2 + k^2 \right] \frac{e^{ikr}}{r} = \left[\frac{1}{r^2} \frac{d}{dr} \left(r^2 \frac{d}{dr} \right) + k^2 \right] \frac{e^{ikr}}{r}$$

$$= \left[\frac{1}{r^2}(ik + (ik)^2 r^2 - ik) + k^2 \right] \frac{e^{ikr}}{r}$$

$$= 0. \tag{6-6.8}$$

But the volume integral over a small sphere centered at the origin approaches a constant as the radius approaches zero:

$$\iiint \left[\nabla^2 + k^2 \right] \frac{e^{ikr}}{r} d^3r = \oiint \nabla \frac{e^{ikr}}{r} \cdot d\mathbf{S} + k^2 \iiint \frac{e^{ikr}}{r} r^2 \sin\theta \, dr \, d\theta \, d\phi$$

$$= 4\pi \left[ikre^{ikr} - e^{ikr} \right] + 4\pi k^2 \int re^{ikr} dr$$

$$\rightarrow -4\pi. \tag{6-6.9}$$

Together, equations (6-6.8) and (6-6.9) define the δ-function; but these are just the essential properties of the Green function.

The Green function for the D'Alembert equation may be found by applying a Fourier transformation to equation (6-6.3). Multiplying by $\exp(ik_j r^j)$ and integrating (by parts) gives:

$$\iiiint_{-\infty}^{\infty} e^{ik_j r^j} \left[-k^j k_j G(k) + 1 \right] d^4k = 0 \tag{6-6.10}$$

from which we can pick off the Fourier transform of the Green function:

$$G(k) = \frac{1}{k^j k_j} = \frac{1}{(\omega/c)^2 - k_1^2 - k_2^2 - k_3^2}. \tag{6-6.11}$$

Then one can find the Green function by taking the inverse transform in four-dimensions. However, we may take a short cut by taking the inverse Fourier transform of the Helmholtz equation Green function:

$$G(x^k) = G(\mathbf{r}, t) = \frac{1}{2\pi} \int_{-\infty}^{\infty} \frac{1}{4\pi} \frac{\exp(i\mathbf{k} \cdot \mathbf{r} - i\omega t)}{r} d\omega$$

$$= \frac{1}{4\pi} \frac{\delta(r - ct)}{r}, \tag{6-6.12}$$

where r is constant in the integral and we have used the relation $ck = \omega$.

The physical interpretation of the δ-function in equation (6-6.12) is that electromagnetic field contributions, detected by the observer at time $t = 0$, come only from charges on the lower half of the light cone where $r = ct$. This Green function is called a *retarded Green function*, because the contribution of each charge will reach the detector at a later time than when it left the source. See Figure 6-18.

In these calculations for the Green function we took the value of k to be positive or zero. However, negative values of k are also valid; consider the consequences of this. Then, following the preceding calculations for

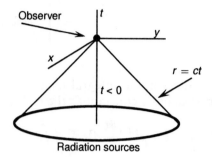

FIGURE 6-18. LIGHT CONE.   Radiation from all sources on the surface, $r = ct$, will reach the observer at the same instant. Causality prevents sources on the forward light cone ($t > 0$) from contributing, i.e., light does not travel backward in time.

negative k, we obtain an *advanced Green function* for the D'Alembertian equation:

$$G(\mathbf{r}, t) = \frac{1}{4\pi} \frac{\delta(r + ct)}{r}. \tag{6-6.13}$$

The existence of two Green functions reflects the fact that for plane waves there are two solutions:

$$\psi = \begin{cases} f(x - ct), \\ f(x + ct), \end{cases} \tag{6-6.14}$$

which may be interpreted as a wave traveling forward and backward in time, respectively.

Because a signal traveling backward in time contradicts causality, we reject the advanced Green function. But rejecting the advanced Green function has implications for the interpretation of equation (6-6.5). The surface of integration is closed, but the forward half of the surface, $t > 0$, does not contribute to the integral. Then the boundary conditions in equation (6-6.5) are effectively Cauchy conditions on an open boundary, which as we saw in Section 3-5, are appropriate for the wave equation. More rigorous calculations of the Green function are given in many advanced textbooks on electromagnetism and mathematical physics, e.g., Barut or Mathews and Walker.

Since the Green function vanishes except at the points where $ct = r$, the retarded Green function may also be written:

$$G(\mathbf{r}, t) = \frac{1}{4\pi} \frac{\delta(ct - r)}{ct}. \tag{6-6.15}$$

This expression refers to a detector at rest and is not obviously invariant; however, we may generalize the denominator for an arbitrarily moving coordinate system by taking the invariant quantity:

$$ct \rightarrow \frac{1}{c} U_j x^j. \tag{6-6.16}$$

The invariant Green function becomes:

$$G(x^j) = \frac{c}{4\pi} \frac{\delta(ct - r)}{U_j x^j} = \frac{1}{4\pi} \frac{\delta(ct - r)}{(1 - \mathbf{v}/c \cdot \mathbf{n}) r}, \tag{6-6.17}$$

where the vector $\mathbf{n}$ is a unit vector parallel to the radius vector $\mathbf{r}$.

Now let us apply our Green function to the wave equation. The homogeneous solution in equation (6-6.5) is:

$$\psi(x_f) = \oiiint \left[\psi(x_s) \partial_j G(x_{fs}) - G(x_{fs}) \partial_j \psi(x_s) \right] dH^j, \tag{6-6.18}$$

$$G(x_{fs}) = \frac{c}{4\pi} \frac{\delta(c(t_f - t_s) - |\mathbf{r}_f - \mathbf{r}_s|)}{U_j(x_f^j - x_s^j)}. \tag{6-6.19}$$

If we write $dH^j = dS^j d(ct)$ and integrate over time, the remaining integral is over the surface dS^j, and the integrand is evaluated at the retarded time $t_s = t_f - |r_f - r_s|/c$. Then the integrand is retarded. The equation becomes:

$$\psi(r_f) = \frac{1}{4\pi} \oiint \left[\psi(r_s) \partial_j \frac{1}{|r_f^j - r_s^j|} - \frac{1}{|r_f^j - r_s^j|} \partial_j \psi(r_s) \right]_{\text{Ret}} dS^j. \tag{6-6.20}$$

This is the *general form of Kirchhoff's theorem*. At small distances where the time delay is negligible, it reduces to the expression for steady-state fields. See equation (5-1.16). This equation is the mathematically rigorous implementation of Huygens' principle. It shows how a wave front is propagated downstream from its initial (boundary) values. Actual integration may be difficult to perform for arbitrary boundary surfaces, but approximations of various kinds are possible, including numerical integration. There are the usual disadvantages to numerical integration, of course.

By limiting the discussion to monochromatic waves, we may use the Helmholtz equation. The solution is:

$$\psi(r_f, k) = \frac{1}{4\pi} \oiint \left[\psi(r_s) \nabla \frac{\exp[i\mathbf{k} \cdot (\mathbf{r}_f - \mathbf{r}_s)]}{|\mathbf{r}_f - \mathbf{r}_s|} \right.$$
$$\left. - \frac{\exp[i\mathbf{k} \cdot (\mathbf{r}_f - \mathbf{r}_s)]}{|\mathbf{r}_f - \mathbf{r}_s|} \nabla \psi(r_s) \right] \cdot d\mathbf{S}, \tag{6-6.21}$$

called the *Helmholtz–Kirchhoff integral*. Because we consider monochromatic light and do not integrate over time, this integrand is not retarded.

In our discussion of Huygens' principle, we have ignored one tricky detail. If a wave front acts as a source of spherical wavelets which overlap and act to form a new wave front, why are there no waves traveling backward? More generally, since we know that waves spread transversely after passing through a slit in a barrier, how does the amplitude depend on the angle of deviation from forward? To illustrate the solution to this problem, consider the following case. Figure 6-19 shows a short section of an incident spherical wave.

Assume the incident wave takes the form (consistent with the spherical Bessel functions we discussed in Section 6-2):

$$\psi_0 = \Psi_0 \frac{\exp(ikR_0)}{R_0}. \tag{6-6.22}$$

By taking a small enough section, it may be treated locally as a plane wave. This wave front is taken to be the Huygens source of waves downstream. Its propagation vector, which can be written, $k\mathbf{n}_0$, is normal to the wave front at each point. In the limit of distances R larger than the wavelength equation (6-6.21) becomes:

$$\psi(x_f, k) = \frac{i\Psi_0 k}{4\pi} \oiint \left[\frac{\exp[ik(R + R_0)]}{RR_0} \right] (\mathbf{n} - \mathbf{n}_0) \cdot d\mathbf{S}. \tag{6-6.23}$$

The direction of the surface element is parallel to the vector $\mathbf{n}_0$; so the scalar product becomes:

$$(\mathbf{n} - \mathbf{n}_0) \cdot d\mathbf{S} = (1 + \cos \chi)\, dS, \tag{6-6.24}$$

where χ is the angle between the normal to the incident (boundary) wave and the direction of the Huygens wavelet. This *inclination factor* eliminates backward propagating waves. When R_0 is large enough to allow the plane wave approximation, it may be convenient to absorb

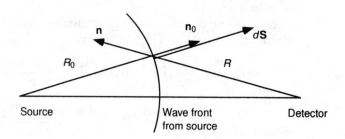

FIGURE 6-19. INCLINATION FACTOR.   The surface element, $d\mathbf{S}$, is parallel to the propagation vector, $k\mathbf{n}_0$. When R_0 is very large (i.e., for plane waves), the propagation vector becomes parallel to the Source–Detector axis.

part of the integrand into the constant:

$$\psi(x_f, k) \rightarrow \frac{i\tilde{\Psi}_0 k}{4\pi} \oiint \left[\frac{\exp(ikR)}{R} \right] dS. \qquad (6\text{-}6.25)$$

Example: Plane Waves

For an incident wave traveling in the z-direction and the surface of integration is the plane, $z_s = 0$, it should be possible to construct the wave function for $z > 0$, by integrating. On the surface, $z_s = 0$, the field has the value $\psi = \Psi_0 \exp(-i\omega t)$. Then the integral (6-6.21) becomes:

$$
\begin{aligned}
\psi &= \frac{1}{4\pi} \iint \left[\Psi_0 e^{-i\omega t} \nabla_s \frac{e^{ikz}}{r_{fs}} - \frac{e^{ikz}}{r_{fs}} \nabla \Psi_0 e^{-i\omega t} \right] \cdot d\mathbf{S}_s \\
&= \frac{1}{4\pi} \Psi_0 e^{i(kz - \omega t)} \iint \nabla \frac{1}{r} \cdot d\mathbf{S} = \frac{1}{4\pi} \Psi_0 e^{i(kz - \omega t)} \int d\Omega \\
&= \Psi_0 e^{i(kz - \omega t)}, \qquad\qquad\qquad\qquad\qquad\qquad\qquad (6\text{-}6.26)
\end{aligned}
$$

which is still a plane wave. That is, a plane wave maintains its shape downstream. It is possible to show that this also holds for spherical waves.

Consider the transmission of a plane wave incident on an aperture in an opaque barrier. The boundary surface is shown in Figure 6-20. The boundary surface has three regions needing different treatment. The active part, aa', is the Huygens source; the shadowed portions, ab, and, $a'b'$, contribute nothing because the field vanishes there. We also do not expect the spherical section, bb', to contribute because the wave is diverging. Another argument is to suppose that the radius of the sphere is so large that the wave has not reached it yet—then the field on the surface is certainly zero. Of course, in this case, we do not strictly have monochromatic waves, but supposing that the coherence of the source of radiation corresponds to a length of only a few centimeters in the radiated field, then this approximation is good enough.

Suppose the location of the surface element dS at the aperture is given by coordinates $(0, \chi, \eta)$ and the position of the detector is (x, y, z). See Figure 6-21. Then the distance from the slit to the detector r and the distance from the element of area dS on the slit R are, respectively:

$$r^2 = x^2 + y^2 + z^2, \qquad (6\text{-}6.27)$$

$$R^2 = (x - \xi)^2 + (y - \eta)^2 + z^2. \qquad (6\text{-}6.28)$$

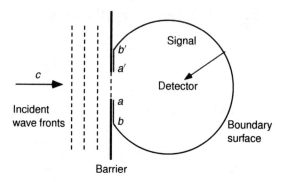

Figure 6-20. Boundary surface.   If the radius of the sphere is sufficiently large, the spherical portion of the boundary surface, bb', does not contribute to the diffracted ray because the signal will not have reached the detector (at the center of the sphere).

At large distances from the aperture, R can be expressed approximately by the first terms in a Taylor expansion:

$$R = r - \frac{x\xi + y\eta}{r} + \cdots,$$

$$= r - m_x\xi - m_y\nu + \cdots,$$

(6-6.29)

where the quantities m_x and m_y are the direction cosines. In this approximation, equation (6-6.23) becomes:

$$\psi(xyz) = -iC \iint e^{-ik(m_x\xi + m_y\eta)} d\xi \, d\eta,$$

(6-6.30)

where C represents the combined constants. Finally, we note that the intensity of the diffraction pattern is proportional to the energy flux, i.e., the square of the field. That is:

$$I = K|\psi|^2.$$

(6-6.31)

EXAMPLE: FRAUNHOFER DIFFRACTION

For a slit of infinite length and width $2d$, find the diffraction pattern on the screen at a very large distance from the slit. For infinite length, the integration over the width of the slit is a constant, giving:

$$\psi(xyz) = -iC_0 \int_{-\infty}^{\infty} \int_{-d}^{d} e^{-ik(m_x\xi + m_y\eta)} d\xi \, d\eta$$

$$= C_1 \int_{-d}^{d} e^{-ik(m_x\xi)} d\xi = \frac{C_1}{km_x d} \left[e^{-ikm_x d} - e^{ikm_x d} \right]$$

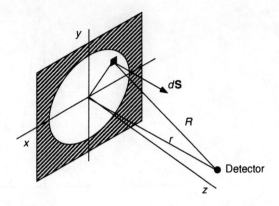

FIGURE 6-21. DIFFRACTION OF PLANE WAVES BY AN APERTURE IN A BARRIER. The field at the detector is obtained by integrating the field values over the boundary surface. At large distances, R approximates a linear function of the location (χ, η) of the element of surface, dS.

$$= C_2 \left[\frac{\sin km_x d}{km_x d} \right].$$

(6-6.32)

The intensity takes the form:

$$I = C^2 \left[\frac{\sin km_x d}{km_x d} \right]^2$$

$$= C^2 \left[\frac{\sin \left[(2\pi d/\lambda) \sin \theta \right]}{(2\pi d/\lambda) \sin \theta} \right]^2,$$

(6-6.33)

and is shown in Figure 6-22. The zeros of the intensity function occur at values:

$$\pi \frac{2d \sin \theta}{\lambda} = \pi \left(m + \frac{1}{2} \right), \qquad m = 0, \pm1, \pm2, \ldots$$

(6-6.34)

in agreement with the Fraunhofer expression.

In the preceding treatment of diffraction, we have used many approximations: scalar fields, spherical wave approximations, large distance approximations, truncated series, and others. The robustness of the theory is shown by the fact that it gives good results in spite of rough treatment. More precise treatments can be found in Born and Wolf and in the technical literature.

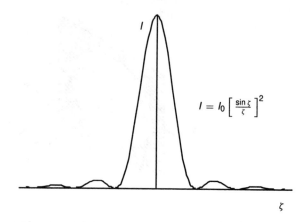

$$I = I_0 \left[\frac{\sin \zeta}{\zeta} \right]^2$$

FIGURE 6-22. FRAUNHOFER SINGLE SLIT DIFFRACTION PATTERN.   The intensity depends on the slit width, $2d$, the wavelength, λ, and the angle, θ, measured from the center of the slit.

6-7   SELECTED BIBLIOGRAPHY

A. Barut, *Electrodynamics and Classical Theory of Fields and Particles*, Macmillan, 1965.

M. Born and E. Wolf, *Principles of Optics*, Macmillan, 1964.

E. Butkov, *Mathematical Physics*, Addison-Wesley, 1968.

Chemical Rubber Publishing Company, *Handbook of Chemistry and Physics*, Chemical Rubber Publishing Co, 1969.

R. Ditchburn, *Light*, Interscience, 2nd ed., 1963.

J. Hamilton, *The Theory of Elementary Particles*, Oxford University Press, 1959.

F. Jenkins and H. White, *Fundamentals of Optics*, McGraw-Hill, 3rd ed., 1957.

C. Kittel, *Introduction to Solid State Physics*, 3rd ed., Wylie, 1967.

J. Kraus, *Electromagnetics*, McGraw-Hill, 1953. This electrical engineering text contains useful information on transmission lines, wave guides, and antennas.

J. Marion, *Classical Electromagnetic Radiation*, Academic Press, 1965.

J. Mathews and R. Walker, *Mathematical Methods of Physics*, 2nd ed., Benjamin, 1970.

C. Misner, K. Thorne, and J. Wheeler, *Gravitation*, Freeman, 1973.

P. Morse and H. Feshbach, *Methods of Theoretical Physics*, McGraw-Hill, 1953.

E. Peat, *Superstrings and the Search for the Theory of Everything*, Contemporary Books, 1988.

J. Reitz, F. Milford, and R. Christy, *Foundations of Electromagnetic Theory*, 3rd ed., Addison-Wesley, 1979.

W. Smythe, *Static and Dynamic Electricity*, 2nd ed, McGraw-Hill, 1950.

I. Sokolnikoff, *Tensor Analysis*, Wiley, 1951.

J. Stratton, *Electromagnetic Theory*, McGraw-Hill, 1941.

T. Wu and T. Ohmura, *Quantum Theory of Scattering*, Prentice-Hall, 1962.

6-8 PROBLEMS

1. A C-shaped magnet has a uniform magnetic field of 5000 Gauss between its pole faces. The pole faces are flat with area 4.0 cm² and separation 1.0 cm.

   A. A. Find the force of attraction of one pole face for the other.
   B. B. Find the energy stored in the gap by the magnetic field.

2. The dielectric strength of air is about 3×10^6 V/m. For a parallel plate capacitor (with area A and separation, (d) calculate:

   A. A. The maximum energy that can be stored per unit area by the capacitor.
   B. B. The maximum force of attraction per unit area, on one plate for the other.

3. A generic small laser has a continuous power of 5 mW, a beam cross section of 1.0 mm², and wavelength 500 nm. Use these numbers to estimate the electric field in the beam.

4. Plasma pinch effect. A long, straight plasma of radius R carries a current I. Suppose that its current density is uniform.

   A. A. Use the stress tensor to calculate the pressure on the plasma as a function of radius, $0 < r < R$.
   B. B. If the plasma is entirely of positive charges, calculate the ratio of inward (pinch) pressure to the outward pressure, due to Coulomb forces, at a given radius.

5. Estimate the penetration depth for sea water for visible light and for radar. ($\rho \approx 25$ ohm-cm, $\varepsilon \approx 80$).

6. Use the generating function to calculate Legendre polynomials P_3 for all allowed m.

7. *Computer application.* Use the method of finite differences (see Chapter 5) to find numerical solutions to the Helmholtz equation. Apply this method to the infinite slit in Section 6-6 to calculate the scalar wave diffraction pattern.

8. Expand a scalar plane wave in spherical coordinates. That is, evaluate the first few coefficients of the spherical expansion by using orthogonality. This technique is useful in evaluating nuclear cross sections.

9. Verify equation (6-1.3) explicitly.

10. Verify equations (6-1.19) and (6-1.20) explicitly.

11. The planet Mars has ice caps at its North and South poles. Model the reflectivity of the planet as a unit step function in spherical coordinates. That is, let the reflectivity be 1.0 for latitudes larger than 80 degrees, North or South and 0.0 otherwise.

12. Show explicitly that the vector expression $\mathbf{E} = \mathbf{r} \times \nabla \psi$ satisfies the wave equation in Cartesian coordinates where $\psi = \psi(\mathbf{k} \cdot \mathbf{r} - \omega t)$.

13. By separation of variables, find the solution of the Helmholtz equation in hyperbolic coordinates. That is, the coordinate set is (u, v, θ) where:

$$u = 2xy,$$

$$v = x^2 - y^2,$$

$$\pi/2 < \theta < \pi/2.$$

(Hint: Find the metric tensor and use it to express the Laplacian in terms of u, v, and θ.)

14. A monochromatic plane wave, traveling in the z-direction, with its magnetic vector, H, polarized in the x-direction, strikes a plane metallic sheet in the x–y plane at $z = 0$. In terms of H, find the direction and magnitude of the current induced at the surface of the metallic sheet. Assume the penetration depth is negligible.

Appendix 6-1　Bessel and Legendre Functions

These functions are found in the solution of the wave equation and the Helmholtz equation by separating variables in spherical coordinates. The Bessel functions give the radial dependence. The wave properties of the wave equation are best represented by the Hankel functions, which are complex linear combinations of the Bessel functions of the first and second kinds. The Legendre functions give the angular dependence. In three spatial dimensions, if the function has no special symmetry, the associated Legendre functions occur as part of the spherical harmonics. In two dimensions, or when the solution has translational symmetry along the z-axis, ordinary Legendre polynomials are used. We assume two dimensions here.

Initialization

```
Remove["Global`@"]
   (* Set highest order functions to be generated *)
lastno = 3;
```

Bessel Functions of Half Order

The Bessel functions of half order give the radial dependence.

Spherical Bessel Functions of the First Kind

```
bessOne = Table[BesselJ[n+1/2,u],{n,0,lastno}];
   Short[bessOne,3]


(* Plot them *)
(* These functions have no singularity. *)
Plot[Evaluate[Table[BesselJ[n+1/2,z],{n,4}]],
{z,0.2,20}, PlotRange->{-.6,.6}]
```

```
                      1
       {Sqrt[2] Sqrt[----] Sin[u],
                    Pi u


       Sqrt[2] <<1>> (-Cos[u] + <<1>>), <<1>>,


                      1              15 Cos[u]    15 Sin[u]
       Sqrt[2] Sqrt[----] (Cos[u] - --------- + --------- -
                    Pi u                2            3
                                       u            u


           6 Sin[u]
           --------)}
              u
```

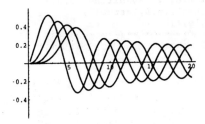

Spherical Bessel Functions of the Second Kind

```
    bessTwo = Table[BesselY[n+1/2,u],{n,0,lastno}];
        Short[bessTwo,3]


    (* Plot them *)
    (* These functions have singularity at the origin. *)
    Plot[Evaluate[Table[BesselY[n+1/2,z],{n,4}]],
    {z,0.2,20}, PlotRange->{-.6,.6}]


                      1
       {-(Sqrt[2] Sqrt[----] Cos[u]), <<2>>,
                      Pi u


                      1      -15 Cos[u]    6 Cos[u]
       Sqrt[2] Sqrt[----] (---------- + -------- + Sin[u] -
                    Pi u        3           u
                               u


           15 Sin[u]
           ---------)}
               2
              u
```

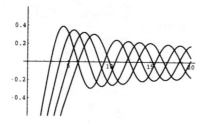

Hankel Functions

The Hankel functions are complex linear combinations of the Bessel functions of the first and second kinds. The Hankel functions bear an analogous relation to the Bessel functions that the exponential functions do to the sine and cosine functions.

```
hank = BesselJ[n+1/2,u]+I*BesselY[n+1/2,u];
hanktable = Table[hank,{n,0,lastno}];
    Short[hanktable,4]
```

```
                       1
{-I Sqrt[2] Sqrt[----] Cos[u] +
                  Pi u


                   1
   Sqrt[2] Sqrt[----] Sin[u], <<2>>,
                Pi u


              1      -15 Cos[u]    6 Cos[u]
  I Sqrt[2] Sqrt[----] (---------- + -------- + Sin[u] -
              Pi u          3          u
                           u


      15 Sin[u]
      ---------) + <<1>>}
          2
         u
```

Legendre Functions

The Legendre polynomials give the angular dependence.

```
Table[LegendreP[n,Cos[q]],{n,0,lastno}]


                        2                            3
             -1 + 3 Cos[q]     -3 Cos[q] + 5 Cos[q]
{1, Cos[q], --------------, ----------------------}
                   2                   2


Plot[Evaluate[Table[LegendreP[n,Cos[q]],{n,0,lastno}]],
{q,0,2*Pi}]
```

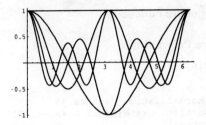

Associated Legendre Functions

Associated Legendre Functions of the First Kind

```
legTable = Table[LegendreP[n,m,q],{n,0,lastno},{m,0,n}];
Short[legTable,3]{ 9}
```

```
                  2
{{1}, {q, -Sqrt[1 - q ]}, <<1>>,
```

```
        3           2                 2
 -3 q + 5 q   3 (1 - 5 q ) Sqrt[1 - q ]            3
{-----------, ----------------------------, 15 (q - q ),
      2                    2
```

```
            2         2
15 Sqrt[1 - q ] (-1 + q )}}
```

Associated Legendre Functions of the Second Kind

```
legTableTwo = Table[LegendreQ[n,m,q],{n,0,2},{m,0,n}];
Short[legTableTwo,3]
```

```
       1 + q                              2     1 + q
  Log[-----]                        (-1 + 3 q ) Log[-----]
       1 - q                 -3 q               1 - q
{{----------}, <<1>>, {---- + ----------------------,
       2                  2               4
```

```
          2
  Sqrt[1 - q ] <<1>>
  ------------------ + <<1>>,
          2
        -1 + q
```

```
                                  2     1 + q
        2           3    3 (1 - q ) Log[-----]
  (1 - q ) (5 q - 3 q )              1 - q
  -------------------- + ----------------------}}
           2 2                       2
        (-1 + q )
```

Orthonormality of Legendre Functions

Normalized Legendre Polynomials

```
lfnct = LegendreP[n,x];
```

```
(* Make table of normalization factors *)
tnorm = Table[1/Sqrt[Integrate[lfnct^2,{x,-1,1}]],
         {n,0,lastno}]
```

```
(* Make table of normalized polynomials *)
norml = Table[lfnct*tnorm[[n+1]],{n,0,lastno}]
```

```
       1          3          5          7
{--------, Sqrt[-], Sqrt[-], Sqrt[-]}
  Sqrt[2]          2          2          2
```

```
                              5              2
                        Sqrt[-]  (-1 + 3 x )
        1          3          2
{--------, Sqrt[-] x, --------------------,
  Sqrt[2]          2              2
```

```
     7                   3
Sqrt[-]  (-3 x + 5 x )
     2
--------------------}
          2
```

Proof of Orthonormality

```
ortho = Table[Integrate[norml[[j]]*norml[[k]],
         {x,-1,1}],
            {j,1,lastno+1},{k,1,lastno+1}];
MatrixForm[ortho]{ 13}
```

```
1   0   0   0
0   1   0   0
0   0   1   0
0   0   0   1
```

Angular Dependent Legendre Functions

The most useful application of the Legendre functions is as part of the solution of partial differential equations in which they give the angular dependence. In these applications the argument of the Legendre polynomial is: $x \to \cos(\theta)$.

```
(* Normalized functions of angular argument *)
normlSin = norml/.x->Cos[q]
```

```
                                  5                2
                            Sqrt[-]  (-1 + 3 Cos[q] )
        1          3              2
{--------, Sqrt[-] Cos[q], ------------------------,
  Sqrt[2]          2                    2
```

$$
\frac{\text{Sqrt}\left[-\frac{7}{2}\right] \ (-3 \ \text{Cos}[q] + 5 \ \text{Cos}[q]^3)}{2}\}
$$

Proof of Orthonormality
The factor sin(q) is required in the integrand, due to the change of variable.

```
proDuct := normlSin[[j]]*normlSin[[k]]
orthoSin = Table[Integrate[proDuct*Sin[q],
       {q,0,Pi}],
       {j,1,lastno+1},{k,1,lastno+1}];
MatrixForm[orthoSin]
```

```
1   0   0   0
0   1   0   0
0   0   1   0
0   0   0   1
```

Legendre Expansion of Function

```
(* Define the function to be represented *)
funct = q^2
```

```
(* Calculate the expansion coefficients *)
coeff = Table[Integrate[funct*normlSin[[n]]*Sin[q],
       {q,0,Pi}],{n,1,lastno+1}]
```

```
(* Calculate the Legendre expansion *)
legendreExpn = coeff.normlSin//N
```

```
(* Plot the function and the expansion *)
Plot[{legendreExpn,q^2},{q,0,Pi}]
```

$$
q^2
$$

$$
\{\frac{-4 + \text{Pi}^2}{\text{Sqrt}[2]}, \ \frac{2 - (\text{Sqrt}[-]^3 \ \text{Pi}^2)}{4}\frac{}{2}, \ \frac{2 \ \text{Sqrt}[10]}{9}, \ \frac{-(\text{Sqrt}[-]^7 \ \text{Pi}^2)}{64}\frac{}{2}\}
$$

$$
2.9348 - 3.7011 \ \text{Cos}[q] + 0.555556 \ (-1. + 3. \ \text{Cos}[q]^2) -
$$

$$
0.269872 \ (-3. \ \text{Cos}[q] + 5. \ \text{Cos}[q]^3)
$$

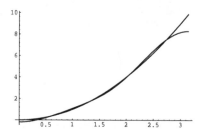

Appendix 6-2   Multipole Radiation Patterns

The solutions to the Helmholtz equation represent a snapshot of the wave equation at a fixed time; that is, the time function has been separated out. The solution of the Helmholtz equation in spherical coordinates is the sum of products of Hankel functions with spherical harmonics of the same order. Because Hankel functions are complex, we take just the imaginary part, that is, Bessel functions of the second kind. These functions have an infinity at the origin, appropriate for a point radiation source.

In order to make the plots easily understandable, we assume two dimensions, where the spherical harmonics reduce to simple Legendre polynomials. The general solution is a superposition of terms.

Initialization

```
Remove["Global`@"]
```

Load the Graphics package for later use.

```
<<Graphics`Graphics`
```

First-Order (Dipole) Pattern

The dipole angular function is the first order Legendre polynomial.

```
n = 1;
```

The Radial and Angular Functions

```
Plot[BesselY[n+1/2,r],{r,0.1,10}, AspectRatio->1/2]
Plot[LegendreP[n,Cos[y]],{y,0,Pi}, AspectRatio->1/2]
```

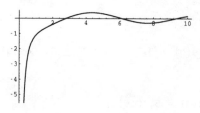

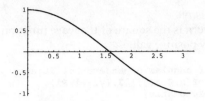

Combined Radial and Angular Functions

For simplicity, only half of the plot is drawn, for θ between $\pm Pi/2$.

```
(* Contour plot showing radial and angular dependence *)
rr = Sqrt[x^2+y^2];
func = BesselY[n+1/2,rr]*LegendreP[n,x/rr];
ContourPlot[func,{x,0.01,15},{y,-10,10},
        AspectRatio->3/2]


(* Surface plot *)
Plot3D[func,{x,1,15},{y,-15,15},
        PlotPoints->30,ClipFill->None]
```

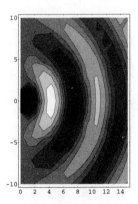

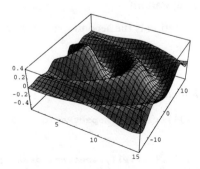

Dipole Radiation Intensity

The intensity flux pattern is the square of the wave function shown above. The intensity takes on only positive values.

```
(* Polar plot of angular dependence of intensity *)
PolarPlot[LegendreP[n,Cos[y]]^2,{y,-Pi,Pi},
    AspectRatio->Automatic]
```

```
(* Surface plot of intensity of dipole radiation *)
```

```
Plot3D[func^2,{x,1,15},{y,-15,15},
   PlotPoints->30,ClipFill->None,
   PlotRange->{0,0.12}]
```

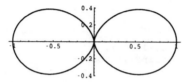

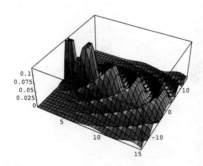

Second-Order (Quadrupole) Pattern

The quadrupole angular function is the second-order Legendre polynomial.

```
n = 2;
```

The Radial and Angular Functions

```
Plot[(BesselY[n+1/2,r]),{r,1,15},AspectRatio->1/2]
```

```
Plot[LegendreP[n,Cos[y]],{y,0,Pi},AspectRatio->1/2]
```

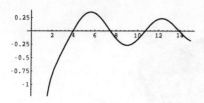

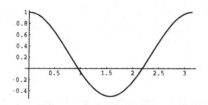

Quadrupole Radiation Function

```
(* Contour plot showing radial and angular dependence *)
rr = Sqrt[x^2+y^2];
func = BesselY[n+1/2,rr]*LegendreP[n,x/rr];
ContourPlot[func,{x,0.01,15},{y,-10,10},
        AspectRatio->3/2]


(* Surface plot *)
Plot3D[func,{x,1,15},{y,-15,15},
        PlotPoints->30,ClipFill->None]
```

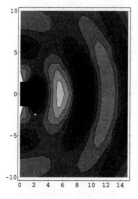

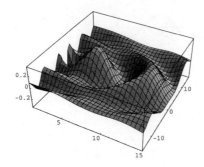

Quadrupole Radiation Intensity

```
(* Polar plot of angular dependence of intensity *)
PolarPlot[LegendreP[n,Cos[y]]^2,{y,-Pi,Pi},
    AspectRatio->Automatic]
```

```
(* Surface plot of intensity of quadrupole radiation *)
Plot3D[func^2,{x,1,15},{y,-15,15},
    PlotPoints->30,ClipFill->None,
    PlotRange->{0,0.1}]
```

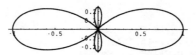

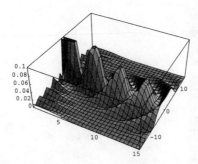

APPENDIX 6-3   SINGLE SLIT DIFFRACTION BY HELMHOLTZ INTEGRAL

Choose Numerical Parameters

```
(* Numerical values are in arbitrary units *)


(* horizontal distance from slit to screen *)
x = 100;


(* wavelength *)
lambda = 0.05;


(* propagation constant *)
k = 2*Pi/lambda//N;


(* slit width *)
d = 0.2;
```

Make Table of Helmholtz Integrals

The table shows the intensity of the diffracted wave at different points on the detecting "screen." Because the pattern is symmetric, it is only necessary to calculate the diffraction pattern for positive values of *y*. We also plot only that half of the pattern.

```
(* Define vertical displacements *)
(* y is vertical height on screen *)
(* y0 is vertical height at slit *)
r = Sqrt[x^2 + (y-y0)^2];


(* Set up Helmholtz integrals *)
(* Evaluation is delayed until table is made*)
func := NIntegrate[E^(I*k*r)/r,{y0,-d/2,d/2}];


(* Make table of Helmholtz integrals *)
(* Intensity is proportional to func squared *)
maketable = Table[Abs[func]^2,{y,0,200,5}];


(* Normalize the table values to one *)
normaltable = maketable/maketable[[1]];
    Short[normaltable,4]

{1., 0.873252, 0.570446, 0.257923, 0.0621512, 0.000890323,
 0.0143946, 0.0370589, 0.0394978, 0.0254923, 0.00960402,
 0.00106238, <<23>>, 0.00205292, 0.00195429, 0.00184382,
 0.00172707, 0.0016083, 0.00149066}
```

Diffraction Pattern Plots

Since the diffraction pattern is symmetric around the origin, we plot only that half of the pattern for the positive axis.

```
(* Linear plot emphasizes central maximum *)
ListPlot[normaltable,PlotRange->{0,1}]


(* Log plot emphasizes secondary maxima *)
(* Data points are normalized to the minimum value *)
logTable = Log[normaltable/Min[normaltable]];
ListPlot[logTable,PlotJoined->True]
```

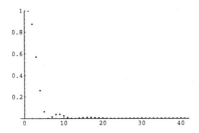

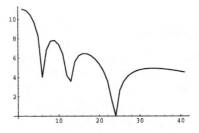

Appendix 6-4   The Electromagnetic Stress Tensor

Define Elementary Tensors

```
(*Field tensor *)
ftensor = {{0,-E1,-E2,-E3},
           {E1,0,-B3,B2},
           {E2,B3,0,-B1},
           {E3,-B2,B1,0}};
ftmat = MatrixForm[ftensor];


(* metric tensor *)
gjk = {{1,0,0,0},
       {0,-1,0,0},
       {0,0,-1,0},
```

```
        {0,0,0,-1}};
gjkmat = MatrixForm[gjk];


(* Kronecker delta tensor *)
delta = IdentityMatrix[4];
demat = MatrixForm[delta];


(* Display the tensors as a List of matrices *)
{ftmat,gjkmat,demat}
```

```
{0    -E1   -E2   -E3, 1    0    0    0 , 1    0    0    0}
 E1    0    -B3   B2   0   -1    0    0   0    1    0    0
 E2    B3    0   -B1   0    0   -1    0   0    0    1    0
 E3   -B2   B1    0    0    0    0   -1   0    0    0    1
```

Calculate the First Term

```
term1 = ftensor.gjk.ftensor.gjk;
        MatrixForm[term1]
```

$$E1^2 + E2^2 + E3^2 \quad -(B3\ E2) + B2\ E3 \quad B3\ E1 - B1\ E3 \quad -(B2\ E1) + B1\ E2$$

$$B3\ E2 - B2\ E3 \quad -B2^2 - B3^2 + E1^2 \quad B1\ B2 + E1\ E2 \quad B1\ B3 + E1\ E3$$

$$-(B3\ E1) + B1\ E3 \quad B1\ B2 + E1\ E2 \quad -B1^2 - B3^2 + E2^2 \quad B2\ B3 + E2\ E3$$

$$B2\ E1 - B1\ E2 \quad B1\ B3 + E1\ E3 \quad B2\ B3 + E2\ E3 \quad -B1^2 - B2^2 + E3^2$$

Calculate the Second Term

The *Mathematica* matrix product sums over the second index of the first matrix, and the first index of the second matrix. Therefore, in calculating the double sum in the expression, FabGab, we take the transpose of Fab to convert the matrix multiplication conventions.

```
innerprod = Transpose[ftensor].gjk.ftensor*gjk;
mytrace := Sum[innerprod[[i,i]],{i,1,4}];
term2 = Simplify[mytrace]*delta/4
```

$$\left\{\left\{\frac{B1^2 + B2^2 + B3^2 - E1^2 - E2^2 - E3^2}{2}, 0, 0, 0\right\},\right.$$

$$\left\{0, \frac{B1^2 + B2^2 + B3^2 - E1^2 - E2^2 - E3^2}{2}, 0, 0\right\},$$

$$\left\{0, 0, \frac{B1^2 + B2^2 + B3^2 - E1^2 - E2^2 - E3^2}{2}, 0\right\},$$

$$\{0,\ 0,\ 0,\ \frac{B1^2 + B2^2 + B3^2 - E1^2 - E2^2 - E3^2}{2}\}\}$$

Make Simplification Rules

```
(* Rules for squares of fields *)
ruleaa = E1^2+E2^2+E3^2->EE;
rulebb = B1^2+B2^2+B3^2->BB;
rulecc = -E1^2-E2^2-E3^2->-EE;
ruledd = -B1^2-B2^2-B3^2->-BB;


(* Rules for other terms *)
ruleee = B1^2+B2^2->BB-B3^2;
ruleff = B3^2+B1^2->BB-B2^2;
rulegg = B2^2+B3^2->BB-B1^2;
negruleee = -B1^2-B2^2->-BB+B3^2;
negruleff = -B3^2-B1^2->-BB+B2^2;
negrulegg = -B2^2-B3^2->-BB+B1^2;


(* Combine rules *)
ruler={ruleaa,rulebb,rulecc,ruledd,
       ruleee,ruleff,rulegg,
       negruleee,negruleff,negrulegg}
```

$\{E1^2 + E2^2 + E3^2 \to EE,\ B1^2 + B2^2 + B3^2 \to BB,$

$-E1^2 - E2^2 - E3^2 \to -EE,\ -B1^2 - B2^2 - B3^2 \to -BB,$

$B1^2 + B2^2 \to BB - B3^2,\ B1^2 + B3^2 \to BB - B2^2,$

$B2^2 + B3^2 \to BB - B1^2,\ -B1^2 - B2^2 \to -BB + B3^2,$

$-B1^2 - B3^2 \to -BB + B2^2,\ -B2^2 - B3^2 \to -BB + B1^2\}$

Apply Rule to Each Term

```
TjkFirst = term1//.ruler;
         MatrixForm[TjkFirst]


TjkSecond = term2//.ruler;
         MatrixForm[TjkSecond]
```

EE	-(B3 E2) + B2 E3	B3 E1 - B1 E3	-(B2 E1) + B1 E2
B3 E2 - B2 E3	$-BB + B1^2 + E1^2$	B1 B2 + E1 E2	B1 B3 + E1 E3

$$-(B3\ E1) + B1\ E3 \quad B1\ B2 + E1\ E2 \quad -BB + B2^2 + E2^2 \quad B2\ B3 + E2\ E3$$

$$B2\ E1 - B1\ E2 \quad B1\ B3 + E1\ E3 \quad B2\ B3 + E2\ E3 \quad -BB + B3^2 + E3^2$$

$$
\begin{array}{cccc}
\dfrac{BB - EE}{2} & 0 & 0 & 0 \\[2mm]
0 & \dfrac{BB - EE}{2} & 0 & 0 \\[2mm]
0 & 0 & \dfrac{BB - EE}{2} & 0 \\[2mm]
0 & 0 & 0 & \dfrac{BB - EE}{2}
\end{array}
$$

Stress Tensor

Mixed Form

```
TjkCombined = TjkFirst + TjkSecond;
TjkMix = Simplify[TjkCombined];
          MatrixForm[TjkMix]
```

$$
\begin{array}{cccc}
\dfrac{BB + EE}{2} & -(B3\ E2) + B2\ E3 & B3\ E1 - B1\ E3 & -(B2\ E1) + B1\ E2 \\[3mm]
B3\ E2 - B2\ E3 & \dfrac{-BB}{2} + B1^2 - \dfrac{EE}{2} + E1^2 & B1\ B2 + E1\ E2 & B1\ B3 + E1\ E3 \\[3mm]
-(B3\ E1) + B1\ E3 & B1\ B2 + E1\ E2 & \dfrac{-BB}{2} + B2^2 - \dfrac{EE}{2} + E2^2 & B2\ B3 + E2\ E3 \\[3mm]
B2\ E1 - B1\ E2 & B1\ B3 + E1\ E3 & B2\ B3 + E2\ E3 & -\dfrac{BB}{2} + B3^2 - \dfrac{EE}{2} + E3^2
\end{array}
$$

Covariant Form

```
Tjk = gjk.TjkMix;
      TableForm[Tjk]
```

$$
\begin{array}{cccc}
\dfrac{BB + EE}{2} & -(B3\ E2) + B2\ E3 & B3\ E1 - B1\ E3 & -(B2\ E1) + B1\ E2 \\[3mm]
-(B3\ E2) + B2\ E3 & \dfrac{BB}{2} - B1^2 + \dfrac{EE}{2} - E1^2 & -(B1\ B2) - E1\ E2 & -(B1\ B3) - E1\ E3
\end{array}
$$

$$B3\ E1\ -\ B1\ E3 \qquad -(B1\ B2)\ -\ E1\ E2 \qquad \frac{BB}{2}\ -\ B2^2\ +\ \frac{EE}{2}\ -\ E2^2 \qquad -(B2\ B3)\ -\ E2\ E3$$

$$-(B2\ E1)\ +\ B1\ E2 \qquad -(B1\ B3)\ -\ E1\ E3 \qquad -(B2\ B3)\ -\ E2\ E3 \qquad \frac{BB}{2}\ -\ B3^2\ +\ \frac{EE}{2}\ -\ E3^2$$

CHAPTER 7

Particle Motion in Electromagnetic Fields

The spirit of the time shall teach me speed.

Shakespeare
King John

I am ... straining at particles of light in the midst of a great darkness.

John Keats
Letter to George Keats

The motion of charged particles moving in an electromagnetic field are determined by the Lorentz force. In this chapter we consider some special applications: uniform fields, electric and magnetic lenses, scattering, and a kinetic field formulation of the equations.

7-1 UNIFORM FIELDS

The equation of motion of a charged particle in electric and magnetic fields is determined by the Lorentz force:

$$\frac{dU^j}{ds} = \frac{q}{mc^2} F^{jk} U_k. \qquad (7\text{-}1.1)$$

Then the antisymmetry of the field tensor ensures the velocity has constant magnitude:

$$U^j U_j = c^2. \tag{7-1.2}$$

When reduced to three-dimensional form, the equation of motion becomes:

$$\frac{d\mathbf{v}}{dt} = \frac{q}{m}(\mathbf{E} + \frac{\mathbf{v}}{c} \times \mathbf{B}). \tag{7-1.3}$$

The simplest application is constant, uniform fields, i.e., the fields are independent of position and time, but have arbitrary strength and direction. Without losing any physical generality, we may choose a co-ordinate system with the z-axis parallel to the magnetic field. In this coordinate system the equations of motion (4-4.10) become:

$$\frac{d}{ds}\begin{bmatrix} U^0 \\ U^1 \\ U^2 \\ U^3 \end{bmatrix} = \frac{q}{mc^2}\begin{bmatrix} 0 & -E_1 & -E_2 & -E_3 \\ E_1 & 0 & -B & 0 \\ E_2 & B & 0 & 0 \\ E_3 & 0 & 0 & 0 \end{bmatrix}\begin{bmatrix} U^0 \\ -U^1 \\ -U^2 \\ -U^3 \end{bmatrix}. \tag{7-1.4}$$

This set of coupled, linear equations with constant coefficients is easily solved by substituting the exponential functions:

$$U^j = U_0^j \exp(k_j s), \qquad \text{no sum on } j, \tag{7-1.5}$$

and solving the resulting algebraic equations for the k_j.

Example

In the nonrelativistic limit, with both $\mathbf{E}$ and $\mathbf{B}$ in the z-direction, the solution becomes:

$$v^1 = v_0 \cos(\omega t + \alpha),$$
$$v^2 = -v_0 \sin(\omega t + \alpha),$$
$$v^3 = \frac{q}{m} Et + v_0^3, \tag{7-1.6}$$

where the angular frequency:

$$\omega = \frac{qB}{mc} \tag{7-1.7}$$

is the *cyclotron frequency*, and where v_0, v_0^3, and α are constants, determined by initial conditions. The trajectory is a helix whose pitch depends on the z-component of the velocity and whose radius is:

$$r = \sqrt{x_1^2 + x_2^2} = \frac{v_0}{\omega}. \tag{7-1.8}$$

In the absence of the magnetic field, the trajectory is a parabola. This occurs, for example, between the deflecting electrodes of a cathode ray tube. Figure 7-1 shows an example of charged particle motion in uniform electric and magnetic fields.

When the electric and magnetic fields are at right angles, it is possible for the forces to cancel, e.g., for an electric field directed along the y-axis, a magnetic field directed along the negative z-axis, and a particle moving with a velocity along the x-axis:

$$v = \frac{E}{B}c \qquad (7\text{-}1.9)$$

will experience no net force. This can be used as a velocity selector for a charged particle beam.

In 1897, J.J. Thomson published the results of experiments that revolutionized the understanding of the nature of matter by showing that all matter is composed, at least in part, of electrically charged particles. The Thomson particles have been given the name *electron*, from the Greek word for amber. When Thomson investigated the properties of electrical discharges in rarified gases, he found that rays emitted from the cathode (i.e., cathode rays) had a specific ratio of charge to mass, *regardless of the composition of the gas or of the cathode*. (By contrast, anode rays now known to be ions, exhibit strong composition dependence). Figure 7-2 is a representation of the Thomson experiment.

The electrons are emitted by cathode, accelerate through a potential difference, ϕ, and then pass through a slit into the main section of the apparatus. As they leave the slit, the electrons have a velocity determined by the energy conservation. In the main section the electric and magnetic fields are adjusted to give no deflection. A simple calculation gives an expression for the ratio of charge to mass in terms of known

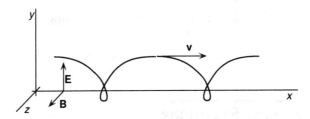

FIGURE 7-1. PARTICLE MOTION IN UNIFORM ELECTRIC AND MAGNETIC FIELDS AT RIGHT ANGLES. The electric and magnetic fields are constant. The exact shape of the cycloid depends on the values of the electric and magnetic fields and on the initial velocity.

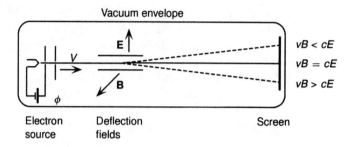

FIGURE 7-2. THE THOMSON EXPERIMENT. The electric field, **E**, and the magnetic field, **B**, and the velocity, **v**, are mutually orthogonal. At a given velocity, the magnetic and electric forces cancel when $v = cE/B$. From this the ratio of the charge to the mass of the particles in the beam can be measured.

quantities, v and ϕ:

$$\frac{e}{m} = \frac{v^2}{2\phi}.$$ (7-1.10)

The electron charge to mass ratio is:

$$\frac{e}{m} = \begin{cases} 5.2728 \times 10^{17} \dfrac{\text{statcoulombs}}{\text{gram}}, & \text{Gaussian,} \\[2ex] \sqrt{4\pi} \times 5.2728 \times 10^{17} \dfrac{\text{statcoulombs}}{\text{gram}}, & \text{Heaviside,} \\[2ex] 1.7588 \times 10^{11} \dfrac{\text{coulombs}}{\text{kilogram}}, & \text{SI.} \end{cases}$$ (7-1.11)

Subsequent experiments by Thomson, Millikan, and others fixed the charge (and mass) of the electron. Its modern value is:

$$e = \begin{cases} 4.8032067991 \times 10^{-10} \text{ statcoulombs,} & \text{Gaussian,} \\[1ex] \sqrt{4\pi} \times 4.8032067991 \times 10^{-10} \text{ statcoulombs,} & \text{Heaviside,} \\[1ex] 1.60217733 \times 10^{-19} \text{ coulombs,} & \text{SI.} \end{cases}$$

(7-1.12)

7-2   NUMERICAL SOLUTIONS

The equations of motion exemplify a common mathematical problem: a set of coupled equations. When the fields are constant the equations have constant coefficients, and there are methods of solution in terms of known functions. For an arbitrary field however, numerical solutions may be more convenient. On the other hand, although numerical solution is very powerful, the solution is expressed only as tables of numbers, and may therefore not give good theoretical insight. A great deal of literature exists for the numerical solution of differential equations.

The first method is Euler's *difference equation method*. It is also called the finite element method. It can be applied to systems of linear or nonlinear equations. The method is based on replacing the continuous variable x in the differential equation, by a set of discrete values, x_n, in a difference equation; the interval between any two adjacent, x_n, is written, δx_n, which is taken to be finite but small. Compare this with Section 5-5. A function of the discrete variable takes discrete values:

$$y_n = f(x_n). \tag{7-2.1}$$

We will suppose all the δx_n to be equal, i.e., the x_n are equally spaced. The derivatives may be approximated by the ratios:

$$\left[\frac{dy}{dx}\right]_n \approx \frac{y_{n+1} - y_n}{x_{n+1} - x_n} = \frac{y_{n+1} - y_n}{\delta x}, \tag{7-2.2}$$

$$\left[\frac{d^2y}{dx^2}\right]_n \approx \frac{\delta(dy/dx)}{\delta x} = \frac{y_{n+2} - 2y_n + y_{n-1}}{\delta x^2}. \tag{7-2.3}$$

Since the first derivative is not symmetric about the point x_n, we may prefer to replace it with the ratio spanning two intervals:

$$\left[\frac{dy}{dx}\right]_n \approx \frac{y_{n+1} - y_{n-1}}{2\delta x}. \tag{7-2.4}$$

Substituting these expressions into a differential equation gives a *recursion relation* between the y_n.

EXAMPLE

The harmonic oscillator equation approximated this way is:

$$\frac{y_{n+1} - 2y_n + y_{n-1}}{\delta t^2} = -\frac{k}{m}y_n = -\omega^2 y_n. \tag{7-2.5}$$

This difference equation yields the recursion formula:

$$y_n = \left[2 - \omega^2 \delta t^2\right] y_{n-1} - y_{n-2}. \tag{7-2.6}$$

If the first two values of y_n were known, the recursion formula would enable calculation of all successive values. But the initial values of y and its first derivative are the initial conditions for a second-order equation. We let them determine the first two y_n:

$$y_0 = y(t_0),$$
(7-2.7)

$$y_1 \approx y_0 + \left[\frac{dy}{dt}\right]_0 \delta t.$$
(7-2.8)

Equations (7-2.6), (7-2.7), and (7-2.8) give the numerical solution to the harmonic oscillator. Appendix 7-1 shows the solution to the harmonic oscillator done with a spreadsheet.

The accuracy of the finite difference method depends on the degree to which derivatives can be approximated by the finite ratios. It is convenient, when discussing accuracy, to write the exact solution as $Y(x_n)$ and the finite approximation values as $y_n(x_n)$. In general, the accuracy of the expression is a function of the interval δx, and we may express it as a Taylor series:

$$\delta Y(x_n) = Y(x_n) - y_n(x_n) = a\,\delta x^2 + b\,\delta x^3 + c\,\delta x^4 + \cdots.$$
(7-2.9)

Then the lowest power occurring in this series is called the order of the approximation. (The zeroth and first-order terms are already accounted for since they are the initial conditions.) Obviously, it is in our interest to have high-order approximations, when possible. This is the purpose of the *Runge–Kutta methods*. We have seen that in the Euler method the derivative can be evaluated at the midpoint by taking an interval $2\delta x$. Then the derivative represents an average value in the interval rather than its value at the beginning of the interval. The basic intent of the Runge–Kutta method is to improve the accuracy by cancelling some of the lower orders in equation (7-2.9). First, note that we can transform a second-order equation to a pair of first-order equations. A single second-order equation:

$$\frac{d^2y}{dx^2} = f\left(\frac{dy}{dx}, y, x\right)$$
(7-2.10)

may be written as two first-order equations:

$$\frac{dz}{dx} = f(z, y, x),$$
$$\frac{dy}{dx} = z,$$
(7-2.11)

which can be solved for y and z simultaneously. Hereafter, we will restrict our discussion to first-order equations.

In order to find solutions to the equation:

$$\frac{dy}{dx} = f(y, x)$$

we define the quantities:

$$P_n = \frac{dy}{dx}\, \delta x = f(x_n, y_n)\, \delta x \qquad (7\text{-}2.12)$$

$$Q_n = f\left(x_n + \frac{\delta x}{2}, y_n + \frac{dy}{dx}\frac{\delta x}{2}\right)\delta x$$

$$= f\left(x_n + \frac{\delta x}{2}, y_n + \frac{P_n}{2}\right)\delta x. \qquad (7\text{-}2.13)$$

In terms of P_n and Q_n the recursion formula for y_{n+1} is written:

$$y_{n+1} = y_n + Q_n + O(\delta x^3). \qquad (7\text{-}2.14)$$

That is, in the quantity Q_n the derivative is evaluated at the midpoint. In these second order Runge–Kutta expressions, the second-order error is cancelled out. It is therefore preferable to the original Euler method; higher-order Runge–Kutta methods give even better accuracy. The fourth-order expressions are:

$$P_n = f(x_n, y_n)\delta x, \qquad (7\text{-}2.15)$$

$$Q_n = f\left(x_n + \frac{\delta x}{2}, y_n + \frac{P_n}{2}\right)\delta x, \qquad (7\text{-}2.16)$$

$$R_n = f\left(x_n + \frac{\delta x}{2}, y_n + \frac{Q_n}{2}\right)\delta x, \qquad (7\text{-}2.17)$$

$$S_n = f\left(x_n + \delta x, y_n + R_n\right)\delta x, \qquad (7\text{-}2.18)$$

$$y_{n+1} = y_n + \frac{P_n}{6} + \frac{Q_n}{3} + \frac{R_n}{3} + \frac{S_n}{6} + O(\delta x^5). \qquad (7\text{-}2.19)$$

In actual applications, the method can be tailored to the specific problem. For example, in regions where the derivatives are large, it may be useful to decrease locally the step size interval δx. Of course, one could use the smaller step size globally, but the computation time would increase proportionally. Press et al. give a set of computer routines for the fourth-order formulas.

A second method of solution of differential equations is based on the Taylor series for functions. A second-order differential equation can be written (using a convenient parentheses notation for derivatives):

$$y^{(2)} = f(y^{(1)}(x), y(x), x). \qquad (7\text{-}2.20)$$

Assume that the functional is analytic in x and differentiable and that the value and slope of $y(x)$ are known at some initial point, x_0. These are Cauchy conditions:

$$y^{(0)}(x_0) = y(x_0) = \text{constant}, \qquad (7\text{-}2.21)$$

$$y^{(1)}(x_0) = \text{constant}. \qquad (7\text{-}2.22)$$

Differentiate equation (7-2.20) to get the third and higher derivatives:

$$y^{(2)} = f(y^{(1)}(x), y(x), x),$$

$$y^{(3)} = \frac{d}{dx} f(y^{(1)}(x), y(x), x) = f(y^{(2)}(x), y^{(1)}(x), y(x), x),$$

$$y^{(4)} = \frac{d^2}{dx^2} f(y^{(1)}(x), y(x), x) = f(y^{(3)}(x), y^{(2)}(x), y^{(1)}(x), y(x), x), \quad (7\text{-}2.23)$$

etc.

But the Taylor series expansion of a function:

$$y(x) = y(x_0) + y^{(1)}(x_0)(x - x_0) + \frac{1}{2!} y^{(2)}(x_0)(x - x_0)^2$$

$$+ \frac{1}{3!} y^{(3)}(x_0)(x - x_0) + \cdots \qquad (7\text{-}2.24)$$

contains the derivatives of the function evaluated at the initial point in its coefficients. Now for each term in (7-2.21), (7-2.22), and (7-2.23), substitute lower derivative expressions into the higher. The result of these multiple substitutions is that each expression is expressed in just $y(x_0)$ and $y^{(1)}(x_0)$. Thus we know (in principle) all the derivatives of the solution of the differential equation, and have its complete solution including the initial conditions.

The advantages of this method is its ease of application and its high accuracy over short ranges. It is very suitable for representing polynomials, and applies to nonlinear equations. Its limitations are related to the covergence of the series. In some cases power series only converge for finite ranges of the variable. Furthermore, we normally must truncate the series, since the general term cannot be determined. Then truncation errors occur, having a strong dependence on x. These problems are usually handled by *analytic continuation*, i.e., at some point x_1, use the series to calculate the function $y(x_1)$ and its derivatives, and then use them as initial values for a second expansion of the function centered at x_1. Further analytic continuations can be made at x_2 and so on. Thus the method is well suited for polynomials and short ranges of x, but accumulated truncation errors may make the method unreliable at large x.

EXAMPLE

The Taylor series solution for the harmonic oscillator is as follows:

$$y^{(2)} = -\omega^2 y. \qquad (7\text{-}2.25)$$

Taking the first six terms (expanded around the point $t_0 = 0$) and substituting lower derivatives as described above gives:

$$y = y_0 + y_0^{(1)}t + \frac{1}{2!}[-\omega^2 y_0]t^2 + \frac{1}{3!}[-\omega^2 y_0^{(1)}]t^3$$

$$+ \frac{1}{4!}[\omega^4 y_0]t^4 + \frac{1}{5!}[\omega^4 y_0^{(1)}]t^5 + \cdots. \qquad (7\text{-}2.26)$$

Then, by rearranging:

$$y = y_0 \left[1 - \frac{1}{2!}[\omega t]^2 + \frac{1}{4!}[\omega t]^4 + \cdots \right]$$

$$+ \frac{y_0^{(1)}}{\omega} \left[\omega t - \frac{1}{3!}[\omega t]^3 + \frac{1}{5!}[\omega t]^5 + \cdots \right] \qquad (7\text{-}2.27)$$

the two series are recognizable functions:

$$y = y_0 \cos[\omega t] + \frac{y_0^{(1)}}{\omega} \sin[\omega t]. \qquad (7\text{-}2.28)$$

This is the well-known exact solution of the harmonic oscillator for initial displacement y_0 and initial velocity $y_0^{(1)}$. Figure 7-3 shows the Taylor series solution for an anharmonic oscillator.

Although it is necessary to truncate the series to a small number of terms, the truncation error associated with this method can be made smaller than the error in the difference equation methods for a given δx. Consider a first-order equation. The overall error over a finite interval δx can be expressed by the integral:

$$\delta y(x + \delta x) = \int_x^{x+\delta x} [y_s' - f(y_s, x)]\, dx, \qquad (7\text{-}2.29)$$

where y_s and y_s' are the series representations for the variable y and its derivative. For the exact solution, the integral vanishes. Then, the series

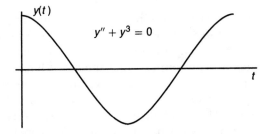

$$y'' + y^3 = 0$$

FIGURE 7-3. ANHARMONIC OSCILLATOR.   This is the solution to the nonlinear equation, $y'' = -y^3$ by the Taylor series method. The solution is similar to a sine function but is more triangular. The truncation error is maximum at the analytic continuation points.

solution and the error can be used as the basis for a *predictor–corrector* solution. In these solutions a value for y is predicted by the Taylor series at some value of x. Then the error at x is determined as above, and a corrected value of y is found by subtracting the error:

$$y_C(x + \delta x) = y_S(x + \delta x) - \delta y(x + \delta x)$$

$$= y_S(x + \delta x) - \int_x^{x+\delta x} \left[y'_S - f(y_S, x) \right] dx. \quad (7\text{-}2.30)$$

The corrected value is now used as the initial (seed) value for a prediction of $y_s(x+\delta x)$. Then the procedure of prediction, error, and correction is continued downstream over the range of x.

In general, predictor–corrector schemes can give very nice fits, and the Taylor series method is particularly appropriate as the predictor method. The time and effort required to set up the computation may be considerably reduced by calculating the derivatives with one of the symbolic mathematics programs. A *Mathematica* routine for Taylor series solutions is given in Appendix 7-2. A discussion of the relative merits of the Runge–Kutta, predictor–corrector, and other numerical methods is given in Press et al.

Equations (7-2.9) and (7-2.29) are local expressions for the error, i.e., they are functions of x. They do not deal with a gradual drift from the exact solution. For example, if the differential error $\delta y(x_n)$ is always positive, then the total error, $\Sigma \delta y(x_n)$ could become quite large for large x. It is partly for this reason that space vehicles make mid-course corrections in their trajectories. This should remind us that there is no substitute for inspecting calculated results for agreement with the basic laws of physics and of mathematics.

7-3   An Example: Particle Optics

Charged particle beams have many important uses including television tubes, electron microscopes, mass spectrometers, and elementary particle experiments. Much of the language describing such beams is taken from optics—one speaks of lenses, aperture stops, focal planes, and chromatic and spherical aberration. We will consider a high-energy particle beam as a representative example.

An external beam of K^- particles is an excellent way to do experiments on particles with the property of strangeness because the particles are created and decay inside the detector (a spark chamber, for example). Such experiments were used to measure the properties of the Λ, Σ, Ξ, Ω, and other particles. The design and construction of a beam optical system to transport particles (having an acceptable momentum range and uncontaminated by other particles) from the accelerator target to the detector is a major undertaking. A diagram of a representative external beam transport system is shown in Figure 7-4. The main components in this beam are: (1) Evacuated drift spaces between the components minimize scattering by air molecules. (2) Bending magnets (B) steer the beam and determine its momentum spread. They are equivalent to prisms in optics. (3) Aperture stops or slits (S) determine the outer envelope of the beam and (with the bending magnets) its momentum spread. (4) Quadrupole magnets (Q) focus the beam; they are equivalent to lenses. (5) Velocity spectrometers (VS) limit the mass of the particles passed through the beam. Together velocity selection and momentum selection determine the mass of the particles. The final beam profile, energy, and composition is determined by the interaction of these components. The total length of the beam between target and detector and the particle lifetime limit the particle flux at the detector. Higher energy particles have the advantage because of time dilation.

First, consider the bending magnets. Although the main functions of the bending or dipole magnets is steering and momentum selection, fringe fields and nonuniformities in the field may cause a certain amount of focusing or defocusing. For our simplified analysis, we will suppose that the magnetic field is uniform in the z-direction, and that fringe fields are negligible. The trajectory of a particle in the uniform field of the bending magnet satisfies the equations:

$$\frac{dU_0}{ds} = 0,$$

$$\frac{dU_1}{ds} = \frac{qB}{mc^2} U_2 = \frac{\omega}{c} U_2,$$

$$\frac{dU_2}{ds} = -\frac{qB}{mc^2} U_1 = -\frac{\omega}{c} U_1,$$

$$\frac{dU_3}{ds} = 0,$$

(7-3.1)

S_1 B_1 S_2 Q_1 Q_2 VS Q_3 Q_4 S_2

Figure 7-4. Representative beam transport system.   The first slit S_1 determines the total flux of the system. The function of the remaining components is to select particles of a given mass and momentum, and to transport as much flux of these particles to the detector as possible.

which can be solved exactly:

$$U_j = \left[\text{const.}, \ A \cos \left(\frac{\omega}{c} s + \alpha \right), \ A \sin \left(\frac{\omega}{c} s + \alpha \right), \ \text{const.} \right]. \quad (7\text{-}3.2)$$

The trajectory is a circular helix of radius r:

$$r = \frac{mU}{qB} c \quad (7\text{-}3.3)$$

proportional to the particle momentum. The deflection angle for a particle of momentum, p, traversing a magnet of length, L, is:

$$\tan \theta = \frac{dy}{dx} = \frac{L}{\sqrt{r^2 - L^2}} = \frac{1}{\sqrt{\left[\frac{c}{qBL} \right]^2 p^2 - 1}}. \quad (7\text{-}3.4)$$

The momentum range of the beam is thereby determined by the width of a downstream slit.

Quadrupole magnets are the particle beam equivalent of lenses. An ideal quadrupole has a magnetic field given by:

$$B_1 = 0,$$
$$B_2 = Gx_3, \quad (7\text{-}3.5)$$
$$B_3 = Gx_2,$$

where we suppose the reference trajectory is in the x-direction and enters the quadrupole at point $y = z = 0$. See equations (5-6.13), (5-6.14), and (5-6.15). In one transverse (e.g., horizontal) plane the quadrupole acts as a converging lens, but in the orthogonal (e.g., vertical) plane it acts as a diverging lens. In this sense it can be said to be astigmatic. However, a pair of quadrupoles with opposite polarity can give convergence in both dimensions.

Consider the trajectory equations of a beam particle in the quadrupole lens, assuming that the transverse velocity component is small compared to the longitudinal component, i.e., in the direction of the beam.

$$\frac{dU_0}{ds} = 0,$$
$$\frac{dU_1}{ds} = \frac{qG}{mc^2} [x_2 U_2 - x_3 U_3] \approx 0,$$
$$\frac{dU_2}{ds} = - \left[\frac{qU_1 G}{mc^2} \right] x_2 = -cK^2 x_2, \quad (7\text{-}3.6)$$
$$\frac{dU_3}{ds} = \left[\frac{qU_1 G}{mc^2} \right] x_3 = cK^2 x_3,$$

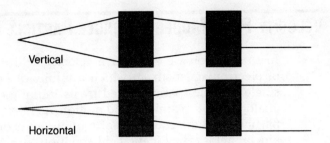

FIGURE 7-5. BEAM PROFILES FOR QUADRUPOLE FIELDS.   A pair of quadrupole magnets can be made to act as an effective convergent lens in both planes even though a single quadrupole has a divergent effect in one plane and a convergent effect in the other. The parallel section of the beam is needed in a velocity spectrometer.

where we define the constant K with units of inverse length. The longitudinal component of the velocity is virtually unchanged, and the two transverse displacements are:

$$x_2 \approx x_{20} \cos[Ks] + \frac{U_{20}}{cK} \sin[Ks],$$

$$x_3 \approx x_{30} \cosh[Ks] + \frac{U_{30}}{cK} \sinh[Ks]. \tag{7-3.7}$$

The quadrupole lens diverges the beam in one transverse axis but converges it in the other. This can be expressed in terms of focal lengths. Let the length of the lens be L, and take a paraxial ray at transverse displacements x_{20} and x_{30}. Then the two focal lengths are:

$$f_2 = \frac{x_{20}}{[dx_2/dx_1]} = \frac{U_1}{U_2} x_{20} = \frac{U_1}{cK \sin(Ks_L)}, \tag{7-3.8}$$

$$f_3 = \frac{x_{30}}{[dx_3/dx_1]} = \frac{U_1}{U_3} x_{30} = \frac{U_1}{cK \sinh(Ks_L)}. \tag{7-3.9}$$

The quantity s_L is the pathlength through the lens, $s_L = cL/U_1$. Since the "constant", K, depends on the beam velocity, U_1, the lens will clearly have chromatic aberration, i.e., its focal length depends on the beam velocity. Figure 7-5 shows the effective convergence due to a pair of quadrupole lenses.

The velocity spectrometers are actually giant Thomson experiments. The important parameter for these devices is the angular deviation of the beam as a function of the velocity of the particles. Their trajectory equations were discussed in Section 7-1.

7-4 Velocity Field Model of Single Particle Kinematics

In this section we consider a velocity field model of the kinematics of electromagnetism, in which a family of possible trajectories of a single particle (parametrized by its initial conditions) is taken to be analogous to streamline fluid motion. A *velocity field* is defined for each particle, and the equations of motion are expressed in terms of these kinetic fields. The kinetic field equations are partial differential equations of particularly simple form. Two-particle interaction becomes an interaction between the two-velocity fields.

Charged particle dynamics deals with the effect of electromagnetic fields on the particle trajectory. In four dimensions a given trajectory is parametrized by a set of eight quantities: the initial position and velocity components. A set of trajectories forms a family of curves which are parametrized by these eight quantities, and whose curvature at each point is determined by interaction with the electromagnetic field. An individual particle follows a trajectory determined by the equation of motion:

$$\frac{dU^j}{ds} = \frac{q}{mc^2} F^{jk} U_k, \qquad (7\text{-}4.1)$$

and its initial position and velocity; each initial state determines a unique trajectory. We are interested here in choosing families of such trajectories that may be parametrized by smooth functions of the initial conditions. In particular, we will choose all the initial positions to lie on some smooth spacelike initial three-dimensional hypersurface (e.g., the hypersurface $t = 0$). Then, since the initial hypersurface is spacelike, the initial velocity may be chosen to be a smooth function of the coordinates of the initial hypersurface, with a component orthogonal to the hypersurface. For such a family of trajectories, the velocity is defined at each point in four-dimensional space–time, parametrized by its initial value on the initial three-dimensional hypersurface. The velocity becomes a field analogous to the velocity field defined in fluid dynamics: each trajectory becomes a streamline to which the velocity at each point is tangent, and the initial conditions become boundary conditions (a specification of the velocity field on the initial spacelike hypersurface). For this reason velocity field models may be referred to as *fluid models*. They are also called *dust models*. The idea of a velocity field is not new since we used it in the current density, $J^k = \rho U^k$. It is thus reasonable to use it in an equation of motion.

Symbolically, we write the transformation from a single trajectory, which is a function of the pathlength parameter s to a velocity field which is a function of four-dimensional space–time:

$$U^k(s) \rightarrow U^k(txyz). \qquad (7\text{-}4.2)$$

In this transformation, the velocity field has no explicit dependence on the parameter s which survives in the fluid model only as the parameter of motion along some particular streamline. As a practical matter, we assume the initial velocity function to be smooth; however, we cannot a priori rule out the possibility that some choices of the velocity boundary function might result in trajectories crossing at certain points. This crossing would imply a multivalued velocity field at those points; this matter will be discussed later, after obtaining the field equations of motion. Finally, the fluid theory implies a density function; normally a Dirac delta function would be used, but for some applications a uniform density might be useful.

Rewriting the equations of motion in terms of the velocity field requires several mathematical adjustments. In the first place, the equation of motion for a particle (7-4.1) contains the *total* derivative of the velocity with respect to the pathlength parameter. We must consider how to express the total derivative in terms of partial derivatives. In general, the total derivative of a function $f(txyz)$ is the directional derivative, that is, the rate of change of $f(txyz)$ along some arbitrary path for which we write:

$$\frac{df}{ds} = \frac{dx_k}{ds}\frac{\partial f}{\partial x_k} = n_k \frac{\partial f}{\partial x_k}. \tag{7-4.3}$$

The vector n_k is tangent to the path; it acquires physical interpretation by relating it to the velocity of an imaginary observer moving along the path of measurement. That is:

$$n_k = \frac{dx_k}{ds} = \frac{1}{c} U_k. \tag{7-4.4}$$

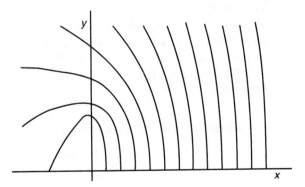

FIGURE 7-6. VELOCITY FIELD AND INITIAL BOUNDARY SURFACE. This represents the two dimensional velocity field $U^j(x, y)$ associated with charged particle scattering. The boundary "surface" is the line $x > 0$. The velocity is tangent to the trajectory at each point.

Then the total derivative of the field depends on the observer's motion and the gradient of the field along the path:

$$\frac{df}{ds} = \frac{1}{c} U_k \, \partial^k f. \tag{7-4.5}$$

An example of this occurs for an airplane flying through the air to measure its temperature (which varies from point to point). The rate of change of temperature seen by the airplane depends on its velocity and on the gradient of the temperature field.

The local differential properties of a field can be expressed by its gradient. For a vector field V^j, the derivative dV^j/ds is that part of the gradient $\partial^k V^j$ that lies parallel to the vector n_k. We call $n_k \partial^k V^j$ the longitudinal part of the gradient; the transverse part is found by subtracting the longitudinal part from the entire gradient. The gradient of a vector field can also be expressed as the sum of a symmetric term and an antisymmetric term; we show that these are identified with the longitudinal and transverse parts. Assume for now that the magnitude of the velocity is not necessarily constant; its directional derivative is:

$$\frac{dU^j}{ds} = \frac{dx_k}{ds} \partial^k U^j = [E^{jk} + R^{jk}]U_k, \tag{7-4.6}$$

where the symmetric and antisymmetric matrices E^{jk} and R^{jk}:

$$R^{jk} = \frac{1}{2c} \left(\partial^k U^j - \partial^j U^k \right), \tag{7-4.7}$$

$$E^{jk} = \frac{1}{2c} \left(\partial^k U^j + \partial^j U^k \right) \tag{7-4.8}$$

are linearly independent. They have simple physical interpretations. The antisymmetric tensor $2cR^{jk}$ is the curl of the velocity, and the trace of the symmetric tensor, $2cE^j_j = \partial_j U^j$, is the divergence of the velocity field. Thus they represent local rotation and stretching of the velocity, respectively. To see explicitly that R^{jk} corresponds to the transverse part, and E^{jk} corresponds to the longitudinal part of the gradient, let R^{jk} or E^{jk} vanish:

1. When the symmetric tensor vanishes:

$$E^{jk} = \frac{1}{2c} \left(\partial^k U^j + \partial^j U^k \right) = 0 \tag{7-4.9}$$

the change in the velocity is pure rotation:

$$\frac{dU^j}{ds} = R^{jk}U_k \tag{7-4.10}$$

since by the antisymmetry of R^{jk} and the symmetry of $U_j U_k$, the velocity is orthogonal (transverse) to its derivative:

$$U_j \frac{dU^j}{ds} = R^{jk} U_j U_k = 0. \qquad (7\text{-}4.11)$$

Therefore the velocity has constant magnitude:

$$U_j U^j = c^2. \qquad (7\text{-}4.12)$$

Thus R^{jk} is the transverse part of the gradient; then by subtraction E^{jk} is the longitudinal part. Because R^{jk} is antisymmetric, equation (7-4.10) is equivalent to the Lorentz force equation (7-4.1).

2. When the curl vanishes:

$$R^{jk} = \frac{1}{2c} \left(\partial^k U^j - \partial^j U^k \right) = 0 \qquad (7\text{-}4.13)$$

the change in the velocity is given by the equation:

$$\frac{dU^j}{ds} = E^{jk} U_k \qquad (7\text{-}4.14)$$

where the velocity is not orthogonal to its derivative:

$$U_j \frac{dU^j}{ds} = E^{jk} U_j U_k \neq 0. \qquad (7\text{-}4.15)$$

Thus E^{jk} represents the change of the velocity field in the direction of the velocity, producing a change in its magnitude, and is consistent with E^{jk} containing the divergence $\partial_j U^j$.

Unless E^{jk} vanishes, the magnitude of the velocity is not constant, but depends on position. But the 4-velocity must have constant magnitude to be invariant under Lorentz transformations. Thus E^{jk} must vanish for electrodynamics to be relativistically invariant.

Now we can find the equation of motion of a charged particle in terms of the velocity field. Applying the differential operator (7-4.3) to the equation of motion (7-4.1) gives:

$$U_k \left[\partial^k U^j - \frac{q}{mc} F^{jk} \right] = 0. \qquad (7\text{-}4.16)$$

By writing $\partial^k U^j$ as a sum of symmetric and antisymmetric terms, this becomes:

$$U_k \left[\frac{1}{2} \left[\partial^k U^j - \partial^j U^k \right] - \frac{q}{mc} F^{jk} \right] = \frac{1}{2} \left[\partial^k U^j + \partial^j U^k \right] U_k,$$

$$= 0 \qquad (7\text{-}4.17)$$

where the right side must vanish by equation (7-4.9) (relativistic invariance). The general solution of this equation is:

$$\frac{1}{2}\left[\partial^k U^j - \partial^j U^k\right] = \frac{q}{mc} F^{jk} + Q^{jk},$$
(7-4.18)

where Q^{jk} is an antisymmetric tensor functional, orthogonal to the velocity:

$$Q^{jk} U_k = 0.$$
(7-4.19)

Since the Lorentz equation is first order in the velocity and linear, equation (7-4.18) must retain these properties. Consider how that restricts the form of Q^{jk}. It can maintain its orthogonality to the velocity only by being dependent, in some way, on the velocity (i.e., as U_k rotates, it must correspondingly transform) or by vanishing. That is, Q^{jk} is a tensor functional, a function of U_k and its derivatives. But, as remarked above, it must not be non-linear in the velocity or its derivatives, or have higher than first-order derivatives of the velocity. Since the linear, antisymmetric first derivative expression already occurs in equation (7-4.18), the expression Q^{jk} is redundant. The best choice is zero.

In the following we let Q^{jk} vanish, subject to the test of consistency with known solutions. Therefore, we have the *kinetic field equations*:

$$\partial^k U^j - \partial^j U^k = \frac{2q}{mc} F^{jk},$$
(7-4.20a)

$$\partial^k U^j + \partial^j U^k = 0$$
(7-4.20b)

consistent with equations (7-4.9) and (7-4.10). The kinetic field equations are relativistically covariant, linear, and first order in U_k as desired. They are also *sufficient conditions* for the Lorentz equation because multiplying equation (7-4.20a) by U_k and using equation (7-4.20b) gives equation (7-4.1).

Let us express the electromagnetic field in terms of the vector potentials:

$$F^{jk} = \partial^j A^k - \partial^k A^j$$
(7-4.21)

and assume that the vector potentials obey the Lorentz gauge:

$$\partial_k A^k = 0.$$
(7-4.22)

Then the kinetic field equation becomes:

$$\partial^k\left[\frac{mc}{2} U^j + qA^j\right] - \partial^j\left[\frac{mc}{2} U^k + qA^k\right] = 0.$$
(7-4.23)

Now define a quantity with units of momentum:

$$G^k \equiv \frac{m}{2} U^k + \frac{q}{c} A^k,$$
(7-4.24)

which is similar to the canonical momentum, but differs by the factor of 1/2. In terms of this quantity, equation (7-4.23) may be expressed by:

$$\partial^k G^j - \partial^j G^k = 0. \tag{7-4.25}$$

Its solution is expressed in terms of an (arbitrary) gauge function, Γ:

$$G^k = \partial^k \Gamma. \tag{7-4.26}$$

Since the vector potential A^k is itself invariant under a gauge transformation, we can eliminate the gauge function Γ by absorbing it into the gauge function for A^k. Let us require both A^k and $A^{k'}$ to satisfy the Lorentz gauge condition. Then the Lorentz gauge condition on the transformed potential:

$$\partial_k A^{k'} = \partial_k A^k - \partial_k \partial^k \Gamma = 0 \tag{7-4.27}$$

requires the gauge function to satisfy the homogeneous wave equation:

$$\partial_k \partial^k \Gamma = 0. \tag{7-4.28}$$

Consequently, the divergence of G^k vanishes:

$$\partial_k G^k = \partial_k \partial^k \Gamma = 0. \tag{7-4.29}$$

A consequence of this and the Lorentz gauge condition on A^k requires the divergence of the velocity field to vanish:

$$\partial_k U^k = \frac{2}{m} \partial_k \partial^k \Gamma$$
$$= 0 \tag{7-4.30}$$

consistent with equation (7-4.9). The physical interpretation is that the lines of velocity flux (i.e., trajectories or streamlines) only begin or end at infinity. This is similar behavior to the lines of flux in charge conservation. This result is not unexpected since it is associated with conservation of particles and therefore with conservation of charge and mass, at least for a nonquantum theory. From conservation of charge and equation (7-4.30), we get a condition on the particle density:

$$\partial_k J^k = \partial_k [\rho U^k] = \rho \partial_k U^k + U^k \partial_k \rho$$
$$= c \frac{d\rho}{ds} = 0. \tag{7-4.31}$$

That is, the particle density is constant along each streamline, although it may have an arbitrary function $\rho_0(xyzt)$ on the initial hypersurface. Introduction of a particle density is characteristic of fluid models. Note that we are considering the kinematics of just a single particle, not a collection of particles. For a classical particle, one would normally assume the particle density to be a Dirac delta function. However, for some purposes (e.g., scattering) it might useful to choose the particle density to be constant on the initial hypersurface. Then equation (7-4.31) determines it downstream.

We can find additional physical and pictorial interpretations to G^k and Γ. The preceding discussion suggests an analogy with electrostatic fields and potentials:

$$\mathbf{E} = -\nabla\varphi \qquad \longleftrightarrow \qquad G^k = \partial^k\Gamma \tag{7-4.32}$$

$$\nabla \times \mathbf{E} = 0 \qquad \longleftrightarrow \qquad \partial^k G^j - \partial^j G^k = 0, \tag{7-4.33}$$

$$\phi = -\int \mathbf{E} \cdot d\mathbf{r} \qquad \longleftrightarrow \qquad \Gamma = \int G^k dx_k. \tag{7-4.34}$$

From these equations and arguments similar to electrostatics in Section 2-2, Γ is a single-valued function of the coordinates. Specifically, since it satisfies equation (7-4.28):

$$\Gamma = F\left(g_n x^n\right),$$
$$g_n g^n = 0, \tag{7-4.35}$$

where the g_n are constants and F is an arbitrary function.

EXAMPLE

In many applications, it is convenient to choose:

$$\Gamma = g_n x^n.$$
$$G^n = g^n. \tag{7-4.36}$$

Thus the components of G^n are conserved.[1] This conservation is useful for scattering problems because it allows us to write:

$$U^k = U_0^k - \frac{2q}{mc}A^k, \tag{7-4.37}$$

where U_0^k is the value on a distant initial hypersurface where A^k vanishes.

In analogy to electrostatics, we can define lines of flux for G^k and equipotential hypersurfaces for Γ. Like electrostatics, the lines of G-flux are everywhere orthogonal to the equipotential surfaces, but now in four-dimensions. By starting from one Γ-hypersurface and using the G^k-vector one can generate a higher-valued Γ-hypersurface. The integral in equation (7-4.34) may be expanded out to give:

$$\Gamma = \int G^k dx_k = \int \left[\frac{m}{2} U^k U_k + \frac{q}{c} A^k U_k\right]\frac{ds}{c}. \tag{7-4.38}$$

---

[1] Equations (7-4.25) and (7-4.30) require both the divergence and curl of G^k to vanish, consistent with conservation.

That is, the gauge function is the *action*, and the integrand becomes the Lagrangian L. It is significant that the factor of one-half is required in this Lagrangian to give the equation of motion using least action. A *Lagrangian density* may be obtained by introducing a particle density ρ. Then the integral becomes:

$$\Gamma = \iiiint \left[\frac{m}{2} U^k U_k + \frac{q}{c} A^k U_k \right] \rho \, dV \frac{ds}{c}, \tag{7-4.39}$$

which can be rewritten as an integral over d^4x. Compare it with equation (4-7.14).[2]

EXAMPLE

In the absence of an electromagnetic field, the kinetic field equations are:

$$\partial^k U^j - \partial^j U^k = 0, \tag{7-4.40}$$

$$\partial^k U^j + \partial^j U^k = 0, \tag{7-4.41}$$

whose solution is constant velocity, and the initial surface is any spacelike hyperplane.

EXAMPLE

Consider the general solution. Take the kinetic field equation (7-4.20a) and integrate along an arbitrary world-line starting from a point x_0^k, using equation (7-4.20b). This gives:

$$U^j(s) = \int \frac{q}{mc} F^{jk} \, dx_k = \int \frac{q}{mc^2} F^{ij} U_k \, ds, \tag{7-4.42}$$

which is just the integral form of equation (7-4.1):

$$\frac{dU^j}{ds} = \frac{q}{mc^2} F^{jk} U_k \tag{7-4.43}$$

as required.

EXAMPLE

Consider a charged particle moving in a constant uniform magnetic field in the z-direction. Let the electromagnetic potentials be given by:

$$A^j = \left[0, \frac{1}{2} By, -\frac{1}{2} Bx, 0 \right]. \tag{7-4.44}$$

---

[2] From the form of this equation, and by analogy with the Hamiltonian (the generating function for time), it is tempting to call G^k the electric generator. Unfortunately that name is already taken! Goldstein discusses generating functions in Chapter 8.

Then the field becomes:

$$
F^{jk} = \begin{bmatrix} 0 & 0 & 0 & 0 \\ 0 & 0 & -B & 0 \\ 0 & B & 0 & 0 \\ 0 & 0 & 0 & 0 \end{bmatrix}.
\tag{7-4.45}
$$

We choose the initial surface to be the x–y plane at $t = 0$. The solution is:

$$
U^j = \frac{q}{mc} \int F^{jk}\, dx_k = \left[U_0^t,\ -\frac{qB}{mc}\, y,\ \frac{qB}{mc}\, x,\ U_0^z \right].
\tag{7-4.46}
$$

That is, the zeroth and third velocity components are constant and the world-lines are concentric helices which rotate in the x–y plane with the cyclotron frequency:

$$
\omega = \frac{qB}{mc}
\tag{7-4.47}
$$

in agreement with standard theory. Alternatively, we could have applied equation (7-4.37) directly. The potentials in (7-4.44) give:

$$
U^j = \begin{bmatrix} \text{constant} \\[4pt] -\dfrac{2q}{mc}\left[\dfrac{1}{2} By \right] \\[8pt] -\dfrac{2q}{mc}\left[-\dfrac{1}{2} Bx \right] \\[8pt] \text{constant} \end{bmatrix} = \begin{bmatrix} U_0^t \\[4pt] -\omega y \\[6pt] \omega x \\[6pt] U_0^z \end{bmatrix}
\tag{7-4.48}
$$

in agreement with the above. Note that the factor of one-half cancels.

The kinetic field equations are a convenient method for treating the scattering of charged particles. We have seen that the velocity field is a linear function of the vector potential:

$$
\frac{m}{2}\, U^k = -\frac{q}{c}\, A^k + \partial^k \Gamma.
\tag{7-4.49}
$$

Applying this, and equation (4-5.17), to the Maxwell field equations:

$$
\partial_j F^{jk} = \frac{4\pi}{c}\, J^k
\tag{7-4.50}
$$

gives a field equation for the velocity:

$$
\partial^j \partial_j U^k = -\frac{2q}{mc}\, \partial^j \partial_j A^k = -\frac{8\pi q}{mc^2}\, J^k.
\tag{7-4.51}
$$

The charge q is the charge of the moving particle, and J^k is an external current source, i.e., velocities associated with J^k do not include the velocity U^k of the particle itself. In other words, we do not consider self-interactions.

Consider the interaction (scattering) of two charged particles. We express J^k in terms of the particle density ρ and write:

$$J^k = q\rho U^k. \tag{7-4.52}$$

Then, writing an equation for each particle gives a set of coupled equations for the two velocity fields:

$$\partial^k \partial_k U_1^j = -\frac{8\pi q_1 q_2}{m_1 c^2} \rho_2 U_2^j, \tag{7-4.53}$$

$$\partial^k \partial_k U_2^j = -\frac{8\pi q_1 q_2}{m_2 c^2} \rho_1 U_1^j, \tag{7-4.54}$$

each subject to the auxiliary conditions on its velocity field.

By using the retarded Green function (see equation (6-6.19)), these equations can be expressed as integral equations:

$$U_1^j(z) = U_{10}^j(z) - \varepsilon_1 \int G(z, x) \rho_2(x) U_2^j(x)\, d^4x, \tag{7-4.55a}$$

$$U_2^j(z) = U_{20}^j(z) - \varepsilon_2 \int G(z, x) \rho_1(x) U_1^j(x)\, d^4x, \tag{7-4.55b}$$

with auxiliary conditions:

$$U^k U_k = c^2, \tag{7-4.55c}$$

$$\partial_k U^k = 0. \tag{7-4.55d}$$

The first terms, U_{10}^j and U_{20}^j, are the homogeneous solutions and physically represent the velocity on the initial hypersurface. The integrals are the particular solutions and are, in fact, the usual expressions for the potential, A^j. Compare equations (7-4.55) with equation (7-4.37). These integral equations are coupled Fredholm equations, which are treated extensively in mathematical physics texts, such as Mathews and Walker. The *coupling constant* is:

$$\varepsilon_n = 8\pi \frac{q_1 q_2}{m_n c^2} = 8\pi R_C, \qquad n = 1, 2, \tag{7-4.56}$$

where n is the particle index and R_C has units of length. For electron–electron scattering, it is the classical electron radius, $R_C = 2.82 \times 10^{-13}$ cm, discussed in Section 6-5.

The discussion shows that it is possible to entirely *eliminate the electromagnetic field from the interaction between charged particles*, expressing it solely in terms of the velocity fields. For classical particles, the densities are delta functions at the particles' positions and the Green function is the retarded form, discussed previously. The expression suggests a ready conversion to a quantized form where the densities are given in terms of the state functions and the velocity becomes an operator.

Equations (7-4.55) can be solved by standard methods such as infinite series. See, for example, Mathews and Walker, Section 11-3. Cross substitution of equations (7-4.55a,b) gives the infinite series solution:

$$U^j_1 = U^j_{10}(z) - \varepsilon_1 \int_{z>x} \rho(x)G(z,x)U^j_{20}(x)\,d^4x$$

$$+ \varepsilon_1\varepsilon_2 \iint_{z>x'>x} \rho(x')\rho(x)G(z,x')G(x',x)U^j_{10}(x)\,d^4x'\,d^4x$$

$$+ \cdots, \tag{7-4.57a}$$

$$U^j_2 = U^j_{20}(z) - \varepsilon_2 \int_{z>x} \rho(x)G(z,x)U^j_{10}(x)\,d^4x$$

$$+ \varepsilon_1\varepsilon_2 \iint_{z>x'>x} \rho(x')\rho(x)G(z,x')G(x',x)U^j_{20}(x)\,d^4x'\,d^4x$$

$$+ \cdots. \tag{7-4.57b}$$

Each equation can be separated into two terms, one containing only U^j_{10}, and the other containing only U^j_{20}. The limits on the integrals refer to time ordering. For example, in the second terms we require $t' > t$. This condition ensures causality when both t' and t are integrated. Since the coupling constant ε is much smaller than 1, the series may normally be truncated after a few terms. However, note that at small distances the Green function blows up and the integrals of all orders diverge. Consequently, it is necessary to limit the calculation to distances larger than R_C by applying a cutoff to the integrals. Similar short-range divergences occur regardless of how the interactions are calculated, including quantized treatments.

By expressing the scattering of two particles in terms of a *scattering matrix*, equations (7-4.57a,b) may be written in compact form:

$$\begin{bmatrix} U^j_1(z) \\ U^j_2(z) \end{bmatrix} = \int_{z>x} M^{mn}(z,x) \begin{bmatrix} U^j_{10}(x) \\ U^j_{20}(x) \end{bmatrix} d^4x. \tag{7-4.58}$$

That is, the scattering matrix transforms the initial velocities into the velocities after scattering. In this equation, the vector contains eight elements, four components for each of two particle velocities. The scattering matrix has four indices, two discrete particle indices, m and n, and two continuous variables, x and z. Then the matrix operation on the vector implies an integration over x as well as the summation over *n*. To third order the scattering matrix is:

$$
M^{mn} = \begin{bmatrix} \delta^4(z-x) & 0 \\ 0 & \delta^4(z-x) \end{bmatrix}
$$

$$
+ \begin{bmatrix} \varepsilon^2 \int G(z,x')\rho(x')G(x',x)\rho(x)d^4x' + \cdots & \varepsilon G(z,x)\rho(x) + \varepsilon^3 \iint + \cdots \\ \varepsilon G(z,x)\rho(x) + \varepsilon^3 \iint + \cdots & \varepsilon^2 \int G(z,x')\rho(x')G(x',x)\rho(x)d^4x' + \cdots \end{bmatrix}.
$$

$$(7\text{-}4.59)$$

The first term produces the initial conditions (undeflected beam), while second term contains the interaction to all orders in ε. It is possible to calculate scattering cross sections directly in terms of a scattering matrix, especially in quantum electrodynamics. Because the coupling constant is so small, a small number of terms is sufficient at available energies.

The solution can be represented graphically. Equations (7-4.57) are represented by the *ladder diagrams* in Figure 7-7. Each vertical line represents a velocity and the short horizontal dashed lines represent the Green function between vertices. Each diagram has a corresponding term in equations (7-4.57), where the number of dashed lines corresponds to the order (in ε) of the term. That is, the first diagram has zero dashed lines and corresponds to the zeroth-order term, the second diagram has one dashed line and corresponds to the first-order term, and so on. In this classical theory, the number of particles (i.e., the number of solid lines in each diagram) is conserved. By contrast, particle creation occurs in quantum electrodynamics, where the analogous diagrams, called *Feynman diagrams*, are much more complex, and there are larger numbers of diagrams at each order.

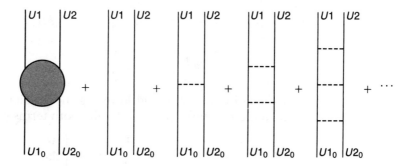

FIGURE 7-7. GRAPHICAL REPRESENTATION OF SOLUTION OF INTEGRAL EQUATIONS OF MOTION. The total interaction is a sum of terms corresponding to the "ladder diagrams." They are reminiscent of Feynman diagrams in quantum electrodynamics.

Consider the conditions for which the velocity field is well defined in the sense that it is a single-valued function of space and time. In general, an arbitrary choice of the velocity function on the initial hypersurface may lead to crossing world-lines downstream and therefore multiple-valued velocity. However, it is possible to form a set of self-consistent boundary conditions as follows: (1) We hypothesize that the vector potential A^k is known and is single valued everywhere in the region of interest. (2) Choose a scalar function Γ with single-valued derivatives everywhere. (3) Choose an appropriate spacelike initial hypersurface. (4) Define, on this hypersurface, the initial velocity U^k consistent with equation (7-4.20).

$$\frac{mc}{2} U^k|_0 \equiv -qA^k|_0 + \partial^k \Gamma|_0. \tag{7-4.60}$$

This choice of the initial velocity is a single-valued solution to the kinetic field equations (7-4.20) everywhere on the initial hypersurface. The solutions are also single valued downstream since we specified that A^k and $\partial^k \Gamma$ be single valued *everywhere* in the region of interest. This self-consistent choice of initial velocity will not *force* multivalued solutions as an arbitrary choice might. In spite of the restriction on the initial velocity, it is still possible to choose a wide range of values for it, because of the arbitrariness of the gauge function. Certainly any individual world-line can be chosen to have essentially any initial position and initial velocity.

In Section 6-5 we investigated the electromagnetic stress tensor T^{jk} formed from the electric and magnetic fields E^j and B^k. That is, the spatial components are:

$$T^{jk} = E^j E^k - \frac{1}{2} \delta^{ik} E^n E_n + B^j B^k - \frac{1}{2} \delta^{jk} B^n B_n. \tag{7-4.61}$$

Our preceding discussion suggests that a similar quantity might be formed from quantities like $G^j G^k$. Calculating its divergence gives:

$$\partial_j(G^j G^k) = (\partial_j G^j)G^k + G^j(\partial_j G^k) = \frac{1}{2} \partial^k(G^j G_j)$$

$$= \partial_j \left(\frac{1}{2} \mu^{jk} G^n G_n \right), \tag{7-4.62}$$

by using equation (7-4.25), and lowering the gradient index with the Minkowski metric tensor μ^{jk}. Combining both terms gives:

$$\partial_j \left[G^j G^k - \frac{1}{2} \mu^{jk} G^n G_n \right] = 0. \tag{7-4.63}$$

We define the second rank tensor:

$$S^{jk} = G^j G^k - \frac{1}{2} \mu^{jk} G^n G_n, \tag{7-4.64}$$

resembling the electromagnetic stress tensor. Then we have:

$$\partial_j S^{jk} = 0, \tag{7-4.65}$$

which is a conservation theorem, analogous to conservation of charge, but with four components.

In the limit where A^k vanishes, the S-tensor is proportional to the free-particle stress tensor, i.e.:

$$S^{jk} \approx \frac{\rho}{2} U^j U^k = T^{jk}_{\text{free}} \tag{7-4.66}$$

where ρ is the mass density. In this limit, equation (7-4.65) implies a constant 4-momentum. When A^k does not vanish, equation (7-4.65) is an equation of motion—this is hardly surprising since the equation of motion went into its making.

7-5 CROSS SECTIONS

The most useful quantities in discussing scattering are the *cross sections* for the various interactions. The cross section for some interaction is defined by the following ideal experiment: Suppose that an incident flux of particles, I_0, interacts with a small object we will call the absorber. The absorber has an area projected into the direction of the incident flux. The projected area is the cross section. If the incident particle hits the absorber it is absorbed by it; if it does not hit the absorber, it passes by unaffected. The rate of absorption, the number of particles absorbed per unit time, of the particle beam is:

$$R = \frac{dN}{dt} \tag{7-5.1}$$

and the incident flux (analogous to current density) is:

$$I_0 = \frac{d^2 N}{dA\,dt} = \rho v_0, \tag{7-5.2}$$

where ρ is the particle density. Then we define the cross section σ by the equation:

$$R = \sigma I_0. \tag{7-5.3}$$

In each case, differential cross sections can be defined for parameters of the interaction such as, for example, energy or scattering angle. Then the total cross section is the integral over all these parameters.

In the preceding example, the absorption cross section is identified with the geometry of the absorber. This simple interpretation does not work well, however, if the cross section is a function of the parameters of the measurement. For example, the cross section may be a function of energy. In general, a cross section is a measure of a certain interaction rate; different interactions will have different cross sections. Thus, the scattering cross section is the total number of particles scattered per second divided by the incident flux. In general, the scattering cross section will not equal the absorption cross section, even at the same energy. In this case, we still *define* the cross sections by equation (7-5.3). The cross section remains an important measure of the interaction, but is not necessarily related to size in any simple way.

Consider the differential cross section for scattering into the solid angle $d\Omega$:

$$\frac{d\sigma}{d\Omega} = f(\theta, \phi). \tag{7-5.4}$$

The scattering angle θ is the asymptotic angle of deviation for the outgoing flux measured from the direction of the incident flux and ϕ is the azimuthal angle. Cylindrical coordinates (r, ϕ, z) are the most natural for scattering with azimuthal symmetry and incident flux in the z-direction. For an incident beam of particles, the *impact parameter* r_0 is its asymptotic r-value at large distances upstream from the scatterer. Since the scattering takes place in a plane, the calculation reduces mainly to finding the relation between the impact parameter and the scattering angle. See Figures 7-8 and 7-9.

A general relation between the cross section and r_0 and θ can be obtained by considering the flux passing through a ring-shaped area of radius r_0 and width dr_0. Thus, in steady state, every particle that passes through the ring, passes through the element of solid angle $d\Omega$ on the side of the cylinder. In this sense, the area of the ring dA_0 equals the differential cross section for that angle:

$$dA_0 = [r_0 \, d\phi \, dr_0]$$

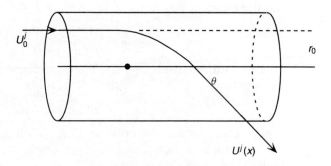

Figure 7-8. Scattering.   Each scattered particle entering the cylinder at some point on the left leaves the cylinder at some corresponding point on the side. There is therefore a functional relation between the scattering angle θ and the impact parameter r_0.

$$= \frac{d\sigma}{d\Omega} d\Omega = -\frac{d\sigma}{d\Omega} \sin\theta \, d\theta \, d\phi. \tag{7-5.5}$$

Then, in terms of the impact parameter and scattering angle, the differential cross section is:

$$\frac{d\sigma}{d\Omega} = -\frac{r_0}{\sin\theta} \frac{dr_0}{d\theta}. \tag{7-5.6}$$

One needs to find the function:

$$r_0 = r_0(\theta) \tag{7-5.7}$$

from the equations of motion of the scattered particle beam.

EXAMPLE

Attractive Coulomb potential at small velocity and small scattering angles. In this case, the force on the scattered particle is:

$$F = \frac{q_1 q_2}{r^2}. \tag{7-5.8}$$

The exact orbit function requires lengthy calculations. We can get some quick results for small deflections by making some approximations: (1) The force is applied during a short time interval, Δt, as the particle passes the scatterer at some average distance r_{avg}. (2) The longitudinal velocity is unchanged (i.e., assume conservation of kinetic energy for the particle). We make the approximations:

$$r_{avg} \approx \frac{r_0}{2},$$

$$\Delta t \approx \frac{r_{avg}}{v_0}, \tag{7-5.9}$$

$$p_z \approx mv_0 = \text{constant},$$

$$p_{r0} \approx 0.$$

Then the momentum changes are:

$$\Delta p_z \approx 0, \tag{7-5.10}$$

$$\Delta p_r = \frac{q_1 q_2}{r_{avg}^2} \Delta t = \frac{q_1 q_2}{r_{avg} v_0}. \tag{7-5.11}$$

Then the tangent of the scattering angle is the ratio of transverse to longitudinal momentum:

$$\tan\theta = \frac{p_r}{p_z} = \frac{\Delta p_r}{mv_0}$$

$$= \frac{q_1 q_2}{T r_0}, \tag{7-5.12}$$

where T is the initial kinetic energy. The exact expression is obtained by solving the equations of motion. This is done in Goldstein, for example, giving:

$$\tan \frac{\theta}{2} = \frac{q_1 q_2}{2 T r_0}, \tag{7-5.13}$$

which agrees with equation (7-5.12) for small deflections, where $\tan \theta = \theta$. Substitution into equation (7-5.6), gives the well-known *Rutherford scattering cross section*:

$$\frac{d\sigma}{d\Omega} = \frac{1}{4} \left[\frac{q_1 q_2}{2T} \right]^2 \frac{1}{\sin^4(\theta/2)}. \tag{7-5.14}$$

The total scattering cross section, obtained by integrating over all angles, is infinite. This is a result of the fact that the Coulomb potential remains comparatively large at large distances, and therefore scattering into small angles is large. Atomic scattering is finite because of the screening effect of the electrons.

The cross section can easily be calculated from the velocity fields (in cylindrical coordinates), $U^j(r\phi zt)$. The incident particle flux is along the z-axis of the cylinder. In the diagram, dA_0 represents the flat area of the left end of the cylinder, while dA_r represents the area of the curved side of the cylinder. A particle striking a particular patch of the cross section dA_0 will be scattered into a corresponding patch of area dA_r on the side of the cylinder.

The rate at which incident particles are scattered is the incident flux times the cross section:

$$dR_0 = I_z \, dA_z = I_0 \, d\sigma, \tag{7-5.15}$$

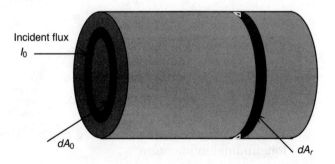

Incident flux
I_0

dA_0

dA_r

Figure 7-9. Details of flux and area elements.   The scattering particles, represented by the flux I_0 enter the end of the cylinder through area $dA_z = dA_0$. All of these particles will then leave the cylinder through the lateral area dA_r. The scattering center is located on the axis of symmetry $r = 0$.

and the rate at which particles leave the cylinder is the product of the radial component of the flux times the element of area of the cylinder:

$$dR_r = I_r \, dA_r. \tag{7-5.16}$$

In the steady state the two rates are equal, $dR_r = dR_0$ (the particles are all scattered through some angle, no matter how small). Then:

$$\frac{d\sigma}{dA_r} = \frac{I_r}{I_0} = \frac{\rho U_r}{\rho_0 U_0}. \tag{7-5.17}$$

But, as we saw in equation (7-4.31), the particle density is constant, i.e., $\rho = \rho_0$. Therefore, the differential scattering cross section can be expressed as the ratio of the radial to incident components of the velocity:

$$\frac{d\sigma}{dA_r} = \frac{U_r}{U_0}. \tag{7-5.18}$$

The differential cross section is more commonly given in terms of the solid angle:

$$\frac{d\sigma}{d\Omega} = \frac{d\sigma}{dA_r}\frac{dA_r}{d\Omega} = \frac{U_r}{U_0}\frac{dA_r}{d\Omega}, \tag{7-5.19}$$

where $dA_r/d\Omega$ is the lateral area of the cylinder subtended per unit solid angle.

7-6   SELECTED BIBLIOGRAPHY

H. Goldstein, *Classical Mechanics*, Addison-Wesley, 1953.

D. Heddle, *Electrostatic Lens Systems*, Adam Hilger, 1991.

E. Konopinski, "What the Electromagnetic Vector Potential Describes," Amer. J. Phys. **46**, (1978), 499.

L. Landau and E. Lifshitz, *The Classical Theory of Fields*, 4th ed., Addison-Wesley, 1975.

J. Mathews and R. Walker, *Mathematical Methods of Physics*, 2nd ed., Benjamin, 1970.

W. Press et al., *Numerical Recipes*, Cambridge University Press, 1986.

W. Rieder and H. Busby, *Introduction to Engineering Modeling*, Wiley, 1986.

F. Rohrlich, *Classical Charged Particles*, Addison-Wesley, 1965.

K. Steffen, *High Energy Beam Optics*, Interscience, 1965.

J. Westgard, "A Fluid Model of Single-Particle Kinematics," Foundations of Physics Letters (2), June 1989.

7-7   PROBLEMS

1. Solve and plot the damped harmonic oscillator by the Taylor method.

2. Solve and plot the damped harmonic oscillator by the Euler method.

3. Write a computer program to compute and plot the trajectory of a particle subject to a Coulomb force using:

   A. Euler method.
   B. Fourth-order Runge–Kutta method.
   C. Predictor–corrector method using Taylor series as the predictor.

   Compare the three methods. Do they give closed orbits?

4. Use the library routines in *Mathematica* for the Runge–Kutta solution of differential equations to compute the trajectory of a particle subject to a Coulomb force.

5. Write a program to calculate and plot the trajectory of a particle in electric and magnetic fields which are uniform but with arbitrary strength and direction.

6. *Project*. Write computer subroutines to calculate and plot the path of a particle through:

   A. field-free regions with length as parameter;
   B. bending magnets with length and strength as parameters;
   C. quadrupole magnets with length and strength as parameters; and
   D. rectangular slits with height and width as parameters.

   Using these routines, plot the beam profile for the particle optical system shown in Figure 7-4.

APPENDIX 7-1   FINITE-ELEMENT SOLUTION OF A DIFFERENTIAL EQUATION USING A SPREADSHEET

The relaxation solution to a differential equation with Cauchy boundary conditions can be solved in simple cases as a spreadsheet calculation. As an example we use the harmonic oscillator since its solutions are well known, and we can easily compare the results of the numerical calculation with the known results. Although the spreadsheet software is relatively slow to execute, entering the data and formulas is extremely easy.

Cauchy Boundary Conditions

We take the harmonic oscillator as our example. For Cauchy boundary conditions the procedure includes the following steps:

Step 1. In the spreadsheet choose a column of desired length and fill its first cell with the initial value $y(0)$, and the second cell with the value:

$$y(1) = \left[\frac{dy}{dt}\right]_0 dt + y(0).$$

Step 2. Fill the rest of the cells with the recursion formula:

$$y(n) = (2 - w^2 dt^2)y(n-1) - y(n-2).$$

For example, if we put the value of $(w\,dt)^2$ in cell D37, then cells B15 and B16 in column B should contain the formulas:

```
B15:  = (2-$D$37)*B14-B13,
B16:  = (2-$D$37)*B15-B14.
```

Some spreadsheets allow formulas to be copied down and automatically adjust the cell addresses. The expression D37 ensures that this cell address is absolute, i.e., does not change when the formula is copied.

Step 3. The spreadsheet will calculate the numerical values of $y(n)$. The calculation can be either automatic or manual.

Step 4. The $y(n)$ can then be plotted by the spreadsheet.

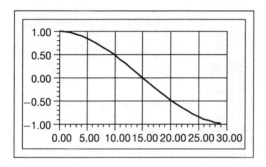

Dirichlet Boundary Conditions

The finite-element method can also be used for Dirichlet boundary conditions. For Dirichlet boundary conditions the procedure becomes:

Step 1. Put the spreadsheet into MANUAL CALCULATION.

Step 2. In the spreadsheet choose a column of desired length and fill its first cell with the initial value $y(0)$, and the last cell with the value $y(N)$.

Step 3. At this point, the spreadsheet must be in MANUAL CALCULATION mode. Fill the intermediate cells with the recursion formula:

$$y(n) = \frac{[y(n+1) + y(n-1)]}{[2 - (w\,dt)^2]}.$$

Note that this formula differs from the Cauchy formula, although it is based on the same recursion relation. For example cell B15 contains:

```
B15:  = (B16 + B14) / (2-(w*dt)^2).
```

Table . Numerical Solution to Harmonic Oscillator.

	A	B	C	D
8	n	y(n)		Comments
9				
10	0	1.000		initial value
11	1	0.988		calculated from initial value and initial slope
12	2	0.966		
13	3	0.935		
14	4	0.894		
15	5	0.844		
16	6	0.786		
17	7	0.719		
18	8	0.646		
19	9	0.566		
20	10	0.481		
21	11	0.391		
22	12	0.297		Cauchy Recursion formula
23	13	0.199		
24	14	0.100		$y(n) = (2 - (wdt)^2)) * y(n-1) - y(n-2)$
25	15	0.000		
26	16	-0.100		
27	17	-0.199		Dirichlet Recursion formula
28	18	-0.296		
29	19	-0.390		$y(n) = (y(n+1) + y(n-1))/(2 - (wdt)^2)$
30	20	-0.481		
31	21	-0.566		
32	22	-0.646		
33	23	-0.719		Recursion parameter
34	24	-0.785		
35	25	-0.844		$(wdt)^2$
36	26	-0.894		
37	27	-0.934		0.01
38	28	-0.966		
39	29	-0.988		
40	30	-1.000		

Step 4. Manually recalculate the spreadsheet untill the values no longer change (to the accuracy shown). The necessary number of recalculations may be quite large, perhaps greater than 100—Dirichlet calculations are not as efficient as Cauchy calculations.

Step 5. The $y(n)$ can then be plotted by the spreadsheet.

If we choose: $y(0) = 1.000$ and $y(30) = -1.000$ for the Dirichlet boundary conditions, the solution turns out to be identical to the solution for Cauchy boundary conditions calculated in the preceding example. Thus, the same table of values and plot apply to both the Cauchy and Dirichlet examples.

APPENDIX 7-2 TAYLOR SERIES SOLUTIONS OF DIFFERENTIAL EQUATIONS

The following is a *Mathematica* implementation of the Taylor series solution of differential equations. The first form, DiffEq, solves equations of any order; the equation need not be linear. A differential equation can be written in the general form in terms of the pth-order functional:

$$F[t, y(t), y'(t), y''(t), \ldots] = 0,$$

where p is the highest derivative in F. We find an approximate solution to this equation by expressing it as a Taylor series truncated to order n, in the variable t.

```
Remove["Global`@"]
```

Mathematica Implementation of the Method

Define the Functional and the Parameters

```
    (* order of the equation *)
p = 2;
(* number of terms in the expansion *)
n = 4;
(* point of expansion *)
q = 0;
(* functional of the differential equation *)
func = x''[t]+x[t]

x[t] + x''[t]
```

Define the Procedure for Applying Taylor Series Solution
Create a set of substitution rules for evaluating a Taylor series expansion of the solution of the differential equation.

```
    (* Define Differential Operation *)
ddt = D[#, t]&


    (* Create list of derivatives of "func".      *)
    (* Evaluate them at the point of expansion, q. *)
    (* This list is a set of recursion formulas.   *)
dlist = NestList[ddt, func, n-p]/.t -> q


    (* Create derivative list of variable x(t)     *)
    (* from the p-th to the n-th derivative.       *)
evallist = Table[Derivative[i][x][q], {i, p, n}]


    (* Create substitution rules from derivatives  *)
    (* of func by solving for variables in evallist *)
ruler = Solve[dlist == 0, evallist]



D[#1, t] &
```

```
                              (3)                      (4)
{x[0] + x''[0], x'[0] + x   [0], x''[0] + x   [0]}
```

```
          (3)        (4)
{x''[0], x   [0], x   [0]}
```

```
     (3)                      (4)
{{x   [0] -> -x'[0], x   [0] -> x[0], x''[0] -> -x[0]}}
```

Create a Taylor series and evaluate its coefficients by using the substitution rules obtained above.

```
(* Create a generic Taylor series in x(t) *)
taylor = Normal[Series[x[t], {t, q, n}]]
```

```
(* Evaluate the coefficients using ruler  *)
solution = taylor//.ruler
```

```
(* Substitute initial values into series  *)
xx = solution[[1]]/.{x[0]->1,x'[0]->0}
```

```
                          2            3 (3)       4 (4)
                         t  x''[0]    t  x   [0]   t  x   [0]
x[0] + t x'[0] + --------- + ----------- + ----------
                         2            6           24
```

```
            2         4                    3
           t  x[0]   t  x[0]              t  x'[0]
{x[0] - ------- + ------- + t x'[0] - --------}
            2         24                   6
```

```
      2    4
     t    t
1 - -- + --
     2    24
```

Plot the Solution

```
Plot[{xx,Cos[t]},{t,0,2}]
```

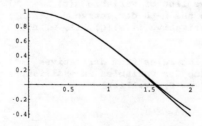

Taylor Series Package

The preceding procedure can be put in the form of a *Mathematica* package.

Define the Package and the Usage

```
DiffEq::usage =
    "DiffEq[F[t,y[t],y'[t],y''[t],...],y[t],n,p,q]
    solves a differential equation of order, p,by
    Taylor series of, n, terms, expanded around
    point,q. Dependent and independent variables
    are y[t] and t, respectively. Default values
    are: n=10,p=2, q=0."
DiffEq[func_, y_[t_], n_:10, p_:2, q_:0] :=
    Block[{ddt,d1,i,evallist,ruler,taylor,soln},
        ddt=D[#,t]&;
        d1=NestList[ddt,func,n-p]/.t->q;
        evallist=Table[Derivative[i][y][q],{i,p,n}];
        ruler=Solve[d1==0,evallist];
        taylor=Normal[Series[y[t],\{t,q,n\}]];
        soln=taylor//.ruler
    ]
```

```
DiffEq[F[t,y[t],y'[t],y''[t],...],y[t],n,p,q] solves a\\
    differential equation of order, p,by Taylor series of,\\
    n, terms, expanded around point,q. Dependent and\\
    independent variables are y[t] and t, respectively.\\
    Default values are: n=10,p=2, q=0.
```

Find the Solution by Using the Package

If not specified, the equation is assumed to be second order, there will be ten terms in the series, and the solution is expanded around the origin $t = 0$, since these are the default values.

```
soln = DiffEq[y''[t]+y[t],y[t]]
```

$$\{y[0] - \frac{t^2 y[0]}{2} + \frac{t^4 y[0]}{24} - \frac{t^6 y[0]}{720} + \frac{t^8 y[0]}{40320} -$$

$$\frac{t^{10} y[0]}{3628800} + t y'[0] - \frac{t^3 y'[0]}{6} + \frac{t^5 y'[0]}{120} - \frac{t^7 y'[0]}{5040} +$$

$$\frac{t^9 y'[0]}{362880}\}$$

This is the expected solution containing both sine and cosine with the initial values.

```
yy = soln[[1]]/.{y[0]->1,y'[0]->0}
```

$$\qquad 2 \qquad 4 \qquad 6 \qquad 8 \qquad \qquad 10$$

```
       t    t     t       t          t
1  -  --  + --  - ---  + -----  -  -------
       2   24    720    40320      3628800
```

Plot the Series and Exact Solution on the Same Axes

```
Plot[{yy, Cos[t]},{t,0,2}]
```

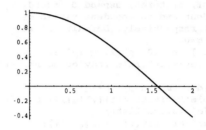

CHAPTER 8

Radiation by Moving Charges

Science is built up with facts as a house is with
stones, but a collection of facts is no more a science than a heap of stones is a house.

Jules Henri Poincaré
La Science et l'Hypothèse

Discovery consists of seeing what everybody has
seen and thinking what nobody has thought.

Albert Szent-Györgyi
The Scientist Speculates

Preceding chapters discussed properties of electromagnetic radiation
in vacuum and dielectrics, without considering the sources of the radiation. In this chapter, we show accelerating electric charge to be the
source of electromagnetic waves, by finding solutions to the inhomogeneous wave equations. Although exact solutions are often difficult,
they are possible in a few important cases. For example, we express the
radiation fields of a single particle undergoing arbitrary acceleration
as a Green function integral. Often it is useful to apply approximation
methods, such as multipole and Fourier series solutions. Not only do
these mathematically powerful methods provide numerical solutions
to specific radiation problems, but provide much physical insight as
well. For example, the Fourier solution gives the frequency distribution of the radiation. Finally, we apply the methods to some important
examples, including radiation from various sources.

8-1   Multipole Expansion

For arbitrary charge and current distributions, it may be impossible to solve the equations in terms of known functions. However, as in Section 5-2, we can try to express the potentials as a multipole expansion. Multipole expansion is useful even when exact integration is possible, since the multipoles carry information about the spatial distribution of the radiating charges and currents. Multipole calculations are not explicitly covariant, since they have a preferred reference frame and treat time unsymmetrically.

First, consider the conditions for the validity of such an expansion. We assume that the size of the charge distribution is small compared with its distance from the field point. Then the (retarded) current density can be expressed as a Taylor series in t:

$$J^k\left(r_s, t_f - \frac{1}{c}|r_{fs}|\right) = J^k\left(r_s, t_f - \frac{1}{c}|r_f|\right) + O\left(\frac{1}{c}|r_s|\,\partial_t J^k\right). \tag{8-1.1}$$

But at sufficiently large distances, the first-order term is much smaller than the zeroth-order term. Thus we write:

$$J^k(r, t) \gg \frac{R_s}{c}\frac{\partial J^k}{\partial t}, \tag{8-1.2}$$

where R_s represents the size of the source. Treating the dependence on time as a Fourier integral, this condition can be expressed in terms of frequency:

$$J^k(r, t) = \frac{1}{2\pi}\int J^k(r, \omega)e^{-i\omega t}d\omega. \tag{8-1.3}$$

Then condition (8-1.2) becomes:

$$J^i(r, \omega) \gg \frac{R_s}{c}\omega J^i(r, \omega). \tag{8-1.4}$$

That is, the wavelength is much larger than the size of the source:

$$\lambda = \frac{2\pi c}{\omega} \gg R_s. \tag{8-1.5}$$

The physical picture associated with this condition is that the period of the radiation is much larger than the time needed for a signal to cross the volume containing the source charges. Consequently, the motions of the source charges can be correlated, and radiation from different parts of the source can have a nonrandom phase relation.

EXAMPLE

Take visible radiation by an atom. In this case:

$$\lambda_{\text{visible}} \sim 10^{-7} \text{ m},$$

$$R_{\text{atom}} \sim 10^{-10} \text{ m},$$

which satisfies the condition quite well.

In the steady-state applications where Poisson's equation applies, the Green function $(1/r)$ is expanded in a Taylor series and truncated after a few terms, giving multipole terms in the expansion. We can use a similar treatment here for radiating fields at large distances, where $r_f \gg r_s$. Thus, since:

$$|\mathbf{r}_f - \mathbf{r}_s| \approx |\mathbf{r}_f| \tag{8-1.6}$$

the potential may be approximated by:[1]

$$A^k \approx \frac{1}{c|\mathbf{r}_f|} \int \left[J^k(r_s, t_s) \right]_{\text{ret}} d^3 r_s. \tag{8-1.7}$$

But since the integrand now depends on time, it now has an implicit dependence on r_f through the time retardation in t_s. It is convenient for our purposes to consider the source current to be a function of two arguments $\mathbf{r}$ and t:

$$J^k = J^k(r_s, t_s) \tag{8-1.8}$$

and expand it as a Taylor series of *the second argument t_s only*. In these terms, the potential becomes:[2]

$$A^k = \frac{1}{c|\mathbf{r}_f|} \int \sum_{n=0}^{\infty} \frac{1}{n!} t_s^n \partial_t^n \left[J^k(\mathbf{r}_s, t_s) \right]_{ct_s = ct_f - r_{fs}} d^3 r_s \tag{8-1.9}$$

showing the time retardation more explicitly. In expanding the integrand we make use of the special properties of the argument. First, let the unit vector $\mathbf{n}$ be:

$$\mathbf{n} = \frac{\mathbf{r}_f - \mathbf{r}_s}{|\mathbf{r}_f - \mathbf{r}_s|} \approx \frac{\mathbf{r}_f}{|\mathbf{r}_f|}. \tag{8-1.10}$$

Now the spatial and time derivatives of the arbitrary function $F(ct - r)$ are related by:

$$\frac{\partial}{\partial x} F(ct - r) = -\frac{\partial r}{\partial x} \frac{\partial F(ct - r)}{\partial r} = -\frac{x}{r} \frac{\partial F(ct - r)}{\partial r} = -\frac{n_x}{c} \frac{\partial}{\partial t} F(ct - r).$$

$$\tag{8-1.11}$$

---

[1] We give a full justification for writing the potential this way in Section 8-5.

[2] The potential can be expanded as a power series of the spatial variable r_f (static multipoles), as well as in time. However, for studying radiation, only the zeroth-order term in r_f is kept.

In three-dimensional vector notation these derivatives can be generalized to include the vector operators gradient, curl and divergence. That is:

$$\nabla \Phi(ct - r) = -\frac{\mathbf{n}}{c}\frac{\partial}{\partial t}\Phi(ct - r), \qquad (8\text{-}1.12)$$

$$\nabla \times \mathbf{F}(ct - r) = -\frac{\mathbf{n}}{c}\frac{\partial}{\partial t} \times \mathbf{F}(ct - r), \qquad (8\text{-}1.13)$$

$$\nabla \cdot \mathbf{F}(ct - r) = -\frac{\mathbf{n}}{c}\frac{\partial}{\partial t} \cdot \mathbf{F}(ct - r). \qquad (8\text{-}1.14)$$

Therefore the potential becomes, after integrating by parts:

$$A^k = \frac{1}{c|\mathbf{r}_f|}\sum_{n=0}^{\infty}\frac{1}{c^n n!}\iiint (-\mathbf{n}\cdot\mathbf{r}_s)^n \left(\partial_t^n J^k(r_s, 0)\right) d^3r_s. \qquad (8\text{-}1.15)$$

We use the dot notation for time derivatives, i.e., $\partial_t Q = \dot{Q}$. To first order in r_s the potential is:

$$A^k = \frac{1}{c|\mathbf{r}_f|}\iiint [J^k]_{\text{ret}}\, d^3r_s$$

$$- \frac{1}{c^2|\mathbf{r}_f|}\iiint [\mathbf{n}\cdot\mathbf{r}_s\mathbf{\dot{J}}^k]_{\text{ret}}\, d^3r_s + \cdots. \qquad (8\text{-}1.16)$$

The first term is the steady-state expression. Following terms represent the time dependence. The integrals may be evaluated using the treatment in Chapters 3 and 5. Thus:

$$\iiint \mathbf{J}\, d^3r = \iiint \rho\mathbf{v}\, d^3r = \mathbf{\dot{p}}, \qquad (8\text{-}1.17)$$

$$\iiint \mathbf{n}\cdot\mathbf{r}\mathbf{\dot{J}}\, d^3r = c\mathbf{\dot{m}} \times \mathbf{n} + \frac{1}{2}\mathbf{\ddot{Q}}\cdot\mathbf{n}, \qquad (8\text{-}1.18)$$

where:

$$\mathbf{\ddot{Q}}\cdot\mathbf{n} = \sum_{k=1}^{3}\ddot{Q}^{jk}n_k. \qquad (8\text{-}1.19)$$

Then to first order the potentials are:

$$\mathbf{A} = \frac{1}{c}\frac{\mathbf{\dot{p}}_{\text{ret}}}{|\mathbf{r}_f|} + \frac{1}{c}\frac{\mathbf{\dot{m}}_{\text{ret}} \times \mathbf{n}}{|\mathbf{r}_f|} + \frac{1}{2c^2}\frac{\mathbf{\ddot{Q}}_{\text{ret}}\cdot\mathbf{n}}{|\mathbf{r}_f|} + \cdots \qquad (8\text{-}1.20)$$

and:

$$\phi = \frac{q}{|\mathbf{r}_f|} + \frac{1}{c}\frac{\mathbf{n}\cdot\mathbf{\dot{p}}_{\text{ret}}}{|\mathbf{r}_f|} + \frac{1}{2c^2}\frac{\sum_{j=1}^{3}\sum_{k=1}^{3}n_j n_k \ddot{Q}_{\text{ret}}^{jk}}{|\mathbf{r}_f|} + \cdots, \qquad (8\text{-}1.21)$$

where **p** is the electric dipole moment, **m** is the magnetic dipole moment, and Q^{jk} is the electric quadrupole moment. The multipole integrals can be evaluated (numerically if necessary) since the integrand contains only the dummy variable r_s. Compare these expressions for the potentials with the static expressions, equations (5-2.11) and (5-2.29). The most significant difference is the strong time dependence in equations (8-1.20) and (8-1.21) manifested by the time derivatives of the multipoles.

We can now calculate the electric and magnetic fields by taking the derivatives of the potentials, using equations (8-1.12)–(8-1.14). Then by showing just the radiation electric and magnetic dipole terms (dropping the static contributions), the fields become:

$$\mathbf{E} = -\nabla\phi - \frac{1}{c}\frac{\partial \mathbf{A}}{\partial t}$$

$$= -\frac{\mathbf{n}}{c}\partial_t\left[\frac{-1}{c}\frac{\mathbf{n}\cdot\dot{\mathbf{p}}}{|\mathbf{r}_f|}+\cdots\right]_{\text{ret}} - \frac{1}{c}\partial_t\left[\frac{1}{c}\frac{\dot{\mathbf{p}}}{|\mathbf{r}_f|}+\frac{1}{c}\frac{\dot{\mathbf{m}}\times\mathbf{n}}{|\mathbf{r}_f|}+\cdots\right]_{\text{ret}}$$

$$= \frac{1}{c^2|\mathbf{r}_f|}[\mathbf{n}(\mathbf{n}\cdot\ddot{\mathbf{p}}) - \mathbf{n}\cdot\mathbf{n}\ddot{\mathbf{p}} + \mathbf{n}\times\ddot{\mathbf{m}} + \cdots]_{\text{ret}}$$

$$= \frac{1}{c^2|\mathbf{r}_f|}[(\ddot{\mathbf{p}}\times\mathbf{n})\times\mathbf{n} + \mathbf{n}\times\ddot{\mathbf{m}} + \cdots]_{\text{ret}}, \qquad (8\text{-}1.22)$$

and:

$$\mathbf{B} = \nabla\times\mathbf{A} = -\frac{\mathbf{n}}{c}\partial_t\times\left[\frac{1}{c}\frac{\dot{\mathbf{p}}}{|\mathbf{r}_f|}+\frac{1}{c}\frac{\dot{\mathbf{m}}\times\mathbf{n}}{|\mathbf{r}_f|}+\cdots\right]_{\text{ret}}$$

$$= \frac{1}{c^2|\mathbf{r}_f|}\left[\ddot{\mathbf{p}}\times\mathbf{n} + [\ddot{\mathbf{m}}\times\mathbf{n}]\times\mathbf{n} + \cdots\right]_{\text{ret}}. \qquad (8\text{-}1.23)$$

We can calculate the Poynting vector and the radiated power for the multipole fields. The quantity $\mathbf{n}\cdot\mathbf{N}$ represents the energy radiated per unit area per unit time into the direction **n**. If we substitute the dipole fields calculated above into the Poynting vector, the energy radiated per unit time per unit solid angle is:

$$\frac{dP}{d\Omega} = \mathbf{n}\cdot\mathbf{N}r^2 = \frac{1}{4\pi c^3}[(\ddot{\mathbf{p}}\times\mathbf{n})\times\mathbf{n} + \ddot{\mathbf{m}}\times\mathbf{n} + \ldots]^2_{\text{ret}}$$

$$= \frac{1}{4\pi c^3}\left[(\ddot{\mathbf{p}}\times\mathbf{n})^2 + (\ddot{\mathbf{m}}\times\mathbf{n})^2 + 2\mathbf{n}\cdot\ddot{\mathbf{p}}\times\ddot{\mathbf{m}} + \ldots\right]_{\text{ret}}, \, (8\text{-}1.24)$$

in terms of the electric dipole, magnetic dipole, and mixed terms, respectively. This expression is useful for calculating radiation patterns from antennas and differential cross sections, for example.

It is possible to show that the mixed term does not contribute to the total energy flux integrated over all solid angles. Thus integrating the mixed term:

$$\oint\!\!\!\oint \mathbf{n} \cdot \ddot{\mathbf{p}} \times \dddot{\mathbf{m}} \, d\Omega = \oint\!\!\!\oint |\ddot{\mathbf{p}}| \, |\dddot{\mathbf{m}}| \sin \psi \cos \theta \sin \theta \, d\theta \, d\phi$$

$$= 2\pi|\ddot{\mathbf{p}}| \, |\dddot{\mathbf{m}}| \sin \psi \int_{-\pi/2}^{\pi/2} \cos \theta \sin \theta \, d\theta$$

$$= 0, \tag{8-1.25}$$

where ψ is the angle between $\ddot{\mathbf{p}}$ and $\dddot{\mathbf{m}}$. Therefore, the total power radiated can be expressed as a sum of the electric and magnetic terms:

$$P = P_E + P_M. \tag{8-1.26}$$

For electric and magnetic dipole radiation we can write for the radiated power:

$$P_E = \frac{1}{4\pi c^3} \oint\!\!\!\oint |\ddot{\mathbf{p}}|^2 \sin^2 \theta \, d\Omega = \frac{2|\ddot{\mathbf{p}}|^2}{3c^3}, \tag{8-1.27}$$

$$P_M = \frac{2|\dddot{\mathbf{m}}|^2}{2c^3}. \tag{8-1.28}$$

We immediately infer two characteristics about radiation from these expressions. First, acceleration of charge is the source of electromagnetic radiation, since second time derivative occurs in these expressions. Second, because of the tiny coefficient c^{-3} the acceleration must be very large, in order for significant amounts of power to be radiated.

EXAMPLE

Consider the scattering of radiation from a free electron. An electron at rest in a beam of radiation will be subject to a force determined by the electric field. The magnetic force is negligible at the small velocities considered here. We assume that the incident radiation is monochromatic, plane polarized, plane waves. Then the electron will have an acceleration given by:

$$\mathbf{a} = \frac{e}{m}\mathbf{E} = \frac{e}{m}\mathbf{E}_0 \sin \omega t. \tag{8-1.29}$$

The charge density of a particle of small radius may be treated as rigid. Then we can calculate the dipole moment and its derivative:

$$\ddot{\mathbf{p}} = \int \rho \ddot{\mathbf{r}} \, d^3 r = e\mathbf{a}$$

$$= \frac{e^2}{m} \mathbf{E}_0 \sin \omega t. \tag{8-1.30}$$

Substitution into the angular distribution formula gives an expression for the energy radiated by the electron. That is, the energy scattered per unit time per unit solid angle is:

$$\frac{dP}{d\Omega} = \frac{\ddot{p}^2}{4\pi c^3} \sin^2\theta = \frac{e^4 \left(E_0^2 \sin^2\omega t\right)}{4\pi m^2 c^3} \sin^2\theta$$

$$= \left[\left(\frac{e^2}{mc^2}\right)^2 \sin^2\theta\right]\left[\frac{c\left(E_0^2 \sin^2\omega t\right)}{4\pi}\right] \qquad (8\text{-}1.31)$$

Recall that the angle θ is the angle between the direction of the electric vector and the direction of the detector of the radiation. The incident flux of radiation is expressed in terms of the incident Poynting vector:

$$I_{\text{inc}} = \mathbf{N}_{\text{inc}} \cdot \mathbf{n} = \frac{c\left(E_0^2 \sin^2\omega t\right)}{4\pi}, \qquad (8\text{-}1.32)$$

where $\mathbf{n}$ represents the direction of the incident flux. Then the differential scattering cross section (see equation (7-5.4)) is the ratio of the power distribution by the incident flux:

$$\frac{d\sigma}{d\Omega} = \left(\frac{e^2}{mc^2}\right)^2 \sin^2\theta$$

$$= R_C^2 \sin^2\theta, \qquad (8\text{-}1.33)$$

where R_C is the classical electron radius. We get the total cross section by integrating over the solid angle:

$$\sigma = \frac{8\pi}{3} R_C^2. \qquad (8\text{-}1.34)$$

This is known as the *Thomson scattering* cross section after J.J. Thomson, who first investigated it. The cross section is 2/3 of the projected area of a sphere of radius, R_C. That does not mean, however, that Thomson scattering can be used to investigate effects at that short distance. Rather, the cross section is the sum of the electric interaction at large distances. Although we have considered monochromatic radiation, the frequency dependence cancels, leaving the cross section independent of the frequency of the incident radiation. A more exact calculation must include quantum effects including the Compton effect and virtual electron–positron pair production.

The preceding calculations apply to plane polarized radiation; we can treat unpolarized, i.e., randomly polarized, radiation by averaging over all polarizations. See Figure 8-1. The light is incident in the z-direction and the electric field is in the x–y plane. Define the three angles: θ is the angle between the direction of the detector and the electric field, ψ is the angle between the direction of the detector and the

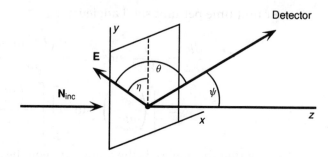

FIGURE 8-1. POLARIZATION ANGLE IN THOMSON SCATTERING. The radiation is incident in the z-direction and the electric field is randomly oriented in the $x-y$ plane. When the radiation received by the detector is averaged over the polarization angle η, the cross section depends only on the angle ψ.

z-axis, and η is the angle between the electric field and the y-axis. It is the angle of polarization and we will sum over it.

There is no loss of generality in letting the detector lie in the $y-z$ plane; then we express θ in terms of ψ and η:

$$\cos \theta = \cos \eta \sin \psi. \tag{8-1.35}$$

Since ψ is independent of η and $\langle \cos^2 \eta \rangle = \frac{1}{2}$, the differential cross section averaged over the polarization angle becomes:

$$\left\langle \frac{d\sigma}{d\Omega} \right\rangle = R_C^2 \langle \sin^2 \theta \rangle = R_C^2 [1 - \langle \cos^2 \eta \rangle \sin^2 \psi]$$

$$= \frac{1}{2} R_C^2 (1 + \cos^2 \psi), \tag{8-1.36}$$

where the angle brackets indicate the average. The angular distribution is shown in Figure 8-2.

EXAMPLE

A small dielectric or conducting sphere scatters incident radiation. The reflection of light by the droplets in a cloud is an example of this. Consider scattering from a dielectric sphere in vacuum or air, in the case where the size of the sphere is smaller than the wavelength of the incident radiation. In that case we may treat the electric field as constant over the sphere, but changing in time. Similarly, we require the relaxation time of the dielectric material to be less than the period of the incident radiation. For visible light the radius must be less than 10^{-5} cm.

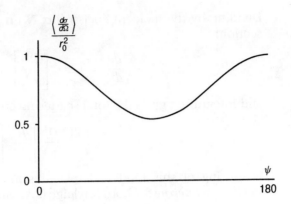

FIGURE 8-2. ANGULAR DISTRIBUTION IN THOMSON CROSS SECTION.

Now take the origin of coordinates at the center of the sphere. Then the potential outside the sphere in spherical coordinates is:

$$\phi_{\text{ext}} = \mathbf{E} \cdot \mathbf{r} = E_{\text{ext}} r \cos \alpha. \tag{8-1.37}$$

The applied uniform electric field induces a dipole moment in the sphere, having the associated dipole potential:

$$\phi_{\text{dipole}} = \frac{pr \cos \alpha}{r^3}. \tag{8-1.38}$$

On the surface of the sphere (at $r = R$) the two expressions are equal, yielding the dipole moment in terms of the applied field:

$$p = R^3 E_{\text{ext}}. \tag{8-1.39}$$

We have assumed that the incident radiation has a wavelength much greater than the radius of the sphere. Then at the sphere the electric field depends only on time:

$$E_{\text{ext}} = E_0 \sin \omega t \tag{8-1.40}$$

and we may calculate the derivatives of the dipole moment:

$$\ddot{p} = -\omega^2 R^3 E_0 \sin \omega t. \tag{8-1.41}$$

The angular distribution of the radiated power becomes:

$$\frac{dP}{d\Omega} = \frac{\ddot{p}^2}{4\pi c^3} \sin^2 \theta = \frac{R^6 \omega^4}{4\pi c^3} E_0^2 \sin^2 \theta \sin^2 \omega t$$

$$= \left[\frac{R^6 \omega^4}{c^4} \sin^2 \theta \right] \left[\frac{c E_0^2 \sin^2 \omega t}{4\pi} \right]. \tag{8-1.42}$$

Dividing by the incident energy flux $\mathbf{N} \cdot \mathbf{n}$ gives the differential cross section:

$$\frac{d\sigma}{d\Omega} = (2\pi)^4 \left[\frac{R^6}{\lambda^4} \right] \sin^2 \theta \qquad (8\text{-}1.43)$$

and integrating gives the total scattering cross section:

$$\sigma = \frac{2(2\pi)^4}{3} \left[\frac{R^6}{\lambda^4} \right]. \qquad (8\text{-}1.44)$$

Scattering of this kind and related scattering processes are called *Rayleigh scattering* for Lord Rayleigh (J.W. Strutt), who first investigated it. Rayleigh scattering is characterized by the inverse fourth power of the wavelength. The related process, scattering due to spontaneous density fluctuations in the atmosphere, is responsible for the blue color of the daytime sky, especially at large angles from the sun. This is a consequence of shorter wavelengths (blue) being scattered much more than longer (red). Conversely, the longer wavelengths are seen closer to the incident light in sunsets, although the source of this scattering is due to dust and aerosols in the atmosphere. The cross section shows very strong dependence on the size of the particle. The Rayleigh scattering cross section can be averaged over the polarization of the incident radiation like the Thomson cross section above.

8-2   A Physical Model

The relation between acceleration and radiation can be understood on an intuitive level with the following model. In Section 4-6 we found the fields due to a linear charge distribution moving with constant velocity in the collinear direction. We reuse that example here, assuming that the line of charge is long enough to ignore edge effects. The fields and potentials have cylindrical symmetry. The contribution of each element of charge is shown in Figure 8-3. Electric field lines diverge radially and the whole thing translates rigidly with velocity v. Magnetic field lines circulate concentrically around the charge distribution, but are not shown here for simplicity.

Consider how the field lines behave when the element of charge accelerates for a brief interval between t_1 and t_2. At the end of the interval, the field lines diverge radially again from the charge *at its new location*, ahead of where it would be if it maintained constant velocity v_1. This causes a dislocation or "tear" in the field line, since at sufficient radial distance the field line will have translated only by the amount

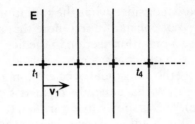

FIGURE 8-3. CONTRIBUTION OF A SINGLE CHARGE IN A LINEAR CHARGE DISTRIBUTION MOVING WITH CONSTANT VELOCITY. The entire system of charge, electric and magnetic field lines, and equipotential surfaces translate rigidly.

determined by the old velocity. This is a direct result of the retardation in the potentials and fields; the change in the position of the source charges is transmitted at the finite velocity c. However, the field lines are continuous—they begin or end only on charges or infinity. Therefore there must be a segment of field line connecting the two pieces, as shown in Figure 8-4. It has a component parallel to the acceleration but travels radially outward with velocity c. This segment of the field line is the *radiation*. The slope of the radiation section is proportional to the acceleration. For very high acceleration, the radiation section is normal to the radial (horizontal in the diagram). Thus, the horizontal component of the field line represents the field strength of the radiation, and the radiation field is proportional to the acceleration. The direction of propagation is normal to the acceleration.

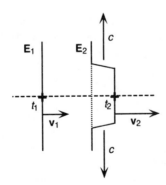

FIGURE 8-4. RADIATION OF AN ACCELERATING CHARGE. The intersection of the two dashed lines is where the LOF would be if the charge had constant velocity. Instead, the acceleration "tears" the LOF. This tear is radiation since it moves with the velocity c away from the charge. Its slope is proportional to the acceleration.

Now consider what the radiation would be like if the acceleration were not constant. First, note that a constant deceleration would also cause a radiation section in the field line, but with opposite slope. Now suppose the charges move periodically. The radiation field must then follow the sinusoidal acceleration, traveling outward with the velocity of light. We have thereby created a radiation antenna (of infinite length here). This is shown in Figure 8-5. It also suggests a mechanical analogy. Imagine holding a rope oriented vertically. Then imagine displacing the end of the rope horizontally; the displacement is transmitted vertically, but becomes small quickly for points up the rope. However, if the rope is given a quick horizontal shake, an impulse climbs the rope with an amplitude that decays only because of viscous forces acting on it. *The acceleration creates a wave.* The requirement for wave motion is a medium in which a displacement travels with a finite velocity, and an accelerated displacement.

We have considered a linear collection of charges above in order to simplify treatment of the field lines. A single point charge undergoing acceleration must also produce radiation but with different geometric distribution. Figure 8-6 indicates the radiation from a moving point charge suddenly brought to rest. The radiation has spherical shape, but nonuniform angular distribution. That is, the strength of the radiation measured by a detector depends on the component of the acceleration normal to the line between the position of the detector, and the (retarded) position of the accelerating charge.

8-3 FREQUENCY ANALYSIS OF RADIATION

One of the most widely applicable ways of representing an arbitrary function is its expansion as a series of orthogonal functions. This

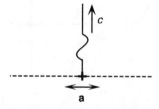

FIGURE 8-5. RADIATION BY ARBITRARY ACCELERATION.   The charge shown here is one of a linear set of charges. The radiation field of each charge is proportional to the acceleration of the charge and travels in the direction shown with velocity *c*. This is the physical basis of transmitting antennas.

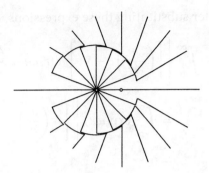

FIGURE 8-6. RADIATION FROM A DECELERATING POINT CHARGE. This illustrates the radiation from a point charge that accelerates for a short time interval. The lines of flux outside the sphere appear to originate from where the charge would be if the charge had constant velocity, shown as an open dot to the right of the center of the sphere. Inside the sphere the lines of flux originate on the actual position of the charge. The radiation is the spherically shaped transition between the two radial configurations of lines of flux. The radiation travels radially outward with velocity c.

method is especially useful for finding solutions to linear differential equations, where it can be used to represent the solution directly, or can be used to determine the Green function. In some cases the individual terms have physical significance of their own, and are therefore of interest even when the solution can be obtained otherwise. An example of this is frequency or *spectral analysis*. In this section we address the problem of finding the frequency distribution of radiation from an arbitrary source. To this end, we find the Fourier transform of the potentials $A^k(\mathbf{r}, \omega)$ just for the variable t alone.[3] Then some further calculations produce the transforms of the electric and magnetic fields and the Poynting vector. Selecting just the one variable, time, prevents the calculations from being covariant.

Consider the particular solutions to the wave equation:

$$\partial^n \partial_n A^j = \frac{4\pi}{c} J^j. \tag{8-3.1}$$

We can express solutions as Fourier series in frequency alone, i.e., the transform is performed on the variable t but not on the spatial variables. The transforms of the potential and current density are:

$$A^j(r, t) = \int_{-\infty}^{\infty} \exp(i\omega t) A^j(r, \omega)\, d\omega, \tag{8-3.2}$$

$$J^j(r, t) = \int_{-\infty}^{\infty} \exp(i\omega t) J^j(r, \omega)\, d\omega. \tag{8-3.3}$$

---

[3] It is possible to extend this to all four dimensions of course (see Section 6.1), but the frequency spectrum of radiation is interesting for its own sake.

After substituting these expressions, equation (8-3.1) becomes:

$$\int_{-\infty}^{\infty} \left[\left[\left(\frac{\partial}{\partial ct} \right)^2 - \nabla^2 \right] A^j(r, \omega) - \frac{4\pi}{c} J^j(r, \omega) \right] \exp(i\omega t) \, d\omega$$

$$= \int_{-\infty}^{\infty} \exp(i\omega t) \left[\left\{ -\left(\frac{\omega}{c} \right)^2 - \nabla^2 \right\} A^j(r, \omega) - \frac{4\pi}{c} J^j(r, \omega) \right] d\omega$$

$$= 0. \tag{8-3.4}$$

Because the exponential functions are orthogonal, the Fourier coefficients in this equation vanish. This gives the Helmholtz equation with its source term:

$$\left[\nabla^2 + \left(\frac{\omega}{c} \right)^2 \right] A^j(r, \omega) = -\frac{4\pi}{c} J^j(r, \omega). \tag{8-3.5}$$

See Section 6-6. The Green function solution is:

$$A^j(r_f, \omega) = \frac{1}{c} \int_{-\infty}^{\infty} \frac{\exp(ik|\mathbf{r}_f - \mathbf{r}_s|)}{|\mathbf{r}_f - \mathbf{r}_s|} J^j(r_s, \omega) \, d^3 r_s, \qquad \omega = ck. \tag{8-3.6}$$

Exact integration of this expression may be difficult. However, if we confine our interest to the radiation spectrum, the integral may be approximated in the radiation zone:

$$r_f \gg r_s. \tag{8-3.7}$$

That is, the radiated wavelengths are much smaller than the distance between the detector and the source:

$$kr_f \gg 1. \tag{8-3.8}$$

Although strictly speaking, that puts finite limits on the Fourier integrals, we will continue to write them as $\pm\infty$. This condition excludes low-frequency radiation, as is reasonable since low frequencies correspond to steady-state fields rather than radiation. Since sufficiently low frequency corresponds to adiabatic conditions, the fields may be treated as quasi-static. In this case, a multipole expansion may be more appropriate. Equation (8-1.5) gives the condition for multipole wavelengths.

For these conditions the displacement $|\mathbf{r}_f - \mathbf{r}_s|$ may be approximated by a truncated power series:

$$|\mathbf{r}_f - \mathbf{r}_s| = r_f - \frac{\mathbf{r}_f \cdot \mathbf{r}_s}{r_f} + \cdots$$

$$\approx r_f - \mathbf{n} \cdot \mathbf{r}_s. \tag{8-3.9}$$

Then the vector potential can be factored into two parts:

$$\mathbf{A}(r_f, \omega) \approx \frac{\exp(ikr_f)}{cr_f} \int_{-\infty}^{\infty} \exp(-ik\mathbf{n} \cdot \mathbf{r}_s)\mathbf{J}(r_s, \omega) \, d^3r_s$$

$$= \frac{\exp(ikr_f)}{cr_f} \mathbf{I}(\mathbf{n}, \omega). \qquad (8\text{-}3.10)$$

This expression is in agreeable conformity with the lowest-order Hankel function, equation (6-2.17), and thus nicely represents a spherical wave.

Because of the Lorentz gauge condition the four components of the vector potential are not independent; thus the zeroth component, $\phi(r, \omega)$, can be expressed in terms of the three-dimensional vector potential, $\mathbf{A}(r, \omega)$. Thus, as above:

$$\frac{1}{c}\phi(r_f, t) + \nabla \cdot \mathbf{A}(r_f, t) = \int_{-\infty}^{\infty} \exp(i\omega t)\left[i\frac{\omega}{c}\phi(r_f, \omega) + \nabla \cdot \mathbf{A}(r_f, \omega)\right] d\omega$$

$$= 0. \qquad (8\text{-}3.11)$$

Since the exponential function is orthogonal, its coefficient vanishes, leaving:

$$i\frac{\omega}{c}\phi(r, \omega) = -\nabla \cdot \mathbf{A}(r, \omega). \qquad (8\text{-}3.12)$$

Now calculate the divergence. Because the integral $\mathbf{I}(\mathbf{n}, \omega)$ depends only on $\mathbf{n}$ which has unit magnitude and no angular components, its derivatives vanish. Therefore the divergence becomes:

$$\nabla \cdot \mathbf{A}(r_f, \omega) = \nabla \frac{\exp(ikr_f)}{cr_f} \cdot \mathbf{I}(\mathbf{n}, \omega) + \frac{\exp(ikr_f)}{cr_f} \nabla \cdot \mathbf{I}(\mathbf{n}, \omega)$$

$$= \frac{\mathbf{n} \cdot \mathbf{I}(\mathbf{n}, \omega)}{c}\left[-\frac{\exp(ikr_f)}{r_f^2} + ik\frac{\exp(ikr_r)}{r_f}\right]. \qquad (8\text{-}3.13)$$

In the radiation zone (equation (8-3.7)) we keep only the second term. Thus:

$$\nabla \cdot \mathbf{A}(r_f, \omega) = ik\mathbf{n} \cdot \frac{\exp(ikr_f)}{cr_f}\mathbf{I}(\mathbf{n}, \omega)$$

$$= ik\mathbf{n} \cdot \mathbf{A}(r_f, \omega). \qquad (8\text{-}3.14)$$

Substituting into equation (8-3.12), the Fourier component of the scalar potential becomes:

$$\phi(r_f, \omega) = -\mathbf{n} \cdot \mathbf{A}(r_f, \omega). \qquad (8\text{-}3.15)$$

Now that we have the Fourier transforms of the potentials, we can calculate the frequency components of the electric and magnetic fields.

First, take the magnetic field:

$$\mathbf{B} - \nabla \times \mathbf{A} = \int_{-\infty}^{\infty} \exp(i\omega t) \left[\mathbf{B}(r_f, \omega) - \nabla \times \mathbf{A}(r_f, \omega) \right] d\omega$$

$$= 0. \tag{8-3.16}$$

As there is no time derivative in the expression, it becomes simply:

$$\mathbf{B}(r, \omega) = \nabla \times \mathbf{A}(r, \omega). \tag{8-3.17}$$

Evaluation of the curl is similar to evaluation of the divergence above:

$$\mathbf{B}(r_f, \omega) = i\mathbf{k} \times \mathbf{A}(r_f, \omega) = \frac{i[\exp(ikr_f)]}{cr_f} \mathbf{k} \times \mathbf{I}(\mathbf{n}, \omega)$$

$$= \frac{i[\exp(ikr_f)]}{cr_f} \mathbf{k} \times \int_{-\infty}^{\infty} \exp(-i\mathbf{k} \cdot \mathbf{r}_s) \mathbf{J}(r_s, \omega) \, d^3 r_s. \tag{8-3.18}$$

It could also be calculated directly from the transform of the current density.

A similar calculation gives the frequency components of the electric field. Thus:

$$\mathbf{E} + \nabla\phi + \frac{\partial \mathbf{A}}{c\partial t} = \int_{-\infty}^{\infty} \exp(i\omega t) \left[\mathbf{E}(r_f, \omega) + \nabla\phi(r_f, \omega) + \frac{\partial \mathbf{A}(r_f, \omega)}{c \, \partial t} \right] d\omega$$

$$= \int_{-\infty}^{\infty} \exp(i\omega t) \left[\mathbf{E}(r_f, \omega) - \nabla\mathbf{n} \cdot \mathbf{A}(r_f, \omega) + ik\mathbf{A}(r_f, \omega) \right] d\omega$$

$$= \int_{-\infty}^{\infty} \exp(i\omega t) \left[\mathbf{E}(r_f, \omega) - i\mathbf{k}(\mathbf{n} \cdot \mathbf{A}(r_f, \omega)) \right.$$

$$\left. + ik(\mathbf{n} \cdot \mathbf{n})\mathbf{A}(r_f, \omega) \right] d\omega. \tag{8-3.19a}$$

Then, continue and set the expression to zero:

$$\mathbf{E} + \nabla\phi + \frac{\partial \mathbf{A}}{c\partial t} = \int_{-\infty}^{\infty} \exp(i\omega t) \left[\mathbf{E}(r_f, \omega) - \mathbf{n} \times (i\mathbf{k} \times \mathbf{A}(r_f, \omega)) \right] d\omega$$

$$= \int_{-\infty}^{\infty} \exp(i\omega t) \left[\mathbf{E}(r_f, \omega) - \mathbf{n} \times \mathbf{B}(r_f, \omega) \right] d\omega$$

$$= 0. \tag{8-3.19b}$$

The vectors, $\mathbf{n}$, $\mathbf{E}(r_f, \omega)$, and $\mathbf{B}(r_f, \omega)$ form a mutually orthogonal set, corresponding to equation (6-1.20):

$$\mathbf{E}(r_f, \omega) = \mathbf{n} \times \mathbf{B}(r_f, \omega). \tag{8-3.20}$$

The magnetic field may therefore replace the electric field in calculations. Because these fields are Fourier transforms, not time-dependent fields, they depend inversely on the first power of distance, not inversely on the square of the distance, as the static fields do.

A problem of great practical interest is the dependence of radiated energy on frequency and direction. Since we now know the frequency distribution of the electric and magnetic fields, we can calculate the Poynting vector. Then by expressing it in terms of the frequency components, we can calculate the desired expressions. The Poynting vector may be written as a double integral:

$$
\begin{aligned}
\mathbf{N} &= \frac{c}{4\pi} \mathbf{E} \times \mathbf{B} \\
&= \frac{c}{4\pi} \int_{-\infty}^{\infty} \exp(i\omega t)\mathbf{E}(r_f, \omega)\, d\omega x \int_{-\infty}^{\infty} \exp(i\omega' t)\mathbf{B}(r_f, \omega')\, d\omega' \\
&= \frac{c}{4\pi} \iint_{-\infty}^{\infty} \exp[i(\omega + \omega')t]\mathbf{E}(r_f, \omega) \times \mathbf{B}(r_f, \omega')\, d\omega\, d\omega' \\
&= \frac{c}{4\pi} \iint_{-\infty}^{\infty} \exp[i(\omega - \omega')t]|\mathbf{B}(r_f, \omega')|^2 \mathbf{n}\, d\omega\, d\omega',
\end{aligned} \tag{8-3.21}
$$

where we have used equation (8-3.20) for the electric field, and the identity:

$$
\begin{aligned}
\mathbf{B}^*(r_f, \omega') &= \frac{1}{2\pi} \int_{-\infty}^{\infty} \exp(-i\omega' t)\, \mathbf{B}(r_f, t)\, dt \\
&= \mathbf{B}(r_f, -\omega').
\end{aligned} \tag{8-3.22}
$$

The integral over time gives the total energy transmitted per unit solid angle in the direction **n**. We define the *angular distribution* of the radiated energy:

$$
\frac{dW(\mathbf{n})}{d\Omega} = r_f^2 \int_{-\infty}^{\infty} \mathbf{n} \cdot \mathbf{N}\, dt \tag{8-3.23}
$$

as the energy radiated per unit solid angle in the direction of the unit vector **n**. We evaluate the integral:

$$
\begin{aligned}
\frac{dW(\mathbf{n})}{d\Omega} &= \frac{cr_f^2}{4\pi} \iiint_{-\infty}^{\infty} \exp[i(\omega - \omega')t]\, |\mathbf{B}(r_f, \omega')|^2 dt\, d\omega\, d\omega' \\
&= 2\pi \frac{cr_f^2}{4\pi} \iint_{-\infty}^{\infty} \delta[(\omega - \omega')t]|\mathbf{B}(r_f, \omega')|^2\, d\omega\, d\omega' \\
&= \frac{cr^2}{2} \int_{-\infty}^{\infty} |\mathbf{B}(r_f, \omega)|^2\, d\omega,
\end{aligned} \tag{8-3.24}
$$

using the representation of the δ-function:

$$
\int_{-\infty}^{\infty} \exp[i(\omega - \omega')t]\, dt = 2\pi\delta(\omega - \omega'). \tag{8-3.25}
$$

Because the integrand is an even function we may change the range of integration to be from zero to infinity, cancelling the factor of 2. Now substitute the magnetic field from equation (8-3.18). Then the angular distribution of the radiated energy is:

$$\frac{dW(\mathbf{n})}{d\Omega} = \frac{1}{c^3} \int_0^\infty \left| \mathbf{n} \times \iiint \exp(-i\mathbf{k} \cdot \mathbf{r}_s)\mathbf{J}(\mathbf{r}_s, \omega) d^3r_s \right|^2 \omega^2 \, d\omega$$

$$= \int_0^\infty W(\mathbf{n}, \omega) \, d\omega, \tag{8-3.26}$$

where the integrand is the *spectral distribution* of the radiated energy:

$$W(\mathbf{n}, \omega) = \frac{\partial^2 W}{\partial \Omega \, \partial \omega}$$

$$= \frac{\omega^2}{c^3} \left| \mathbf{n} \times \iiint \exp\left(-1\frac{\omega}{c}\mathbf{n} \cdot \mathbf{r}_s\right) \mathbf{J}(\mathbf{r}_s, \omega) d^3r_s \right|^2 \tag{8-3.27}$$

that is, the energy radiated per unit solid angle per unit frequency. In principle, the energy radiated is for all time since it is the Fourier transform. However, the expression can be extended to slowly varying (adiabatic) currents.

Expression (8-3.27) can be evaluated in terms of the current density $\mathbf{J}(\mathbf{r}, t)$, instead of its Fourier component $\mathbf{J}(r, \omega)$. Thus:

$$W(\mathbf{n}, \omega)$$

$$= \frac{\omega^2}{c^3} \left| \mathbf{n} \times \iiint \exp\left(-i\frac{\omega}{c}\mathbf{n} \cdot \mathbf{r}_s\right) \int \frac{\exp(i\omega t)}{2\pi} \mathbf{J}(\mathbf{r}_s, t) \, dt \, d^3r_s \right|^2$$

$$= \frac{\omega^2}{4\pi^2 c^3} \left| \mathbf{n} \times \iiiint \exp i(\omega t - \mathbf{k} \cdot \mathbf{r}_s)\mathbf{J}(\mathbf{r}_s, t) d^3r_s \, dt \right|^2. \tag{8-3.28}$$

The integral in this expression is the four-dimensional Fourier transform. We can calculate the *total radiated energy* by integrating over all solid angles and all frequencies. That is:

$$W = \int_0^{4\pi} \int_0^\infty W(\mathbf{n}, \omega) \, d\omega \, d\Omega, \tag{8-3.29}$$

where integration over the angles Ω includes integration over the direction vector $\mathbf{n}$.

EXAMPLE

Consider the forced harmonic oscillations of a bound electron due to an applied electric field. For small displacements from its equilibrium position, an electron is bound in a specific atomic or molecular state with an approximately linear restoring force. Within this approximation, we may treat the motion of the electron as harmonic, with perturbations due to higher-order (nonlinear) terms. This model is useful for treating resonant scattering of radiation in material media and for treating the frequency dependence of dispersive media. In Section 6-3 we found a dispersion relation for the dielectric constant with this model.

For a classical charged particle subject to a linear restoring force, a viscous force proportional to its velocity, and an external sinusoidal electric field, the equation of motion is:

$$m\ddot{x} + b\dot{x} + kx = e|E| \exp(i\omega t). \tag{8-3.30}$$

We will use the following notation for the natural frequency of the oscillator, and the decay constant:

$$\omega_0^2 = \frac{k}{m}, \tag{8-3.31}$$

$$\Gamma = \frac{b}{m}. \tag{8-3.32}$$

For a sinusoidal forced oscillation with frequency ω, the displacement is:

$$x = |x_0| \exp(i\omega t). \tag{8-3.33}$$

After substitution and cancelling the exponential function, the expression for the (complex) amplitude of the displacement x_0 is:

$$-\omega^2 m x_0 + i\omega b x_0 + kx_0 = eE_0. \tag{8-3.34}$$

The amplitude of the acceleration is found by differentiating the displacement. It is:

$$a_0 = \frac{eE_0}{m} \frac{-\omega^2}{\omega_0^2 - \omega^2 + i\omega\Gamma}. \tag{8-3.35}$$

Using this acceleration and the expression for the electric dipole radiation, equation (8-1.27), we can find the rate of radiation by a charged oscillator. Thus:

$$P_E = \frac{2e^2 a^2}{3c^3} = \left[\frac{2e^4 E_0^2}{3c^3 m^2} \frac{\omega^4}{(\omega_0^2 - \omega^2)^2 + \omega^2\Gamma^2} \right] \sin^2 \omega t$$

$$= \left[\frac{8\pi e^4}{3m^2 c^4} \frac{\omega^4}{(\omega_0^2 - \omega^2)^2 + \omega^2\Gamma^2} \right] \left[\frac{cE_0^2}{4\pi} \sin^2 \omega t \right]. \tag{8-3.36}$$

Then, for an incident beam of light, the total scattering cross section is:

$$\sigma = \sigma_{\text{Thomson}} \frac{\omega^4}{(\omega_0^2 - \omega^2)^2 + \omega^2\Gamma^2}. \tag{8-3.37}$$

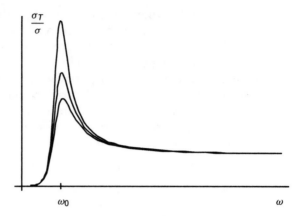

FIGURE 8-7. RESONANCE CROSS SECTIONS.   The height and width of the peak depend on the value of the parameter Γ. At high frequency, classical theory predicts the cross section will become the Thomson cross section, which is constant. The actual cross section decreases at high frequency (i.e., high energy) due to quantum effects.

At low frequencies the cross section rises as ω^4, characteristic of Rayleigh scattering. The height of the resonance depends strongly on the parameter Γ and on the value of the natural frequency of the oscillator ω_0. At high frequency, this expression for the cross section approaches the Thomson cross section. Experimentally, it falls at very high frequency due to quantum effects. The value of the resonance frequency depends on the specific electronic properties of the dielectric material. Figure 8-7 shows the cross section for a range of values of Γ.

8-4   CALCULATING WITH DELTA FUNCTIONS

Green functions and Dirac delta functions provide a very powerful tool for finding solutions to differential equations, as we have seen in the preceding chapters. We now want to apply them to the present problem. First, however, it is necessary to consider some further properties of delta functions. Our discussion is limited to a relatively nonrigorous discussion of results, since these functions are very singular, and do not satisfy many of the theorems that apply to analytical functions. More systematic treatments are given in mathematical physics texts such as Morse and Feshbach, Mathews and Walker, or Butkov.

In this section and the next, the four-dimensional vector x^k (or just x) will usually represent the position of a particle, but when we want to distinguish spatial components from time, the three-dimensional radius vector **r** will represent the position of the particle. In this notation **r**(t) represents the trajectory of the particle.

Consider the delta function with a function as its argument. For this we can write:

$$\delta\left(f(x)\right) = \frac{\delta(x)}{|df/dx|} = \frac{\delta(x)}{|f'|}. \tag{8-4.1}$$

The justification for this is as follows. First, for this equation to be valid, one must take the absolute value of the derivative, since the delta function is positive or zero. Second, an integral containing a delta function only needs to be evaluated in an infinitesmal interval around the singularity where the derivative may be considered constant. Now suppose the function $f(x)$ is zero at x_0, and its derivative is not; then for very small x:

$$f(x) \approx f(x_0) + f'(x_0)x \approx f'(x_0)x. \tag{8-4.2}$$

The integral can be evaluated by reducing the limits of integration to such a small interval around x_0 that $f(x)$ can be replaced by $f(x_0)$, and changing variables. Thus:

$$\int_{-\infty}^{\infty} \delta\left(f(x)\right) g(x)\, dx = \int_{x_0-\varepsilon}^{x_0+\varepsilon} \frac{\delta\left(f'(x_0)x\right) g(x_0)}{|f'(x_0)|}\, d\left(|f'(x_0)|x\right)$$

$$= \frac{g(x_0)}{|f'(x_0)|} \int_{x_0-\varepsilon}^{x_0+\varepsilon} \delta\left(f'(x_0)x\right) d\left(|f'(x_0)|x\right)$$

$$= \frac{g(x_0)}{|f'(x_0)|}, \tag{8-4.3}$$

which is what is required by equation (8-4.1). If the function $f(x)$ has several zeros (i.e., $f(x) = 0$ at $x = x_1, x_2, \ldots$), then the integral becomes a sum over the singularities:

$$\int_{-\infty}^{\infty} \delta\left(f(x)\right) g(x)\, dx = \int_{x_1-\varepsilon}^{x_1+\varepsilon} \frac{\delta(x)g(x)}{|f'|}\, dx + \int_{x_2-\varepsilon}^{x_2+\varepsilon} \frac{\delta(x)g(x)}{|f'|}\, dx + \cdots$$

$$= \left.\frac{g(x)}{|f'|}\right|_{x=x_1} + \left.\frac{g(x)}{|f'|}\right|_{x=x_2} + \cdots, \tag{8-4.4}$$

with a term for each singularity of the function in the range of integration.

It is also possible to define derivatives of the delta functions—in terms of their effect on an integral. Thus:

$$\int \frac{d\delta(x)}{dx} f(x)\, dx = - \int \delta(x) \frac{df(x)}{dx}\, dx = - \left.\frac{df(x)}{dx}\right|_{x=x_0} \tag{8-4.5}$$

by integrating by parts. In general, we can write:

$$\int f(x) \frac{d^n \delta(x - x_0)}{dx^n} \, dx = (-1)^n \left. \frac{d^n f}{dx^n} \right|_{x=x_0} \qquad (8\text{-}4.6)$$

assuming the boundary terms vanish.

Now consider the application of these expressions to electromagnetism. We have expressed the trajectory of a particle in terms of the source time t_s. Let us now express the trajectory as a function of the invariant pathlength parameter s:

$$x_s^k(t_s) \rightarrow x^k(s), \qquad (8\text{-}4.7)$$

where the symbol (s) conveniently signifies both pathlength and source variables. In this notation, the spatial displacement from the position of a particle at $\mathbf{r}(s)$ to the detector at the field point $\mathbf{r}_f$ is:

$$\mathbf{r}_{fs} = \mathbf{r}_f - \mathbf{r}(s) \qquad (8\text{-}4.8)$$

and the 4-velocity is:

$$U^k = \frac{dx^k(s)}{ds} c. \qquad (8\text{-}4.9)$$

We want to find a relativistic expression for the Green function. To this end, we choose a symbol for the Minkowski interval. Let:

$$T = x^n x_n$$
$$= c^2 (t_f - t(s))^2 - |\mathbf{r}_f - \mathbf{r}(s)|^2. \qquad (8\text{-}4.10)$$

There are three derivatives involving T which we will use later:

$$\frac{\partial T}{\partial x^k} = \frac{\partial x^n x_n}{\partial x^k} = 2x_k, \qquad (8\text{-}4.11)$$

$$\frac{dT}{ds} = 2x_n \frac{dx^n}{ds} = \frac{2}{c} x_n U^n, \qquad (8\text{-}4.12)$$

$$\frac{ds}{dT} = \left(\frac{dT}{ds} \right)^{-1}. \qquad (8\text{-}4.13)$$

We now find the delta function of T which, by using equations (8-4.1) and (8-4.12), becomes:

$$\delta(T) = \delta(x^n x_n) = \frac{\delta(ct \pm r)}{dx^n x_n / ds}$$
$$= \frac{c\delta(ct - r)}{2x_n U^n} + \frac{c\delta(ct + r)}{2x_n U^n}, \qquad (8\text{-}4.14)$$

where two terms occur because the argument of the delta function has two zeros. Now consider the interpretation of the two terms. In an alternate development, the delta function can be written:

$$\delta(c^2t^2 - r^2) = \delta((ct - r)(ct + r)) = \frac{\delta(ct - r)}{2r} + \frac{\delta(ct + r)}{2r}. \quad (8\text{-}4.15)$$

But these two terms are just the retarded and advanced Green functions with singularities at $t = +r/c$ and $t = -r/c$. That is:

$$2\delta(c^2t^2 - r^2) = 4\pi(G_{\text{ret}} + G_{\text{adv}}). \quad (8\text{-}4.16)$$

We can pick off the retarded Green function by inspection. If necessary, for mathematical manipulations, we can also pick it off by multiplication with the unit step function:

$$4\pi G_{\text{ret}} = 2\delta(c^2t^2 - r^2)u(ct - r)$$
$$= 2\delta(x^n x_n)u(ct - r), \quad (8\text{-}4.17)$$

$$u(t) = \begin{cases} 1, & t > 0, \\ 0, & t < 0. \end{cases} \quad (8\text{-}4.18)$$

Therefore, the retarded Green function is:

$$G\left(x_f - x(s)\right) = \frac{c}{4\pi} \frac{\delta\left(c\left(t_f - t(s)\right) - |\mathbf{r}_f - \mathbf{r}(s)|\right)}{U_k\left(x_f^k - x^k(s)\right)}$$

$$= \frac{1}{4\pi} \frac{\delta\left(c\left(t_f - t(s)\right) - |\mathbf{r}_{fs}|\right)}{(1 - \mathbf{n} \cdot \boldsymbol{\beta}(s)) |\mathbf{r}_{fs}|}, \quad (8\text{-}4.19)$$

where $\mathbf{n}$ is the unit vector:

$$\mathbf{n} = \frac{\mathbf{r}_{fs}}{|\mathbf{r}_{fs}|} = \frac{\mathbf{r}_f - \mathbf{r}_s}{|\mathbf{r}_f - \mathbf{r}_s|}. \quad (8\text{-}4.20)$$

For a coordinate system at rest, it becomes:

$$G\left(x_f - x(s)\right) = \frac{1}{4\pi} \frac{1}{|\mathbf{r}_{fs}|} \delta\left(c\left(t_f - t(s)\right) - |\mathbf{r}_{fs}|\right). \quad (8\text{-}4.21)$$

Compare this Green function with equation (6-6.12).

8-5 Radiation by Charged Particles

In Section 4-5 we derived the wave equation for the vector potential:

$$\partial_j \partial^j A^k = \frac{4\pi}{c} J^k. \quad (8\text{-}5.1)$$

The solution for the steady-state (Poisson's equation) was found by using a Green function. Similarly, the particular solution to the general problem can be expressed by using the retarded Green function.

The particular solution of equation (8-5.1) becomes:

$$A^k(x_f) = \frac{4\pi}{c} \iiiint G(x_f - x_s) J^k(x_s) \, d^4 x_s. \qquad (8\text{-}5.2)$$

For a single particle the trajectory $\mathbf{r}(s)$ is a function of its four-dimensional pathlength parameter s and the charge and current densities are proportional to the delta function:

$$\rho(x_s) = e\delta^3\left(\mathbf{r}_s - \mathbf{r}(s)\right), \qquad (8\text{-}5.3)$$

$$J^k(x_s) = e\delta^3\left(\mathbf{r}_s - \mathbf{r}(s)\right) U^k(s). \qquad (8\text{-}5.4)$$

With these delta functions the integration is elementary; the potential for the point charge is known as the *Liénard–Wiechert potential*:

$$A^k(x_f) = \frac{e}{c} \left[\frac{U^k(t_s)}{U_n \left(r_f^n - r_s^n (t_s) \right)} \right]_{\text{ret}}$$

$$= \begin{cases} \dfrac{e}{c} \left[\dfrac{c}{(1 - \mathbf{n} \cdot \boldsymbol{\beta}(s)) |\mathbf{r}_{fs}|} \right]_{ct_s = ct_f - r_{fs}}, & k = 0, \\[4mm] \dfrac{e}{c} \left[\dfrac{v^k(t_s)}{(1 - \mathbf{n} \cdot \boldsymbol{\beta}(s)) |\mathbf{r}_{fs}|} \right]_{ct_s = ct_f - r_{fs}}, & k = 1, 2, 3. \end{cases} \qquad (8\text{-}5.5)$$

At small velocities (i.e., where $v_s \ll c$) the Liénard–Wiechert potential becomes identical to the steady-state potential, except for retaining the time delay associated with retardation. Thus:

$$A^k(x_f) \rightarrow \begin{cases} \left[\dfrac{e}{|\mathbf{r}_{fs}|} \right]_{ct_s = ct_f - r_{fs}}, & k = 0, \\[4mm] \dfrac{1}{c} \left[\dfrac{e v^k(t_s)}{|\mathbf{r}_{fs}|} \right]_{ct_s = ct_f - r_{fs}}, & k = 1, 2, 3. \end{cases} \qquad (8\text{-}5.6)$$

At short distances, the time delay is also negligible.

At higher velocities, more interpretation is needed. The signal travels from the source to the field point at the speed of light c. The result is a delay given by the expression $c \, \Delta t = |\mathbf{r}_{fs}|$ between the time t_s when the source is at its location $\mathbf{r}(t_s)$ and the time t_f when the potential is observed at the field point $\mathbf{r}_f(t_f)$. Therefore, by the time the potential

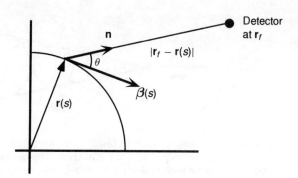

FIGURE 8-8. TRAJECTORY OF SOURCE CHARGE. The signal is received at the detector at a time delayed by an amount depending upon the distance between the source and detector. If the source travels toward the detector at nearly light speed, the detected signal may be bunched up."

is observed, the source may have moved from its initial location. Suppose that the source is moving at nearly the speed of light toward the observer, and consider a sequence of observed potentials due to a sequence of locations of the source. The source nearly catches up to the signal; consequently, the signals (potentials) received from the source are bunched up at the observer and received nearly simultaneously. The overlap of received signal increases the amount of the potential above what a static source would produce; in the extreme case, a charged particle traveling from infinity toward the observer would be perceived as a large electromagnetic pulse, followed immediately by the particle itself. This bunching of the signal is reminiscent of a "sonic boom."

A particle traveling faster than the speed of light in a dielectric medium produces a cone of radiation (like a "photonic boom") called *Cerenkov* radiation. This radiation can be understood in terms of the Huygens wavelets and the direction factor defined above. That is, as the particle travels through the medium, its potential is propagated through the medium with the velocity of light in the medium. However, since the particle itself is moving, the potential is represented by a set of Huygens wavelets forming a cone. We define the *directional factor*:

$$Q = 1 - \mathbf{n} \cdot \boldsymbol{\beta}(s), \qquad (8\text{-}5.7)$$

in which we express the source velocity $\boldsymbol{\beta}(s)$ as a vector. The apex angle of the cone is related to the angle in the direction factor:

$$1 - \mathbf{n} \cdot \boldsymbol{\beta}_m = 1 - \frac{v(s)}{c_m} \sin\left(\frac{\psi}{2}\right) = 0. \qquad (8\text{-}5.8)$$

The Cerenkov cone is analogous to the wake following a moving ship.

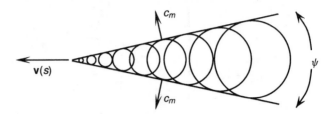

Figure 8-9. Cerenkov radiation cone. The wave front of the radiation travels outward in a cone at the velocity of light in the medium. The circles represent the Huygens wavelets generated by the particle along its trajectory. The apex angle is determined by the ratio of the velocity of light in the medium to the particle velocity.

Detectors based on the Cerenkov effect are very useful as velocity discriminators for high-energy particles. They are frequently used in conjunction with other detectors in logic circuit arrays, to trigger spark chambers for specific event types. Cerenkov detectors are made from an appropriate dielectric and photomultiplier tubes placed to register a flash of light. If a light pulse is produced, the particle velocity is greater than the velocity of light in the dielectric. The intensity of the signal is proportional to the particle's pathlength in the dielectric. By using two Cerenkov detectors in tandem, one can discriminate particles with a certain range of velocity Δv. More sophisticated detectors determine the velocity by measuring the angle of incidence of the light.

Now consider the electric and magnetic fields. The electromagnetic field tensor may be found by differentiating the potentials in equation (8-5.5). However, it is also possible to solve directly from equation (4-4.35):

$$\partial_j \partial^j F^{ki} = \frac{4\pi}{c} \left[\partial^k J^i - \partial^i J^k \right] \tag{8-5.9}$$

and use the Green function explicitly. The field tensor can be written as a sum of two antisymmetric terms; then for simplicity we show the calculations on only the first:

$$F^{ki}(x_f) = \iiiint G(x_f, x_s) \frac{4\pi}{c} \left(\partial^k J_s^i - \partial^i J_s^k \right) d^4 x_s$$

$$= I_1 - I_2. \tag{8-5.10}$$

We will also suppress the subscripts, f and s, until later. Integration by parts gives:

$$F^{ki} = \frac{1}{c} \iiiint 2\delta(T) \, \partial^k J^i \, d^4 x - I_2$$

$$= -\frac{2}{c} \iiiint J^i \partial^k \delta(T) \, d^4 x - I_2. \tag{8-5.11}$$

The pathlength element ds is parallel to the velocity U^n and is timelike. We define a hypersurface element dH^n having units of volume; its projection in the U^n direction represents components of the hypervolume orthogonal to ds, i.e., spacelike. Then the hypervolume element can be written for convenience:

$$d^4x = dH_n \, dx^n = dH_n \frac{U^n}{c} ds = d^3r \, ds. \qquad (8\text{-}5.12)$$

(Recall that U^n/c has unit magnitude.) For a point particle with trajectory given by the parametric equations $x^j(s) = (r(s), t(s))$, the charge density is expressed by delta functions:

$$J^i(r, t) = eU^i \, \delta^3 \left(r - r(s) \right). \qquad (8\text{-}5.13)$$

The volume integration becomes elementary and the field reduces to the single integral:

$$F^{ki} = -\frac{2e}{c} \int U^i \frac{\partial \delta(T)}{\partial x_k} \, ds - I_2$$

$$= -\frac{2e}{c} \int U^i \frac{d\delta(T)}{dT} \frac{\partial T}{\partial x_k} \, ds - I_2. \qquad (8\text{-}5.14)$$

Now we evaluate the delta function of a function:

$$F^{ki} = -\frac{2e}{c} \int U^i \frac{\partial T}{\partial x_k} \frac{d}{dT} \left(\frac{\delta(s)}{dT/ds} \right) ds - I_2. \qquad (8\text{-}5.15)$$

The integral will be evaluated at its two zeros, one advanced and one retarded. Now apply equation (8-4.13), change the variable of the derivative, and integrate by parts:

$$F^{ki} = -\frac{2e}{c} \int U^i \frac{\partial T}{\partial x_k} \frac{ds}{dT} \frac{d}{ds} \left(\delta(s) \frac{ds}{dT} \right) ds - I_2$$

$$= \frac{2e}{c} \int \left(\delta(s) \frac{ds}{dT} \right) \frac{d}{ds} \left(U^i \frac{\partial T}{\partial x_k} \frac{ds}{dT} \right) ds - I_2. \qquad (8\text{-}5.16)$$

Integrate using the delta function:

$$F^{ki} = \frac{2e}{c} \left[\frac{ds}{dT} \frac{d}{ds} \left(U^i \frac{\partial T}{\partial x_k} \frac{ds}{dT} \right) \right]_{T=0} - I_2. \qquad (8\text{-}5.17)$$

Finally, recombine the two terms I_1 and I_2, and evaluate using equations (8-4.11), (8-4.12), and (8-4.13).

$$F^{ki} = \left[\frac{ce}{x_n U^n} \frac{d}{ds} \left(\frac{U^i x^k - U^k x^i}{x_n U^n} \right) \right]_{\text{ret}} + \left[\frac{ce}{x_n U^n} \frac{d}{ds} \left(\frac{U^i x^k - U^k x^i}{x_n U^n} \right) \right]_{\text{adv}}.$$

$$(8\text{-}5.18)$$

So far we have carried both retarded and advanced terms, in order to keep the calculations as general as possible. However, at this point, we drop the advanced term, since fields propagating back from the future do not seem to have physical significance.

The relativistic acceleration was defined in the Lorentz equation (4-4.10).

$$a^k = \frac{dU^k}{ds} c = \gamma \frac{d}{dt} \left(\gamma \frac{dx^k}{dt} \right) \qquad (8\text{-}5.19)$$

Then, calculation of the electromagnetic field tensor only requires taking the derivative:

$$F^{ki} = e \left[\frac{a^i x^k - a^k x^i}{(x_n U^n)^2} + \frac{(U^i x^k - U^k x^i)(c^2 + a^n x_n)}{(x_n U^n)^3} \right]_{\text{ret}}. \qquad (8\text{-}5.20)$$

This expression contains the acceleration, unlike the more familiar steady-state expression. The electromagnetic field tensor can be written as a sum of two terms, one containing only velocity and the other containing velocity and acceleration. They are called, respectively, the *induction field*:

$$F_I^{ki} = ec^2 \left[\frac{(U^i x^k - U^k x^i)}{(x_n U^n)^3} \right]_{\text{ret}} \qquad (8\text{-}5.21)$$

and the *radiation field*:

$$F_R^{ki} = e \left[\left(\frac{a^i x^k - a^k x^i}{(x_n U^n)^2} \right) + \frac{a^n x_n (U^i x^k - U^k x^i)}{(x_n U^n)^3} \right]_{\text{ret}}. \qquad (8\text{-}5.22)$$

In order to compare this expression with experiment and conveniently use it, we need to express the electric and magnetic fields explicitly in three-dimensional vector notation. This is a straightforward calculation, but lengthy. We will do it in detail, but split the calculations into four parts: the electric and magnetic induction fields, and the electric and magnetic radiation fields. Before we begin, it is convenient to express the time retarded product $x_n U^n$ in terms of the vector $\boldsymbol{\beta} = \mathbf{v}/c$, and the directional factor Q. That is:

$$x_n U^n \big|_{\text{ret}} = ct\, c\gamma - c\gamma \boldsymbol{\beta} \cdot \mathbf{r} = rc\gamma - c\gamma \boldsymbol{\beta} \cdot \mathbf{r} = c\gamma r (1 - \boldsymbol{\beta} \cdot \mathbf{n})$$

$$= c\gamma r Q, \tag{8-5.23}$$

$$\dot{Q} = \frac{dQ}{dt} = -\dot{\mathbf{n}} \cdot \boldsymbol{\beta} - \mathbf{n} \cdot \dot{\boldsymbol{\beta}}, \tag{8-5.24}$$

where retardation justifies substitution of r for ct. We also evaluate the time derivative of the unit vector:

$$\dot{\mathbf{n}} = \frac{c(\mathbf{n} \times (\mathbf{n} \times \boldsymbol{\beta}))}{r}. \tag{8-5.25}$$

The proof is easy and is left for the problems. We can now find the induction fields in detail and convert to three-dimensional vector notation by writing (see equation (4-4.11)):

$$\begin{aligned} E^i &= -F^{0i}, \\ B^j &= -F^{ki}, \end{aligned} \qquad i, j, k \text{ in cyclical order.} \tag{8-5.26}$$

Now we can evaluate the induction fields. The electric field is:

$$\begin{aligned} E_I^i &= -ec^2 \frac{(U^i x^0 - U^0 x^i)}{(x_n U^n)^3} = \frac{ec^3 \gamma(r^i - \beta^i|r|)}{(c\gamma r)^3 (1 - \mathbf{n} \cdot \boldsymbol{\beta})^3} \\ &= \frac{e}{r^2} \frac{[\mathbf{n} - \boldsymbol{\beta}]^i}{\gamma^2 (1 - \mathbf{n} \cdot \boldsymbol{\beta})^3} \end{aligned} \tag{8-5.27}$$

replacing r for ct again. Let the indices i, j, and k be in cyclical order; then the induction magnetic field is:

$$\begin{aligned} B_I^j &= -ec^2 \frac{U^i x^k - U^k x^i}{(x_n U^n)^3} = -ec^2 \frac{(c\gamma)[\mathbf{r} \times \boldsymbol{\beta}]^j}{(c\gamma)^3 r^3 (1 - \mathbf{n} \cdot \boldsymbol{\beta})^3} \\ &= \frac{e}{r^2} \frac{[\mathbf{n} \times (\mathbf{n} - \boldsymbol{\beta})]^j}{\gamma^2 (1 - \mathbf{n} \cdot \boldsymbol{\beta})^3}. \end{aligned} \tag{8-5.28}$$

The electric and magnetic induction fields are clearly orthogonal.

The calculations for the radiation fields are equally straightforward, but more lengthy because of the inclusion of the acceleration. The procedure we take is to evaluate equation (8-5.18) in a particular coordinate system containing $\mathbf{r}$ and t. Of course, that hides its manifest covariance, but that is inevitable when expressing the fields as separate electric and magnetic fields. We replace the derivative d/ds by d/dt and express the velocities explicitly in terms of β and γ. That is:

$$\begin{aligned} F^{ki} &= \frac{ec}{x_n U^n} \frac{d}{ds}\left(\frac{U^i x^k - U^k x^i}{x_n U^n}\right) = \frac{ec}{c\gamma r Q} \frac{\gamma}{c} \frac{d}{dt}\left(c\gamma \frac{\beta^i x^k - \beta^k x^i}{c\gamma r Q}\right) \\ &= \frac{e}{cr Q} \frac{d}{dt}\left(\frac{\beta^i x^k - \beta^k x^i}{r Q}\right). \end{aligned} \tag{8-5.29}$$

Now consider the electric and magnetic components of this field (i, j, k are in cyclical order). Then:

$$F^{ki} = \begin{cases} \dfrac{-e}{crQ}\dfrac{d}{dt}\left(\dfrac{x^i - \beta^i ct}{rQ}\right) \\[2mm] \dfrac{-e}{crQ}\dfrac{d}{dt}\left(\dfrac{\beta \times \mathbf{n}}{Q}\right)^j \end{cases}$$

$$= \begin{cases} \dfrac{-e}{crQ}\dfrac{d}{dt}\left(\dfrac{n^i - \beta^i}{Q}\right), & k = 0, \\[3mm] \dfrac{-e}{crQ}\dfrac{d}{dt}\left(\dfrac{\beta \times \mathbf{n}}{Q}\right)^j, & i, j, k = 1, 2, 3. \end{cases} \qquad (8\text{-}5.30)$$

Taking the derivative, the electric field becomes:

$$F^{0i} = \frac{-e}{crQ}\left(\frac{\dot{n}^i - \dot{\beta}^i}{Q} - \frac{n^i - \beta^i}{Q^2}\dot{Q}\right)$$

$$= \frac{-e}{cr}\frac{Q(\dot{n}^i - \dot{\beta}^i) - \dot{Q}\left(n^i - \beta^i\right)}{Q^3}. \qquad (8\text{-}5.31)$$

Inspection of the numerator shows that it can be split into terms containing acceleration and terms containing only velocity:

$$Q(\dot{n}^i - \dot{\beta}^i) - \dot{Q}(n^i - \beta^i) = [(\mathbf{n}\cdot\beta)(n^i - \beta^i) - Q\dot{\beta}^i]$$
$$+ [(\dot{\mathbf{n}}\cdot\beta)(n^i - \beta^i) + Q\dot{n}^i]. \quad (8\text{-}5.32)$$

However, as we have seen, the pure velocity term is the induction field, which was evaluated above. Then, the electric radiation field becomes:

$$\mathbf{E}_R = \frac{e}{cr}\frac{(\mathbf{n}\cdot\dot{\beta}(\mathbf{n} - \beta) - (1 - \mathbf{n}\cdot\beta)\dot{\beta}}{Q^3}$$

$$= \frac{e}{cr}\frac{(\mathbf{n}\cdot\dot{\beta})(\mathbf{n} - \beta) - \{\mathbf{n}\cdot(\mathbf{n} - \beta)\}\dot{\beta}}{(1 - \mathbf{n}\cdot\beta)^3}$$

$$= \frac{e}{cr}\frac{\mathbf{n}\times\{(\mathbf{n} - \beta)\times\dot{\beta}\}}{(1 - \mathbf{n}\cdot\beta)^3}. \qquad (8\text{-}5.33)$$

Now calculate the magnetic field from the spatial components of the electromagnetic field tensor. In three-dimensional vector notation it is:

$$\mathbf{B}_R = \frac{e}{crQ}\frac{d}{dt}\left(\frac{\beta \times \mathbf{n}}{Q}\right) = \frac{e}{cr}\frac{Q\dot{\beta}\times\mathbf{n} + Q\beta\times\dot{\mathbf{n}} - \dot{Q}\beta\times\mathbf{n}}{Q^3}. \qquad (8\text{-}5.34)$$

Evaluate Q and its derivative, and split the expression into velocity and acceleration terms:

$$\mathbf{B}_R = \frac{e}{cr} \frac{(1 - \mathbf{n} \cdot \boldsymbol{\beta})\dot{\boldsymbol{\beta}} \times \mathbf{n} + (1 - \mathbf{n} \cdot \boldsymbol{\beta})\boldsymbol{\beta} \times \dot{\mathbf{n}} + (\dot{\mathbf{n}} \cdot \boldsymbol{\beta} + \mathbf{n} \cdot \dot{\boldsymbol{\beta}})\boldsymbol{\beta} \times \mathbf{n}}{(1 - \mathbf{n} \cdot \boldsymbol{\beta})^3}$$

$$= \frac{e}{cr} \frac{(1 - \mathbf{n} \cdot \boldsymbol{\beta})\dot{\boldsymbol{\beta}} \times \mathbf{n} + (\mathbf{n} \cdot \dot{\boldsymbol{\beta}})\boldsymbol{\beta} \times \mathbf{n}}{(1 - \mathbf{n} \cdot \boldsymbol{\beta})^3}$$

$$+ \frac{e}{cr} \frac{(1 - \mathbf{n} \cdot \boldsymbol{\beta})\boldsymbol{\beta} \times \dot{\mathbf{n}} + (\dot{\mathbf{n}} \cdot \boldsymbol{\beta})\boldsymbol{\beta} \times \mathbf{n}}{(1 - \mathbf{n} \cdot \boldsymbol{\beta})^3}. \tag{8-5.35}$$

The second term is the induction field; thus the radiation magnetic field is:

$$\mathbf{B}_R = \frac{e}{cr} \frac{(1 - \mathbf{n} \cdot \boldsymbol{\beta})\dot{\boldsymbol{\beta}} + (\mathbf{n} \cdot \dot{\boldsymbol{\beta}})\boldsymbol{\beta}}{(1 - \mathbf{n} \cdot \boldsymbol{\beta})^3} \times \mathbf{n} = \frac{e}{cr} \frac{\mathbf{n} \cdot (\mathbf{n} - \boldsymbol{\beta})\dot{\boldsymbol{\beta}} + (\mathbf{n} \cdot \dot{\boldsymbol{\beta}})\boldsymbol{\beta}}{(1 - \mathbf{n} \cdot \boldsymbol{\beta})^3} \times \mathbf{n}$$

$$= \frac{e}{cr} \frac{\mathbf{n} \cdot (\mathbf{n} - \boldsymbol{\beta})\dot{\boldsymbol{\beta}} - (\mathbf{n} \cdot \dot{\boldsymbol{\beta}})(\mathbf{n} - \boldsymbol{\beta})}{(1 - \mathbf{n} \cdot \boldsymbol{\beta})^3} \times \mathbf{n} = \frac{e}{cr} \frac{\mathbf{n} \times \{\dot{\boldsymbol{\beta}} \times (\mathbf{n} - \boldsymbol{\beta})\}}{(1 - \mathbf{n} \cdot \boldsymbol{\beta})^3} \times \mathbf{n}$$

$$= \frac{e}{cr} \frac{\mathbf{n} \times [\mathbf{n} \times \{(\mathbf{n} - \boldsymbol{\beta}) \times \dot{\boldsymbol{\beta}}\}]}{(1 - \mathbf{n} \cdot \boldsymbol{\beta})^3}. \tag{8-5.36}$$

It is obviously orthogonal to the electric radiation field.

We now summarize the results. Written in full, the expressions for displacement, velocity, and acceleration are, respectively: $\mathbf{r} = \mathbf{r}_{fs}$, $\boldsymbol{\beta} = \boldsymbol{\beta}(s)$, and $\mathbf{a} = \mathbf{a}(s)$. Thus:

$$\mathbf{E}(x_f) = \left[\frac{e}{r_{fs}^2} \frac{\left(1 - \beta^2(s)\right)(\mathbf{n} - \boldsymbol{\beta}(s))}{\left(1 - \mathbf{n} \cdot \boldsymbol{\beta}(s)\right)^3} \right.$$

$$\left. + \frac{e}{c^2 r_{fs}} \frac{\mathbf{n} \times \left[(\mathbf{n} - \boldsymbol{\beta}(s)) \times \mathbf{a}(s)\right]}{\left(1 - \mathbf{n} \cdot \boldsymbol{\beta}(s)\right)^3} \right]_{\text{ret}} \tag{8-5.37}$$

$$\mathbf{B}(x_f) = \left[\frac{e}{r_{fs}^2} \frac{\left(1 - \beta^2(s)\right)\mathbf{n} \times (\mathbf{n} - \boldsymbol{\beta}(s))}{\left(1 - \mathbf{n} \cdot \boldsymbol{\beta}(s)\right)^3} \right.$$

$$\left. + \frac{e}{c^2 r_{fs}} \frac{\mathbf{n} \times \left[\mathbf{n} \times \left[(\mathbf{n} - \boldsymbol{\beta}(s)) \times \mathbf{a}(s)\right]\right]}{\left(1 - \mathbf{n} \cdot \boldsymbol{\beta}(s)\right)^3} \right]_{\text{ret}}. \tag{8-5.38}$$

The electric and magnetic radiation fields, and induction fields, are orthogonal to each other and to the vector $\mathbf{n}$:

$$\mathbf{B} = \mathbf{n} \times \mathbf{E}. \tag{8-5.39}$$

The names, induction field and radiation field, refer to the fact that at short distances and low frequencies, the induction field is larger than the radiation field. The induction field is responsible for many important practical applications. At sufficiently large distances, the radiation field dominates because it is inversely proportional to the displacement r_{fs}. To produce a given amount of radiation, relatively large accelerations are needed because of the inverse factor c^2.

For small velocities, $v \ll c$, the fields reduce to:

$$\mathbf{E} = e \left[\frac{1}{r_{fs}^2} \mathbf{n} \right]_{ret} + \frac{e}{c^2} \left[\frac{\mathbf{n} \times (\mathbf{n} \times \mathbf{a})}{|\mathbf{r}_{fs}|} \right]_{ret}, \qquad (8\text{-}5.40)$$

$$\mathbf{B} = \frac{e}{c} \left[\frac{\mathbf{v} \times \mathbf{n}}{r_{fs}^2} \right]_{ret} + \frac{e}{c^2} \left[\frac{\mathbf{a} \times \mathbf{n}}{|\mathbf{r}_{fs}|} \right]_{ret}. \qquad (8\text{-}5.41)$$

The lowest-order terms agree with those in Chapter 2 for steady-state conditions (except for retardation). For a charge at rest, the electric field becomes the Coulomb field, and the magnetic field vanishes. Inspection of equation (8-5.40) shows that the radiation is plane polarized, with the electric field in the $\mathbf{n}$–$\mathbf{a}$ plane. Its magnitude is proportional to the sine of the angle of the observer and the acceleration:

$$|\mathbf{E}| = \frac{ea}{c^2 r_{fs}} \sin \theta. \qquad (8\text{-}5.42)$$

Thus, an observer looking at the radiating charge in the $-\mathbf{n}$ direction will observe the component of $\mathbf{E}$ orthogonal to the direction of $\mathbf{n}$. It follows that when the acceleration is directly toward the observer, no radiation will be observed.

The Poynting vector can be calculated for these fields. In general, it will contain three terms: radiation terms, induction terms, and mixed terms. We need to consider the energy and momentum carried by each of these three terms. First, because of equation (8-5.39), we can express the Poynting vector just in terms of the electric field:

$$\mathbf{N} = \frac{c}{4\pi} \mathbf{E} \times \mathbf{B}$$

$$= \frac{cE^2}{4\pi} \mathbf{n}. \qquad (8\text{-}5.43)$$

Then the radiation, induction, and mixed terms are, respectively:

$$N_R = \frac{e^2}{4\pi c^3} \left| \frac{\mathbf{n} \times [(\mathbf{n} - \boldsymbol{\beta}(s)) \times \mathbf{a}]}{(1 - \mathbf{n} \cdot \boldsymbol{\beta}(s))^3 r_{fs}} \right|^2 \to \frac{1}{4\pi c^3} \frac{e^2 a^2}{r_{fs}^2} \sin^2 \theta, \qquad (8\text{-}5.44)$$

$$N_I = \frac{ce^2}{4\pi} \left| \frac{\left(1 - \beta^2(s)\right)\left(\mathbf{n} - \boldsymbol{\beta}(s)\right)}{\left(1 - \mathbf{n} \cdot \boldsymbol{\beta}(s)\right)^3 r_{fs}^2} \right|^2 \rightarrow \frac{c}{4\pi} \frac{e^2}{r_{fs}^4}, \tag{8-5.45}$$

$$N_M \rightarrow \frac{1}{4\pi c^2} \frac{e^2 a}{r_{fs}^3} \sin\theta, \qquad v \ll c, \tag{8-5.46}$$

and we have indicated the limits as the velocity becomes small. Because the direction factor appears in the Poynting vector to the sixth power, the dependence of the radiation distribution on direction is very strong at high velocity. Figure 8-10 shows Q^{-2} between the angles $\pm\pi$, at the value $\beta = 0.8$.

As we saw in Chapter 6, the Poynting vector is interpreted as the energy radiated per unit time, per unit area. It is, in fact, the exact analog of the current density $\mathbf{J}$:

$$\mathbf{J}\left(\frac{\text{statcoulombs}}{\text{s cm}^2}\right) \leftrightarrow \mathbf{N}\left(\frac{\text{ergs}}{\text{s cm}^2}\right). \tag{8-5.47}$$

By relating the area to the corresponding solid angle, one can calculate the angular distribution of radiated power:

$$\frac{dP_R}{d\Omega} = \mathbf{N}_R \cdot \mathbf{n} r_{fs}^2$$

$$= \frac{e^2}{4\pi c^3} \frac{|\mathbf{n} \times (\mathbf{n} \times \mathbf{a}) - \mathbf{n} \times (\boldsymbol{\beta}(s) \times \mathbf{a})|^2}{\left(1 - \mathbf{n} \cdot \boldsymbol{\beta}(s)\right)^6}. \tag{8-5.48}$$

It divides nicely into nonrelativistic and relativistic terms. At relativistic velocities, the rate of radiation into a given direction may not equal the rate at which that radiation is received. This is a consequence

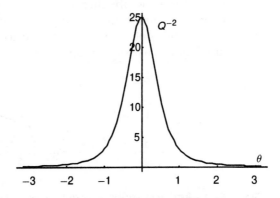

Figure 8-10. Direction factor as a function of angle. This plot is for a representative velocity $v = 0.8c$. Note that it does not quite vanish at the end points $\pm\pi$.

of the fact that the velocity of the observer differs from that of the radiator, and that time dilation applies. We can find the relation between the two rates from the retardation expression. That is:

$$dt_f = dt_s + \frac{1}{c} d|r_f - r_s| = dt_s + \frac{1}{c} \nabla_s |r_f - r_s| \cdot \mathbf{v}_s \, dt_s$$

$$= dt_s - \frac{1}{c} \mathbf{n} \cdot \mathbf{v}_s \, dt_s$$

$$= (1 - \mathbf{n} \cdot \boldsymbol{\beta}_s) \, dt_s, \tag{8-5.49}$$

which contains the direction factor. In these terms, the amount of energy radiated per unit time, per unit solid angle, in the frame of reference of the radiator is:

$$\left. \frac{dP}{d\Omega} \right|_s = \frac{dW}{d\Omega \, dt_f} \frac{dt_f}{dt_s}$$

$$= \frac{e^2}{4\pi c^3} \frac{\left| \mathbf{n} \times \left[(\mathbf{n} - \boldsymbol{\beta}(s)) \times \mathbf{a} \right] \right|^2}{\left(1 - \mathbf{n} \cdot \boldsymbol{\beta}(s) \right)^5}. \tag{8-5.50}$$

We can picture this by considering a particle moving at high velocity in a curved path. See Figure 8-11. It will be accelerating and emitting radiation. However, since it is moving at nearly the speed of light, it will almost catch up to the radiation it emits in the direction of its velocity. The radiation in this direction will be compressed. By this, we mean that radiation emitted during time Δt_s will be detected during time Δt_f where $\Delta t_f < \Delta t_s$. Similarly, radiation in the direction opposite to the direction of the velocity will be spread out. At intermediate angles, the amount of compression depends on the projection of the particle velocity into the direction of $\mathbf{n}$, i.e., the radiation. The compression factor is just the direction factor. We have not considered the frequency spectrum of the radiation, however, it is clear that a Doppler shift occurs and that the shift is related to the direction factor as it is in the analogous case of sound waves.

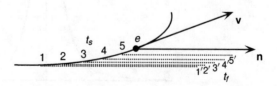

Figure 8-11. Rate compression in observed radiation. The radiation may be supposed to be emitted at times 1–5, and detected at times 1'–5'. Because of the high speed of the radiator, it nearly catches up to the radiation. Thus the radiated rate is slower than the detected rate.

At low velocity, the total radiated power is given by the *Larmor power formula*:

$$P_R = \oiint \mathbf{N}_R \cdot \mathbf{n} r_{fs}^2 \, d\Omega = \frac{e^2 a^2}{4\pi c^3} \int_0^{2\pi} \int_{-\pi/2}^{\pi/2} \sin^3 \theta \, d\theta \, d\varphi$$

$$= \frac{2e^2}{3c^3} a^2, \tag{8-5.51}$$

i.e., the rate of energy radiated through a sphere of large radius $r_{fs} = r_f = r$. Imagine a short pulse of radiation. Then the energy of this pulse will travel radially outward in a shell of radius r growing at the speed of light, and negligible thickness. Since the Larmor formula is independent of the radius, and the total energy is conserved, the energy of the pulse is spread over the increasing area of the sphere. Therefore, the attenuation of energy transmitted into an element of solid angle depends on r^{-2}. The induction and mixed contributions vanish for sufficiently large radius since they vary as r^{-4} and r^{-3}, respectively:

$$P_I = \frac{ce^2}{4\pi r_f^2} \int_0^{4\pi} d\Omega = ce^2 \frac{1}{r_f^2} \rightarrow 0, \tag{8-5.52}$$

$$P_M = \frac{e^2 a}{4\pi c^2 r_f} \int_0^{4\pi} \sin^2 \theta \, d\Omega \rightarrow 0. \tag{8-5.53}$$

Nevertheless, at short distances induction is important, since it is the energy transfer mechanism for many kinds of electronic devices, including transformers. (It's not called induction for nothing!)

The Larmor formula calculation was done at low velocity. However, the radiated power depends only on the square of the acceleration, and the square of a 4-vector is invariant. Power is also invariant since energy and time are each the zeroth component of their respective 4-vectors. They therefore transform identically, leaving the derivative of energy with respect to time invariant. This suggests that Larmor's formula is the low-velocity limit of a general expression. The four-dimensional acceleration can be expressed in three-dimensional terms:

$$a^k = c\gamma \frac{d}{dt}(\gamma \beta^k) = \begin{cases} c\gamma^4 (\boldsymbol{\beta} \cdot \dot{\boldsymbol{\beta}}), \\ c\gamma \left(\gamma \dot{\beta}^k + \gamma^3 (\boldsymbol{\beta} \cdot \dot{\boldsymbol{\beta}}) \beta^k \right). \end{cases} \tag{8-5.54}$$

Then, taking the inner product of a^k with itself, the *relativistic Larmor formula* becomes:

$$P_R = \frac{2e^2}{3c} \left[\beta^2 - \left(\boldsymbol{\beta} \times \dot{\boldsymbol{\beta}} \right)^2 \right] \gamma^6. \tag{8-5.55}$$

The calculation is straightforward and is left to the problems.

EXAMPLE

Consider the radiation due to a single charge in simple harmonic motion. Assume the velocity and acceleration are parallel to the z–axis and $v \ll c$. Then the velocity of the particle and its acceleration are:

$$v = v_0 \cos \omega t, \qquad (8\text{-}5.56)$$

$$a = -\omega v_0 \sin \omega t. \qquad (8\text{-}5.57)$$

The radiation is plane polarized in the plane defined by the vector **n** and the z–axis. The angular distribution and radiated power are:

$$\frac{dP}{d\Omega} = \frac{e^2 a^2 \sin^2 \theta}{4\pi c^3} = \frac{e^2 \omega^2 v_0^2 \sin^2 \omega t \sin^2 \theta}{4\pi c^3}, \qquad (8\text{-}5.58)$$

$$P = \frac{dW}{dt} = \frac{2e^2}{3c^3} a^2 = \frac{2e^2 \omega^2 v_0^2 \sin^2 \omega t}{3c^3}. \qquad (8\text{-}5.59)$$

Figure 8-12 shows the radiated power as a function of time.

However, since the charge radiates energy, the amplitude should decrease. We can calculate its rate of decrease as follows. The average kinetic energy of the particle is:

$$W_{\text{avg}} = m \left[v_0^2 \sin^2 \omega t \right]_{\text{avg}} = \frac{m}{2} v_{\text{avg}}^2, \qquad (8\text{-}5.60)$$

where $\sin^2 \omega t$ is averaged over any integer number of periods. Now apply this to equation (8-5.59) which becomes an (approximate) equation for the decay of energy of the system:

$$\frac{dW}{dt} = -\frac{2e^2 \omega^2}{3c^3} \frac{W_{\text{avg}}}{m} = -\frac{1}{T_{\text{decay}}} W_{\text{avg}}. \qquad (8\text{-}5.61)$$

Solution to this equation shows that the total energy of the oscillator decreases exponentially with the indicated characteristic time. The va-

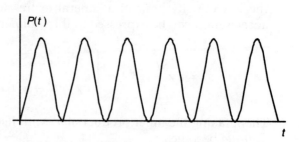

FIGURE 8-12. RATE OF RADIATION FROM A HARMONIC OSCILLATOR.   The rate of radiation is proportional to the square of the acceleration. It is thus always positive, and vanishes in each oscillation when the charge passes the point of equilibrium. Although it is not shown here, the amplitude of the radiation rate gradually decreases.

lidity condition for the approximation is for the decay time, T_{decay}, to be much longer than the period of the oscillator T. That is:

$$\frac{T_{\text{decay}}}{T} = \frac{3mc^3}{2e^2\omega^2} \frac{\omega}{2\pi} = \frac{3/4}{(e^2/mc^2)} \frac{c}{\omega} = \frac{\lambda}{(4\pi/3)R_C} \gg 1. \qquad (8\text{-}5.62)$$

This is well satisfied for optical wavelengths, $\lambda = 5 \times 10^{-7}$ m. The *characteristic time*:

$$\tau = \frac{2e^2}{3mc^3} = \frac{2}{3}\frac{R_C}{c} = 6.26 \times 10^{-24} \text{ s} \qquad (8\text{-}5.63)$$

is the time required for a signal traveling at the speed of light to cross a distance R_C, the classical electron radius.

EXAMPLE

An elliptical orbit may be considered to be the superposition of two harmonic oscillations. The preceding calculation for a decaying oscillator allows us to estimate the decay time of a charged particle in an elliptical orbit. According to this calculation, the electron in a hydrogen atom will spiral inward as it loses energy to radiation. The decay time will be given approximately by T_{decay}, calculated above. Take the following approximate values for the hydrogen atom:

$$r = 10^{-8} \text{ cm,}$$

$$v = \sqrt{\frac{e^2}{mr}} = 10^8 \frac{\text{cm}}{\text{s}} \ll c, \qquad (8\text{-}5.64)$$

$$\omega = \frac{v}{r} = 10^{16} \text{ Hz.}$$

Then the decay time for the orbit is:

$$T_{\text{decay}} = \frac{3}{2}\frac{mc^3}{e^2\omega^2} = \frac{1}{\tau\omega^2} \approx 10^{-8} \text{ s.} \qquad (8\text{-}5.65)$$

This is a serious problem for classical electrodynamics, since it implies that a hydrogen atom is unstable, with a very short lifetime. The solution to this problem does not lie in better approximation, but rather in quantum effects, where discrete orbits govern calculations of radiation and orbital decay. Classical theory is no longer sufficient at this distance.

Radiation patterns from a single charge moving in different directions can be interpreted as transformations of the radiation pattern from a charge that is momentarily at rest, but accelerating. Equation (8-5.48) gives the actual intensity distribution; it can be seen that the effect of high velocity is to throw the direction of maximum radiation

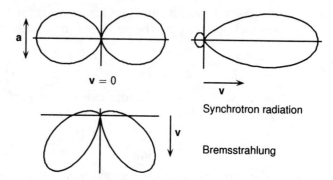

FIGURE 8-13. RADIATION PATTERNS FOR VELOCITIES IN DIFFERENT DIRECTIONS.   The angular distribution pattern at high velocity β is obtained by a transformation of the radiation pattern for the particle (nearly) at rest, into the direction of the velocity.

forward into the direction of the velocity. The transformed radiation patterns are illustrated by Figure 8-13.

Bremsstrahlung is a German word meaning "breaking radiation." It refers to a charged particle undergoing deceleration; however, much of what we say also applies to an accelerating particle. Suppose that a charged particle is decelerating in a straight line. Then the radiation Poynting vector becomes:

$$N_R = \frac{e^2}{4\pi c^3} \left[\frac{\mathbf{n} \times [(\mathbf{n} - \boldsymbol{\beta}) \times \mathbf{a}]}{(1 - \mathbf{n} \cdot \boldsymbol{\beta})^3 r} \right]^2$$

$$= \frac{e^2}{4\pi c^3 r^2} \frac{[\mathbf{a} - (\mathbf{n} \cdot \mathbf{a})\mathbf{n}]^2}{(1 - \beta \cos\theta)^6}, \tag{8-5.66}$$

where θ is the angle between $\boldsymbol{\beta}$ and $\mathbf{n}$, the direction of the detector. The quantity in square brackets is just the component of the acceleration orthogonal to $\mathbf{n}$, since $\boldsymbol{\beta}$ and $\mathbf{a}$ are (anti)parallel. The angular distribution in a plane containing the velocity is:

$$\frac{dP}{d\Omega} = \frac{e^2 a^2}{4\pi c^3} \frac{\sin^2\theta}{(1 - \beta \cos\theta)^5}. \tag{8-5.67}$$

This formula gives the symmetric double-lobed angular distribution illustrated in Figure 8-13. At the velocity increases, the direction of the maximum intensity is rotated forward due to the effect of the direction factor. Figure 8-15 shows the radiation pattern in three dimensions.

Equation (8-5.67) can be integrated to give the total radiated power:

$$P = \frac{e^2 a^2}{4\pi c^3} \frac{8\pi}{3(1 - \beta^2)^3}$$

$$= \frac{2}{3} \frac{e^2 a^2}{c^3} \gamma^6, \tag{8-5.68}$$

which reduces to the Larmor expression at small velocity.

Bremsstrahlung is responsible for the continuous spectrum observed in the production of x-rays. That is, electrons striking the anode of an x-ray tube undergo atomic collisions. The observed radiation is more complex than simple bremsstrahlung, because of the other processes taking place. For example, multiple collisions scatter the incident electrons and smear the radiation distribution. Furthermore, the struck atoms become excited or ionized, and radiate x-rays with wavelength distributions that are characteristic of the material of the target anode.

The charged particles in a synchrotron accelerator move in a circular orbit and radiate as the consequence of the centripetal acceleration. Typically the tangential acceleration takes place in a few short segments of the ring. The particles then drift the rest of the way around the ring, under the influence of the magnetic field, where the acceleration is entirely centripetal. Let us calculate the radiated power in the plane of the ring. Equation (8-5.50) becomes:

$$\frac{dP}{d\Omega} = \frac{e^2}{4\pi c^3} \frac{[\mathbf{n} \times [(\mathbf{n} - \boldsymbol{\beta}) \times \mathbf{a}]]^2}{(1 - \mathbf{n} \cdot \boldsymbol{\beta})^5}$$

$$\rightarrow \frac{e^2 a^2}{4\pi c^3} \frac{[\beta - \cos\theta]^2}{(1 - \beta\cos\theta)^5}, \tag{8-5.69}$$

where the origin of the coordinate system is at the location of the particle. The unit vector $\mathbf{n}$ is directed toward the detector. At high velocity it is strongly peaked in the direction of the velocity, with a small trailing lobe. See Figure 8.14. The radiation pattern in three dimensions is shown by Figure 8-15.

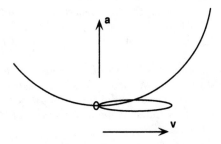

Figure 8-14. Synchrotron radiation. The centripetal acceleration at very high velocity produces a strongly peaked radiation pattern tangent to the orbit.

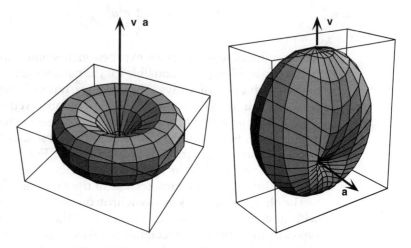

Figure 8-15. Angular distributions for bremsstrahlung and synchrotron radiation. The centripetal acceleration produces a strongly directed radiation pattern tangent to the orbit.

8-6   Selected Bibliography

M. Abraham and R. Becker, *The Classical Electricity and Magnetism*, Vols. 1 and 2, Stechert, 1932.

A. Barut, *Electrodynamics and Classical Theory of Fields and Particles*, Macmillan, 1964.

M. Born and E. Wolf, *Principles of Optics*, Macmillan (Pergamon), 1964. Standard reference on optics.

E. Butkov, *Mathematical Physics*, Addison-Wesley, 1970.

B. DiBartolo, *Classical Theory of Electromagnetism*, Prentice-Hall, 1991. Contains very detailed derivation of the retarded fields from the Liénard–Wiechert potentials.

H. Hertz, *Electric Waves*, Macmillan (London), 1893.

J. Jackson, *Classical Electrodynamics*, Wiley, 1962.

E. Konopinski, *Electromagnetic Fields and Relativistic Particles*, McGraw-Hill, 1981. Discusses self-energy of the electron and radiation reaction.

L. Landau and E. Lifshitz, *The Classical Theory of Fields*, 2nd ed., Pergamon, 1975.

H. Lorentz, *The Theory of Electrons*, 2nd ed., Dover, 1952.

J. Marion and M. Heald, *Classical Electromagnetic Radiation*, Academic Press, 1980. Discusses the Hertz vector potential method.

J. Mathews and R. Walker, *Mathematical Methods of Physics*, 2nd ed., Addison-Wesley, 1968.

J. Maxwell, *A Treatise on Electricity and Magnetism*, 3rd ed., Oxford University Press, 1904.

P. Morse and H. Feshbach, *Theoretical Physics*, McGraw-Hill, 1953.

W. Panofsky and M. Phillips, *Classical Electricity and Magnetism*, Addison-Wesley, 1955.

E. Purcell, *Electricity and Magnetism*, 2nd ed., McGraw-Hill, 1985.

F. Rohrlich, *Classical Charged Particles*, Addison-Wesley, 1965.

E. Segre, *Nuclei and Particles*, Benjamin, 1964. Like many introductory texts on nuclear physics this book discusses the passage of high-energy particles through matter.

8-7 PROBLEMS

1. Discuss the effect of retarded potentials on Newton's third law.

2. The integral in equation (8-3.28) contains $\mathbf{J}(\mathbf{r}, t)$ which is a function of velocity, not acceleration. Show that it predicts no radiation from a charge moving with constant velocity.

3. An electron has a constant acceleration in the opposite direction from its velocity. For $\beta = 0.8$, at what angle (relative to the velocity) does the maximum in the radiated power distribution $dW(\mathbf{n})/dt$ occur?

4. Find the spectral distribution $W(\mathbf{n}, \omega)$ radiated by a current loop (a loop antenna) of radius R, with the current given by:

$$I = I_0 e^{-t/T} e^{i\omega_0 t},$$

where $\omega_0 T > 1$. You will have to use a delta function in the current density. Sketch the directional radiation pattern and a graph of its frequency distribution. Estimate the values of the parameters R, ω_0, and T needed to radiate 1.0 microJoule into all angles.

5. Use equation (8-5.48) to calculate the directional radiation pattern of the loop antenna in Problem 4. Note: This problem is nonrelativistic.

6. Find the electric dipole, magnetic dipole, and electric quadrupole moments for a single charged particle moving with a constant velocity v around a circular loop of radius R. You will have to express the charge and current densities in terms of a delta function.

7. Use the magnetic dipole moment to calculate the directional radiation pattern of the loop antenna in Problem 4.

8. Do the calculations explicitly in equations (8-5.37) and (8-5.38) showing that for a radiating charge at small velocities:

$$\mathbf{E} = \frac{e}{c^2} \left[\frac{\mathbf{n} \times (\mathbf{n} \times \mathbf{a})}{|\mathbf{r}_{fs}|} \right]_{\text{ret}},$$

$$\mathbf{B} = \frac{e}{c^2} \left[\frac{\mathbf{a} \times \mathbf{n}}{|\mathbf{r}_{fs}|} \right]_{\text{ret}}.$$

9. Calculate the stress tensor explicitly for the fields equation given in Problem 8.

10. Prove the formula for the time derivative of the unit vector:

$$\frac{d\mathbf{n}}{dt} = \frac{c(\mathbf{n} \times (\mathbf{n} \times \boldsymbol{\beta}))}{r}.$$

(Hint: The vector $\mathbf{n}$ is a unit vector, orthogonal to $d\mathbf{n}$.)

11. Prove the relativistic Larmor formula (equation (8-5.55)):

$$P_R = \frac{2e^2}{3c} \left[\dot{\boldsymbol{\beta}}^2 - (\boldsymbol{\beta} \times \dot{\boldsymbol{\beta}})^2 \right] \gamma^6.$$

12. A proton is moving with velocity $b = 0.999$ in a circular orbit of 1.0 kms. Find the ratio of its backward tangential radiation intensity to its forward tangential radiation intensity. Also find the ratio of its radiation intensity in the radial direction to its forward radiation intensity.

APPENDIX 8-1   RADIATION BY A FAST CHARGED PARTICLE

This notebook investigates the radiation distributions of charged particles accelerated to velocities comparable to the speed of light. There are two cases of particular interest: *bremsstrahlung* and *synchrotron radiation*. Bremsstrahlung occurs during the deceleration of charged particles during passage through a material medium, where the acceleration is in the same (or opposite) direction as the velocity. It is the familiar source of x-rays. Synchrotron radiation occurs for charged particles moving in circular orbits, where the acceleration is orthogonal to the velocity. It is an excellent coherent light source.

Initialization

```
Remove["Global`@"]


Needs["Graphics`Graphics`"]
Needs["Calculus`VectorAnalysis`"]
Needs["Graphics`ParametricPlot3D`"]
```

Define Vectors

```
(*   acceleration *)
   aa = {ax,ay,az};
(*   velocity given as bayta = v/c *)
   bb = {bx,by,bz};
(*   direction of detector *)
   nn = {Sin[q]*Sin[f],Sin[q]*Cos[f],Cos[q]};
```

Calculate the Angular Distribution Function

```
(*  angular distribution function *)
(*  in acceleration\emdash velocity plane *)
   innr = CrossProduct[(nn-bb),aa];
   outr = CrossProduct[nn,innr];
   numer = outr.outr;
   denom = (1-nn.bb)^5;
(*  general angular distribution *)
   distrib = numer/denom;
   Short[distrib,4]
(*  distribution in a-v plane where f=0 *)
   distribav = distrib/.f->0;
   Short[distribav,4]
```

$$(\text{Power}[-(az\ bx\ \text{Cos}[q]) + ax\ bz\ \text{Cos}[q] - ax\ \text{Cos}[q]^2 - $$

$$ay\ bx\ \text{Cos}[f]\ \text{Sin}[q] + ax\ by\ \text{Cos}[f]\ \text{Sin}[q] + <<1>> - $$

$$ax\ \text{Cos}[f]^2\ \text{Sin}[q]^2 + ay\ \text{Cos}[f]\ \text{Sin}[f]\ \text{Sin}[q]^2\ ,\ 2] + <<2>>)\backslash\backslash$$

$$/\ (1 - bz\ \text{Cos}[q] - by\ \text{Cos}[f]\ \text{Sin}[q] - bx\ \text{Sin}[f]\ \text{Sin}[q])^5$$

$$(\text{Power}[-(az\ by\ \text{Cos}[q]) + ay\ bz\ \text{Cos}[q] - ay\ \text{Cos}[q]^2 + $$

$$az\ \text{Cos}[q]\ \text{Sin}[q],\ 2] + <<1>> + $$

$$\text{Power}[az\ by\ \text{Sin}[q] - ay\ bz\ \text{Sin}[q] + ay\ \text{Cos}[q]\ \text{Sin}[q] - $$

$$az\ \text{Sin}[q]^2\ ,\ 2]) \ /\ (1 - bz\ \text{Cos}[q] - by\ \text{Sin}[q])^5$$

Radiation Patterns

Bremsstrahlung Radiation Patterns

In bremsstrahlung the acceleration is in the same (or opposite) direction as the velocity. As the velocity increases to the speed of light, the total radiation increases, and its distribution is thrown forward. This is shown clearly in the following diagrams:

Cartesian Coordinates

```
(*  acceleration is is z-direction *)
(*  velocity is in z-direction *)
(*  detector is in y-z plane *)


(*  Velocity increment *)
bstep = 0.3;
```

```
(*  Number of drawings calculated in plottable *)
picCount = 2;
(*  The product bstep*picCount is the final velocity *)
(*  It must be less than one *)


values = {ax->0,ay->0,az->1,bx->0,by->0,bz->bstep*j};
func = distribav/.values;
seefunc = func/.j->picCount
doPlot = Plot[Evaluate[Table[func,{j,0,picCount}]],
          {q,-Pi,Pi},PlotRange->All]
```

$$
\frac{Cos[q]^2\ Sin[q]^2\ +\ Sin[q]^4}{(1\ -\ 0.6\ Cos[q])^5}
$$

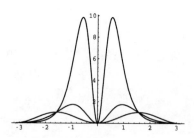

Polar Coordinates

```
bstep = 0.15;
picCount = 3;
values = {ax->0,ay->0,az->1,bx->0,by->0,bz->bstep*j};
func = distribav/.values;


doPlot = PolarPlot[Evaluate[Table[func,{j,0,picCount}]],
          {q,-Pi,Pi},Ticks->None]
```

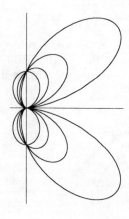

Pattern in Three Dimensions
This plot is split open to show the connection with the polar plot.

```
bayta = 0.65;
values = {ax->0,ay->0,az->1,bx->0,by->0,bz->bayta};
func = distrib/.values
brems = SphericalPlot3D[func,{q,0,Pi},{f,-3,3},
        Ticks->None,PlotRange->All,
        PlotPoints->30,
        ViewPoint->{-5,-4,2}]
```

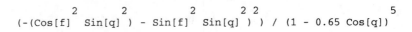

$$(\text{Cos}[f]^2\ \text{Cos}[q]^2\ \text{Sin}[q]^2 + \text{Cos}[q]^2\ \text{Sin}[f]^2\ \text{Sin}[q]^2 +$$

$$(-(\text{Cos}[f]^2\ \text{Sin}[q]^2) - \text{Sin}[f]^2\ \text{Sin}[q]^2)^2\ ) / (1 - 0.65\ \text{Cos}[q])^5$$

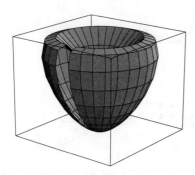

Synchrotron Radiation Patterns
In synchrotron radiation the acceleration is at right angles to the velocity. As in the case of bremsstrahlung, increasing the velocity causes the radiation pattern to become more peaked in the forward direction, that is, toward the direction of the velocity. Its total intensity also increases. The backward lobe, which is the mirror image of the forward lobe at low velocity, becomes negligibly small at high velocities.

Cartesian Coordinates

```
(*  acceleration is is y-direction *)
(*  velocity is in z-direction *)
(*  detector is in y-z plane *)


(*  Velocity increment *)
bstep = 0.2;
(*  Number of drawings calculated in plottable *)
picCount = 2;
(*  The product bstep*picCount is the final velocity *)
(*  It must be less than one *)


values = {ax->0,ay->1,az->0,bx->0,by->0,bz->j*bstep};
func = distribav/.values;
```

```
seefunc = func/.j->picCount
doPlot = Plot[Evaluate[Table[func,{j,0,picCount}]],
              {q,-Pi,Pi},PlotRange->All]
```

$$((0.4\ Cos[q]\ -\ Cos[q]^2)^2\ +$$

$$(-0.4\ Sin[q]\ +\ Cos[q]\ Sin[q])^2)\ /\ (1\ -\ 0.4\ Cos[q])^5$$

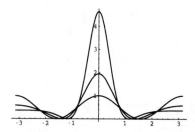

Polar Coordinates

```
(*  acceleration is is y-direction *)
(*  velocity increases in z-direction *)


(*  Velocity increment *)
bstep = 0.2;
(*  Number of drawings calculated in plottable *)
picCount = 2;
(*  The product bstep*picCount is the final velocity *)
(*  It must be less than one *)


values = {ax->0,ay->1,az->0,bx->0,by->0,bz->bstep*j};
func = distribav/.values;
plotfn := PolarPlot[func,{q,0,2*Pi},Ticks->None];
plottable = Table[plotfn,{j,0,picCount}]
```

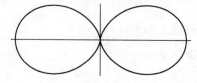

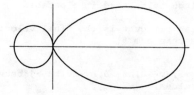

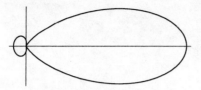

Patterns in Three Dimensions

The second plot is split in half to show the connection with the polar plot.

```
bayta = 0.2;
values = {ax->0,ay->1,az->0,bx->0,by->0,bz->bayta};
func = distrib/.values
synch = SphericalPlot3D[func,{q,0,Pi},{f,0,2*Pi},
        Ticks->None,
        ViewPoint->{5,-4,2}]
synchhalf = SphericalPlot3D[func,{q,0,Pi},{f,0,Pi},
            Ticks->None,
            ViewPoint->{5,-4,2}]
```

$$\left(\text{Cos}[f]^2 \; \text{Sin}[f]^2 \; \text{Sin}[q]^4 + \right.$$

$$\left(-0.2 \; \text{Cos}[f] \; \text{Sin}[q] + \text{Cos}[f] \; \text{Cos}[q] \; \text{Sin}[q] \right)^2 +$$

$$\left. \left(0.2 \; \text{Cos}[q] - \text{Cos}[q]^2 - \text{Sin}[f]^2 \; \text{Sin}[q]^2 \right) \right) /$$

$$\left(1 - 0.2 \; \text{Cos}[q] \right)^5$$

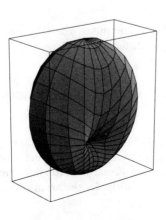

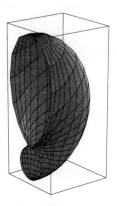

Appendix 8-2   Tensor Potentials

We have seen that electromagnetic theory can be expressed conveniently in terms of the vector potentials, A^k. There are, in fact, other potential-like quantities that can be useful in certain circumstances. In treating radiation, we may introduce a set of vector potentials for the A^k, called the *Hertz potentials*. Thus:

$$A^0 = -\nabla \cdot \Pi_E, \tag{1}$$

$$A = \frac{1}{c} \frac{\partial \Pi_E}{\partial t} + \nabla \times \Pi_M. \tag{2}$$

The discussion of radiation by multipoles is simplified by introducing these quantities. Useful accounts of the Hertz potentials may be found in Panofsky and Phillips, Marion and Heald, and Stratton.

We will consider a relativistic treatment which generalizes the method. Let the vector potentials be written in terms of the tensor potential:

$$A^i = \partial_j \Pi^{ij}. \tag{3}$$

By convention, we sum over the second index. Then the tensor potential satisfies a generalized Lorentz gauge condition:

$$\partial_i \partial_j \Pi^{ij} = 0, \tag{4}$$

which is an identity when the tensor potential is antisymmetric:

$$\Pi^{ij} = -\Pi^{ji}. \tag{5}$$

Then the tensor potential has six independent elements, corresponding to the six components of the two three-dimensional Hertz vector potentials. Replacing a vector by a tensor may not seem especially useful, but whenever the radiation source is a magnetic or electric multipole, the solutions can be expressed in simple form.

Define the *relativistic dipole moment* at point x_f due to source currents at points x_s:

$$m^{ij}(x_f) = \iiint \frac{1}{3c} \left[x_{fs}^i J^j(x_s) - x_{fs}^j J^i(x_s) \right] d^3 r_s. \tag{6}$$

This expression differs from the magnetic dipole moment defined in Chapter 3 in being four dimensional, and thereby includes both magnetic and electric dipole moments. Note that a factor of 3 in the relativistic form replaces a factor of 2 in the magnetic moment. Since:

$$dQ = \rho \, d^3 r, \tag{7}$$

is invariant, the relativistic dipole moment is covariant under Lorentz and rotation transformations. The integrand, M^{ij}, is the relativistic dipole moment density:

$$m^{ij}(x_f) = \iiint M^{ij} \, d^3 r_s, \tag{8}$$

corresponding to polarization and magnetization.

Now find the corresponding current densities. By analogy with Section 3-2, we find the four-dimensional divergence of the dipole moment density, where the derivatives are with respect to the field points x_f. Thus:

$$\frac{\partial}{\partial x_f^j} M^{ij} = \frac{1}{3c} \left[\frac{\partial x_{fs}^i}{\partial x_f^j} J_s^j - \frac{\partial x_{fs}^j}{\partial x_f^j} J_s^i + x_{fs}^i \frac{\partial J_s^j}{\partial x_f^j} - x_{fs}^j \frac{\partial J_s^i}{\partial x_f^j} \right]_{\text{avg}} = \frac{1}{3c} \left[\delta_j^i J_s^j - 4 J_s^i \right]_{\text{avg}}$$

$$= \frac{1}{c} J^i, \tag{9}$$

where, like Section 3-2, the average is over a very small region around the field point. The reason for the factor 3 in the relativistic form is apparent. In terms of the temporal and spatial components, equation (9) can be written:

$$\nabla \cdot \mathbf{P} = \frac{1}{c} J^0, \tag{10}$$

$$\nabla \times \mathbf{M} + \frac{1}{c} \dot{\mathbf{P}} = \frac{1}{c} \mathbf{J}, \tag{11}$$

which are isomorphic to equations (3-1.7), (3-2.3), and (3-2.12) for dielectric and magnetic materials. A similar isomorphism holds for Maxwell's field equations in three dimensions:

$$\nabla \cdot \mathbf{E} = \frac{4\pi}{c} \rho c, \tag{12}$$

$$\nabla \times \mathbf{B} - \frac{1}{c} \dot{\mathbf{E}} = \frac{4\pi}{c} \mathbf{J}, \tag{13}$$

and the corresponding single field equation in four dimensions:

$$\partial_j F^{ij} = -\frac{4\pi}{c} J^i. \tag{14}$$

Now apply the expressions for the current densities in terms of the dipole moments to the electromagnetic field equations. Then the equations for the vector potential, A^i, may be written in terms of the tensor potential, Π^{ik}:

$$\partial_j \partial^j A^i = \partial_k \left[\partial_j \partial^j \Pi^{ik} \right] = \frac{4\pi}{c} J^i = 4\pi \partial_k M^{ik}. \tag{15}$$

From this we can pick off an equation for the tensor potential:

$$\partial_j \partial^j \Pi^{ik} = 4\pi M^{ik}, \tag{16}$$

which is subject to the gauge transformation:

$$\Pi^{ik} \longleftrightarrow \Pi^{ik} + \gamma^i \Gamma^k, \tag{17}$$

with the condition on the gauge function:

$$\partial_n \partial^n \Gamma^k = 0. \tag{18}$$

This tensor potential represents the field of a (space–time-dependent) elementary dipole current distribution.

Since the method is successful for dipoles, we are encouraged to generalize the method to higher multipoles. Define higher-rank tensor potentials obeying generalizations of the Lorentz gauge condition:

$$
\begin{aligned}
\Pi^i = A^i = \partial_j \Pi^{ij}, && \partial_i \Pi^i = \partial_i A^i = 0, \\
\Pi^{ij} = \partial_k \Pi^{ijk}, && \partial_i \partial_j \Pi^{ij} = 0, \\
\Pi^{ijk} = \partial_m \partial_k \Pi^{ijkm}, && \partial_i \partial_j \partial_k \Pi^{ijk} = 0, \\
\text{etc.} && \text{etc.}
\end{aligned}
\tag{19}
$$

The corresponding gauge fields $\Gamma^{i\cdots z}$ satisfy generalizations of equation (18). The higher multipoles satisfy corresponding conditions based on the continuity equation:

$$
\begin{aligned}
M^i = \frac{1}{c} J^i = \partial_j M^{ij}, && \partial_i M^i = \frac{1}{c} \partial_i J^i = 0, \\
M^{ij} = \partial_k M^{ijk}, && \partial_i \partial_j M^{ij} = 0, \\
M^{ijk} = \partial_m M^{ijkm}, && \partial_i \partial_j \partial_k M^{ijk} = 0, \\
\text{etc.} && \text{etc.}
\end{aligned}
\tag{20}
$$

The most general expression for the vector potential is the sum over the tensor potentials:

$$A^i = \Pi^i + \partial_j \Pi^{ij} + \partial_j \partial_k \Pi^{ijk} + \cdots. \tag{21}$$

Similarly, the current density can be expressed as a sum of multipole terms:

$$-\frac{1}{c} J^i = M^i + \partial_j M^{ij} + \partial_j \partial_k M^{ijk} + \cdots. \tag{22}$$

Then if these are substituted into the equation for the potential, equation 4-5.17, we get an infinite regression:

$$\left[\partial_n \partial^n \Pi^i + 4\pi M^i + \partial_j \left[\partial_n \partial^n \Pi^{ij} + 4\pi M^{ij} + \partial_k \left[\partial_n \partial^n \Pi^{ijk} + 4\pi M^{ijk} + \cdots \right] \right] \right] = 0. \tag{23}$$

A convenient solution of this equation is obtained by setting equal rank terms to zero. That is, let the tensor potentials satisfy the equations:

$$\partial_n \partial^n \Pi^i = 4\pi M^i \quad \text{(monopole)},$$

$$\partial_n \partial^n \Pi^{ij} = 4\pi M^{ij} \quad \text{(dipole)},$$

$$\partial_n \partial^n \Pi^{ijk} = 4\pi M^{ijk} \quad \text{(quadrupole)},$$

etc.

(24)

Then each of the tensor potentials represents the contribution of a pure multipole moment.

The solutions may be expressed in terms of the retarded Green function:

$$\Pi^{i\cdots z}(x_f) = \iiiint G(x_f, x_s) M^{i\cdots z}(x_s) \, d^4 x_s.$$

(25)

We can now express the electromagnetic field tensor as a tensor potential series. Thus:

$$F^{ik} = \left(\partial^i \Pi^k - \partial^k \Pi^i \right) + \partial_n \left(\partial^i \Pi^{kn} - \partial^k \Pi^{in} \right)$$

$$+ \partial_n \partial_m \left(\partial^i \Pi^{kmn} - \partial^k \Pi^{imn} \right) + \cdots.$$

(26)

Because the method is general and relativistically invariant, it applies to both steady-state and radiation fields. Compare it to the multipole expansions in Sections 5-2 and 8-1.

CHAPTER 9

Beyond the Classical Theory

We have found it of paramount importance that in order to progress we must recognize our ignorance and leave room for doubt. Scientific knowledge is a body of statements of varying degrees of certainty—some most unsure, some nearly sure, but none absolutely certain.

Richard Feynman
The Value of Science

Mathematical discoveries, small or great . . . are never born of spontaneous generation. They always presuppose a soil seeded with preliminary knowledge and well prepared by labour, both conscious and subconscious.

Jules Henri Poincaré
La Science et l'Hypothèse

The discovery of the electron was a mixed blessing for classical physics. It led directly to atomic and nuclear physics and a physical foundation for chemistry. The electron was the first elementary particle to be discovered, and the practical applications of electronics are virtually innumerable. On the other hand, investigations of its properties and interactions was a major factor in the demise of classical physics and its replacement by quantum physics.

The electron has, of course, a fixed quantity of charge. But is the particle truly elementary, or does it have a structure? Is it possible to create a self-consistent electromagnetic theory which includes a point electron? If the electron has microstructure, what are its interactions with electromagnetic fields at this level? What is the relationship between the charge of the electron and its mass? Can it be split into still smaller particles? Is it unique, or are there other charged particles with different masses and charges, perhaps forming a family of particles? These questions and others were implicit right from the initial discovery of the electron. In fact, these questions are asked of any new particle, even today.

In the preceding chapters, classical electromagnetic theory was developed to describe the interactions between point charges at relatively large distances. Following Lorentz and Abraham, we now apply the theory to the particle itself and its self-fields. This purely classical investigation of the properties of an elementary particle leads to fascinating results, and to certain problems and inconsistencies. Since radiation carries energy and momentum, a radiating particle ought to experience a force of reaction to the radiation. We consider how this reaction force can be incorporated into the equations of motion. A second problem concerns the distribution of the electron's charge. If the electron is truly a point particle, its charge density is a delta function. Then the potential energy due to this charge density is infinitely large, as is its mass, because of the relativistic mass-energy relation. On the other hand, a finite charge distribution would also have observable effects on the dynamics of the particle.

These questions become most important at high energy and short distances, just the region where quantum effects dominate. Thus one cannot reasonably neglect quantization in discussing the microscopic properties of the electron; in fact, many of the problems with the classical theory can only be resolved by a quantized treatment. Recently, a unified theory of electrodynamics and weak interactions has been found. Although details of the unification cannot be discussed in a purely classical context, students should know something about this critical event which closes the story of electromagnetism as an independent branch of physics. For these reasons we give a short overview of how electrodynamics was quantized, and an essentially qualitative discussion of the unified theory.

9-1   Radiation Reaction

We have seen that electromagnetic radiation carries energy and momentum. Then, because of conservation of momentum, a particle emitting radiation ought to experience a reaction force associated with the momentum of the radiation–quite apart from external forces acting on the particle. It is convenient for us to distinguish between the external force and the *radiation reaction force*. We write:

$$\mathbf{F} = \mathbf{F}_{ext} + \mathbf{F}_{rad}.$$ (9-1.1)

Then the equation of motion is:

$$m\frac{d\mathbf{v}}{dt} = \mathbf{F}_{ext} + \mathbf{F}_{rad}.$$ (9-1.2)

Now we can try to find an expression for the radiation reaction force using the Larmor expression, equation (8-5.51), for the radiated power:

$$P_{rad} = \frac{2e^2a^2}{3c^3}.$$ (9-1.3)

That is, integrate the radiated power and set it equal to the work done by a force acting on a moving particle:

$$\int_0^T \frac{2e^2}{3c^3}\dot{\mathbf{v}}\cdot\dot{\mathbf{v}}\,dt = \int_0^T \frac{d}{dt}\left(\frac{2e^2}{3c^3}\dot{\mathbf{v}}\cdot\mathbf{v}\right)dt - \int_0^T \frac{2e^2}{3c^3}\ddot{\mathbf{v}}\cdot\mathbf{v}\,dt$$

$$= \Delta W_{rad} = -\int_0^T \mathbf{F}_{rad}\cdot\mathbf{v}\,dt.$$ (9-1.4)

Radiation decreases the particle's kinetic energy. Thus the minus sign in the second line indicates work done on the moving charge by radiation energy loss, i.e., work due to radiation reaction forces. We can extract an expression for this force from the integral by making two strong assumptions: the exact integral vanishes over the time interval, and the integrands are equal if the integrals are equal. Then the radiation reaction force might be written:

$$\mathbf{F}_{rad} = \frac{2e^2}{3c^3}\ddot{\mathbf{v}} = m\tau\ddot{\mathbf{v}}.$$ (9-1.5)

The first assumption is correct for periodic motion, or when the acceleration vanishes at the end points of the interval, or for other special cases. The second assumption is reasonable, but not mathematically rigorous. We must consider its consequences. The characteristic time in this expression:

$$\tau = \frac{2e^2}{3mc^3} = \frac{2}{3}\frac{R_C}{c}$$ (9-1.6)

represents the approximate time for a signal traveling at the speed of light to cross the classical electron radius. Its small value, 2.62×10^{-24} s, suggests relatively small contributions of radiation reaction to forces on the electron, except at very high frequency.

Example

The equation of motion for an electron in an external field becomes:

$$m\dot{\mathbf{v}} - m\tau\ddot{\mathbf{v}} = \mathbf{F}_{ext}, \tag{9-1.7}$$

known as the *Lorentz–Abraham equation*. It is distinguished from the usual equation of motion by being of third order. In the absence of external forces, i.e., for a free electron, the solution of the equation is:

$$\mathbf{v} = \mathbf{v}_1 + \mathbf{v}_2 \exp\left(t/\tau\right), \tag{9-1.8}$$

where $\mathbf{v}_1$ and $\mathbf{v}_2$ are constants. The first term with constant velocity is correct for free particles; however, the second term, called the *runaway solution*, is absurd since it implies the particle accelerates exponentially with extremely short characteristic time in an arbitrary direction. Since the particle's energy becomes larger than its initial value, the conditions under which the reaction force was determined (conservation of energy) are contradicted. It is tempting to think that since we obtained the reaction force from the square of the acceleration, we might change the sign of the reaction force. Then the second solution becomes a damped exponential function of time. However, in that case, a free electron would accelerate from its initial velocity, radiate energy, and reach a different steady velocity after about 10^{-23} s. But, since they are still free particles, the cycle would continue indefinitely. This is clearly unphysical.

Because of these disastrous results, it might seem best just to call the runaway solution unphysical and ignore it by setting $\mathbf{v}_2 = 0$. However, the equation of motion itself ought to be more reliable, since it is based on conservation of energy. Therefore, we will consider it further by partially integrating it. Multiply by the exponential function and get:

$$m(\tau\ddot{\mathbf{v}} - \dot{\mathbf{v}})\exp(-t/\tau) = m\tau\frac{d}{dt}[\dot{\mathbf{v}}\exp(-t/\tau)] = -\mathbf{F}_{ext}\exp(-t/\tau). \tag{9-1.9}$$

Integration gives an integrodifferential equation of motion:

$$m\dot{\mathbf{v}} = -\exp(t/\tau)\int_0^t \mathbf{F}_{ext}(t')\exp(-t'/\tau)\,d(t'/\tau). \tag{9-1.10}$$

Consider the properties of the solutions. Since it contains the time integral of the force, the acceleration depends not only on the force at the same time as the acceleration, but on the force at earlier, and later times as well. Since it takes time for a signal to travel from one point in space to another, this is a *nonlocal interaction*. Furthermore, there are questions about causality here. Consider the special case shown in Figure 9-1, where the external force is a step function applied at time $t = t_0$. The acceleration rises exponentially to the classical value and becomes constant after $t = t_0$. But, the problem is, the acceleration anticipates

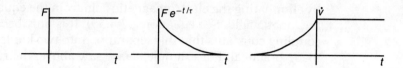

FIGURE 9-1. PREACCELERATION OF THE CHARGED PARTICLE. The acceleration shown in the third curve is a solution of equations (9-1.9) and (9-1.10). Since the particle accelerates before the force is applied, it is known as preacceleration.

the change in the force; it is said to *preaccelerate*. The preacceleration violates our expectations of causality between force and acceleration. It is perhaps not unreasonable to take it to be the effect of an advanced potential; but then one is inclined to ask why advanced fields act only at short range, when retarded fields have unlimited range. In any case, the range is the classical electron radius, i.e., much less than the electron Compton length, which we may let represent the onset of the quantum domain.

9-2 CLASSICAL MODELS OF THE ELECTRON

In the preceding discussion of electrodynamics we distinguished for the most part between two complementary types of problems. If we know the sources of the fields, i.e., the current densities, then we determine the potentials, the fields, and the energy and momentum radiated by the sources. That is, we solve the Maxwell equations, subject to certain auxiliary conditions. On the other hand, we can consider the inverse problem, in which the fields are the given quantities; then we solve the Lorentz force equations for the trajectory of the particle. This dichotomy between the two types of problems is artificial in principle, since a particle moving in an external field creates its own field, bearing its attendant energy and momentum. Charges, having velocities determined by the fields in which they move, become the source of (retarded) secondary fields, thus modifying the velocities. In mathematical terms, the two sets of equations

$$\partial_n \partial^n F^{ik} = \frac{4\pi}{c} (\partial^i J^k - \partial^k J^i), \qquad (9\text{-}2.1)$$

$$\frac{dU^i}{ds} = \frac{e}{mc^2} F^{ik} U_k, \qquad (9\text{-}2.2)$$

are a system of simultaneous equations for the velocities and fields to be solved simultaneously. This was done in Section 7-4, for example, by eliminating the electromagnetic field from the equations in terms of the kinetic field. See equations (7-4.55). It is only because of the small coupling constants that this separation into two kinds problems can be made practical. For example, as we saw above, the radiation reaction has a negligible effect except at extremely short times and distances.

We have heretofore assumed an electron to be a point particle with mass and charge determined by experiment, and considered its interaction with external fields. The point model of the electron has had great success for calculating the interactions between electrons. As far as is known empirically, electrons are elementary particles, in the sense that they are not composite. High-energy scattering experiments indicate that the electron charge density is zero down to 10^{-15} cm, i.e., two orders of magnitude smaller than the classical electron radius. The maximum range of any charge distribution must therefore be smaller than this value. However, if we apply the classical theory to the electron itself, there are some problems. In particular, if we use the stress tensor to calculate the potential energy due to a point charge, the result is infinite. It is conceivable then that the physical electron has microscopic structures, such as charge or mass density, differing from a delta function at short range.

The intrinsic properties of the electron are of great interest in themselves. For example, it should be possible to calculate an electromagnetic contribution to the mass from the charge distribution. Further, an extended electron would have an angular momentum and a magnetic moment that depend on the charge and mass distributions. This is certainly true for nuclei, in which electromagnetic mass contributes about 0.1% to the nuclear mass, in good agreement with calculation. However, nuclei are composites, with charge distributions resulting from their nucleon distributions. In fact, the nucleons themselves are apparently also composites with properties dependent on the distributions of their constituent quarks. Figure 9-2 illustrates some model charge distributions for a classical electron.

We must therefore consider the interactions between one part of the charge distribution and another, i.e., its self-interaction. For a rigid charge distribution, the self-interaction is most clearly manifested as the self-energy, i.e., the energy needed to assemble the charge into its final distribution. However, relativistic systems can never really be rigid, if only because of a Lorentz contraction along the direction of motion. A realistic model of an extended electron should include short-range forces on the charge, binding it to the electron. In other words, we take a nuclear model of the electron. Then one can consider deformations of the charge distribution under interaction with external fields, and possible oscillations.

In the following we consider the interactions of a particle with a distributed charge. We will distinguish between purely external forces

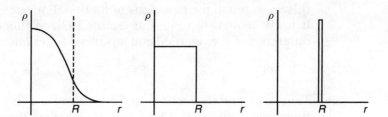

FIGURE 9-2. MODEL CHARGE DISTRIBUTIONS.   The exact shape of the charge density is unknown, but each has a characteristic length, R, which represents the range of the distribution. A true point charge has a delta function with its singularity at $r = 0$, but its self-energy would be infinite. The third distribution represents a shell of charge with radius, R. The three diagrams are not to scale, but each of the three distributions gives the same total charge.

and currents and the self-forces and currents. That is:

$$F^k = F^k_{ext} + F^k_{self}, \tag{9-2.3}$$

$$J^k = J^k_{ext} + J^k_{self}. \tag{9-2.4}$$

Then the force on an electron due to an external field is the integral of the force over all of its distributed charge. Figure 9-3 illustrates the force on an electron due to another electron separated by a distance r_{ext}. At small velocities the force becomes the Coulomb force, analogous to the moon's gravitational force on the Earth, but at higher velocities magnetic effects become important.

The total force acting on the particle is due to all the external fields and any self-force contributions. Then the equation of motion can be expressed in the usual form of forces, masses and acceleration and the damping term obtained above. We can calculate the forces acting on an extended particle with the intention of investigating the radiation reaction. First, we calculate the potentials and the field tensor. From

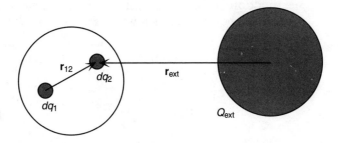

FIGURE 9-3. THE TOTAL FORCE ACTING ON A CHARGE ELEMENT.   The total force on dq_2 is the sum of the internal forces between different parts of the charge density, and external forces, illustrated here by a second electron.

this we can find the equations of motion. For purposes of calculation, it is convenient to write the retarded Green function as a sum and difference of the retarded and advanced functions:

$$G_{\text{ret}} = \frac{1}{2}(G_{\text{ret}} - G_{\text{adv}}) + \frac{1}{2}(G_{\text{ret}} + G_{\text{adv}}) = G_\alpha + G_\sigma. \qquad (9\text{-}2.5)$$

That is, G_α and G_σ are time–antisymmetric and time–symmetric Green functions, respectively. Now we can calculate the antisymmetric and symmetric contributions to the potential and field separately and then recombine them later.

First consider the time–symmetric contribution to the self-force. For this part of the calculation we use equation (4-4.35) for the field:

$$\partial_n \partial^n F^{jk}_{\text{self}} = \frac{4\pi}{c}(\partial^j J^k_{\text{self}} - \partial^k J^j_{\text{self}}). \qquad (9\text{-}2.6)$$

In the following we suppress the designations, self-force and self-current. Now solve in terms of the time-symmetric Green function:

$$F^{jk}_\sigma = \frac{4\pi}{c}\int (\partial^j J^k - \partial^k J^j)\frac{1}{2}(G_{\text{ret}} + G_{\text{adv}})\,d^4x. \qquad (9\text{-}2.7)$$

In Section 8-4 we saw that the sum of the two Green functions can be expressed as the delta function:

$$4\pi(G_{\text{ret}} + G_{\text{adv}}) = 2\delta(T), \qquad (9\text{-}2.8)$$

where T is a shorthand notation for the square of the interval $T = |x_2 - x_1|^2$ between points x_1 and x_2 in 4-space. The force is calculated by integrating the Lorentz force density:

$$F^k = \frac{e}{c}F^{ki}U_i \rightarrow \frac{1}{c}\int F^{ki}J_i\,d^3r. \qquad (9\text{-}2.9)$$

Most of the calculations for the field are already done in Section 8-5, and we can use the results. Rewrite equation (8-5.15) for a continuous distribution of the charge of the electron. The electromagnetic field is a function of the point x_2, summed over the contributions from charge at all the points x_1, as shown in Figure 9-3. Then:

$$F^{ki}_\sigma(x_2) = \frac{-1}{c}\int\int \rho(x_1)\left(U^i(x_1)\frac{\partial T}{\partial x_k} - U^k(x_1)\frac{\partial T}{\partial x_i}\right)\frac{d}{dT}\left(\frac{\delta(T_0)}{dT/ds}\right)ds\,d^3r_1$$

$$= \frac{-1}{c}\int\int \rho(x_1)\left(U^i(x_1)\frac{\partial s}{\partial x_k} - U^k(x_1)\frac{\partial s}{\partial x_i}\right)\frac{d}{ds}\left(\frac{\delta(T_0)}{dT/ds}\right)ds\,d^3r_1.$$

$$(9\text{-}2.10)$$

Note that the evaluation of the delta function of a function is already calculated. The expression $\delta(T_0)$ is a shorthand notation for $\delta(ct_{21} \pm$

$|r_{21}| = 0$), i.e., for advanced and retarded values of x. Now, multiply by $J_i(x_2)/c$ and integrate; then the force is:

$$F_\sigma^k = \frac{1}{c^2} \iiint \rho(x_1)\,\rho(x_2) \frac{ds}{dT} \frac{d}{ds}\left[U^i(x_1)U_i(x_2)\frac{\partial s}{\partial x_k}\right] \delta(T_0)\, ds\, d^3r_1\, d^3r_2$$

$$-\frac{1}{c^2} \iiint \rho(x_1)\,\rho(x_2) \frac{ds}{dT} \frac{d}{ds}\left[U^k(x_1)U_i(x_2)\frac{\partial s}{\partial x_i}\right] \delta(T_0)\, ds\, d^3r_1\, d^3r_2,$$

$$(9\text{-}2.11)$$

where we have used $d\rho/ds = 0$. Because the expression $\delta s/\partial x_k$ is anti-symmetric under the exchange $x_1 \longleftrightarrow x_2$, the first term vanishes (i.e., $F = -F = 0$).

Now evaluate the integrals using the expressions:

$$\frac{\partial s}{\partial x_i} U_i(x_2) = \frac{\partial s}{\partial x_i}\frac{dx_i}{ds}c = c, \qquad (9\text{-}2.12)$$

$$a^k = \frac{dU^k}{ds}c, \qquad (9\text{-}2.13)$$

$$\frac{ds}{dT} = \frac{c}{2x_n U^n}$$

$$\rightarrow \frac{1}{2|r|}, \qquad v \ll c. \qquad (9\text{-}2.14)$$

The time–symmetric contribution to the force becomes:

$$F_\sigma^k = \frac{-1}{c} \iiint \rho(x_1)\rho(x_2) \frac{ds}{dT} \frac{dU^k(x_1)}{ds} \delta(T_0)\, ds\, d^3r_1\, d^3r_2$$

$$= \frac{-1}{c} \iiint \left[\frac{\rho(x_1)\rho(x_2)}{2U^n x_n} c\delta(T_0)\, d^3r_2\right] \frac{a^k(x_1)}{c} ds\, d^3r_1$$

$$= -\iint mW(x_1)a^k(x_1)\, ds\, d^3r_1$$

$$= -m\langle a^k \rangle, \qquad (9\text{-}2.15)$$

which we express as a weighted average. The weight function is the retarded energy density for the self-field, normalized to the rest energy of the electron:

$$W(r_1) = \frac{1}{2mc^2} \int \frac{\rho(x_2)\rho(x_1)}{|x_n U^n|} c\,\delta(T_0)\, d^3r_2. \qquad (9\text{-}2.16)$$

Integrating the energy density gives the *electromagnetic self-energy*:

$$E_{\text{EM}} = \frac{1}{2} \iint \left.\frac{\rho(x_1)\rho(x_2)}{|r_2 - r_1|}\right|_{\text{ret}} d^3r_1\, d^3r_2. \qquad (9\text{-}2.17)$$

Note that the differential charge element is invariant, even though it has only three dimensions:

$$\rho \, d^3r = \frac{dQ}{d^3r} \, d^3r = dQ. \tag{9-2.18}$$

Thus the electromagnetic self-energy is invariant, and the *electromagnetic mass* of the particle is the relativistic mass associated with its self-energy:

$$m_{\text{EM}} = \frac{1}{c^2} E_{\text{EM}}. \tag{9-2.19}$$

We should not be too surprised that the self-energy appears, since there is precedent for it in nuclear masses, where it represents the energy necessary to assemble the nucleus against the repulsive forces between the protons. Thus the electromagnetic mass of the electron ought to contribute to its observed mass also. Exact integration of the electromagnetic energy is not possible unless the charge density is known. However, the charge density is not given by the classical theory but found empirically. When the acceleration depends only on the path length s, then the integral can be evaluated by using the time-retarded delta function, and the symmetric contribution to the force becomes:

$$F_\alpha^k = -m_{\text{EM}} a^k. \tag{9-2.20}$$

Now consider the time-antisymmetric contribution to the force. In this calculation we first find the time-antisymmetric vector potential, and then obtain the field and the force from it. Taking the time-antisymmetric Green function, the vector potential is:

$$A_\alpha^k(x_2) = \frac{4\pi}{c} \int \frac{1}{2} \, (G_{\text{ret}}^k(x_{21}) - G_{\text{adv}}^k(x_{21})) J^k(x_1) \, d^4x_1$$

$$= \frac{1}{c} \int \frac{1}{2} \left(\frac{J_{\text{ret}}^k(x_1) - J_{\text{adv}}^k(x_1)}{|r_{21}|} \right) d^3r_1. \tag{9-2.21}$$

Now express the currents as Taylor expansions in the variable t. That is:

$$J_{\text{ret}}^k = J_0^k - \frac{r_{21}}{c} \frac{\partial J_0^k}{\partial t} + \frac{r_{21}^2}{2c^2} \frac{\partial^2 J_0^k}{\partial t^2} - \frac{r_{21}^3}{6c^3} \frac{\partial^3 J_0^k}{\partial t^3} + \cdots, \tag{9-2.22}$$

$$J_{\text{adv}}^k = J_0^k + \frac{r_{21}}{c} \frac{\partial J_0^k}{\partial t} + \frac{r_{21}^2}{2c^2} \frac{\partial^2 J_0^k}{\partial t^2} + \frac{r_{21}^3}{6c^3} \frac{\partial^3 J_0^k}{\partial t^3} + \cdots. \tag{9-2.23}$$

Because of the signs, the even terms cancel.

To be useful, the series must converge; thus each term will be successively smaller. We take a ratio test as an auxiliary condition and represent time derivatives by their Fourier components. That is:

$$\left| \frac{(r_1^5/c^5)(\partial^5 J/\partial t^5)}{(r_1^3/c^3)(\partial^3 J/\partial t^3)} \right| \approx \left| \frac{(r_1^5/c^5)\omega^5 J}{(r_1^3/c^3)\omega^3 J} \right| = \frac{r_1^2}{\lambda^2} \ll 1. \tag{9-2.24}$$

The wavelength must be much larger than the range of the charge distribution. If we let the classical electron radius represent the range, then the wavelength must satisfy the condition:

$$\lambda \gg R_C, \tag{9-2.25}$$

which is satisfactory from the static case to gamma ray wavelengths. Although we give up relativistic invariance by taking the Taylor expansion, our results should still be valid over a wide energy range.

Assume the velocity v_1 to be a function of time only. Then, to lowest nonvanishing order, the time-antisymmetric potential is:

$$\mathbf{A}_\alpha = \frac{-1}{c^2} \int \frac{\partial \mathbf{J}_1}{\partial t} \, d^3 r_1 = \frac{-1}{c^2} \int (\dot{\mathbf{v}}\rho + \mathbf{v}\dot{\rho}) \, d^3 r_1$$

$$= \frac{-1}{c^2} \int (\dot{\mathbf{v}}\rho - \mathbf{v}\nabla \cdot \mathbf{J}_1) \, d^3 r_1$$

$$= \frac{-e}{c^2} \dot{\mathbf{v}} \tag{9-2.26}$$

for the spatial components. The second term vanishes after integrating by parts, since the gradient of the velocity vanishes.

The zeroth component (the scalar potential) is:

$$A_\alpha^0 = -\int \left(\frac{1}{c} \dot{\rho} + \frac{\mathbf{r}_{21}^2}{6c^3} \dddot{\rho} \right) d^3 r_1 = -\int \left(\frac{-1}{c} \nabla \cdot \mathbf{J}_1 + \frac{r_{21}^2}{6c^3} \dddot{\rho} \right) d^3 r_1 \tag{9-2.27}$$

using $\mathbf{J}_1 = \mathbf{J}(x_1)$. The first term vanishes since it can be expressed as an integral over a surface where the current density vanishes. Then the second term becomes:

$$A_\alpha^0 = \frac{-1}{6c^3} \frac{\partial^2}{\partial t^2} \int r_{21}^2 \nabla \cdot \mathbf{J}_1 \, d^3 r_1 = \frac{1}{6c^3} \frac{\partial^2}{\partial t^2} \int \mathbf{J}_1 \cdot \nabla r_{21}^2 \, d^3 r_1$$

$$= \frac{-1}{3c^3} \frac{\partial^2}{\partial t^2} \int \mathbf{J}_1 \cdot \mathbf{r}_{21} \, d^3 r_1. \tag{9-2.28}$$

Now calculate the electric field.

$$\mathbf{E}_\alpha = -\frac{1}{c} \dot{\mathbf{A}}_\alpha - \nabla A_\alpha^0 = \frac{e}{c^3} \ddot{\mathbf{v}} + \frac{1}{3c^3} \frac{\partial^2}{\partial t^2} \int \nabla_2 (\mathbf{J}_1 \cdot \mathbf{r}_{21}) \, d^3 r_1. \tag{9-2.29}$$

Then, using equation (8-1.17), we get:

$$\mathbf{E}_\alpha = \frac{e}{c^3} \ddot{\mathbf{v}} - \frac{1}{3c^3} \frac{\partial^2}{\partial t^2} \int \mathbf{J}_1 \, d^3 r_1$$

$$= \frac{2}{3} \frac{e}{c^3} \dot{\mathbf{a}}. \tag{9-2.30}$$

Since the vector potential is a function of time only, the magnetic field vanishes:

$$\mathbf{B}_\alpha = \nabla \times \mathbf{A}_\alpha(t) = 0. \tag{9-2.31}$$

Now let us find the force on the particle due to this field acting on its own charge distribution, i.e., the self-field. Integrating over the charge density gives:

$$\mathbf{F}_\alpha = \int \rho \, \mathbf{E}_\alpha \, d^3 r$$

$$= \frac{2}{3} \frac{e^2}{c^3} \dot{\mathbf{a}}, \tag{9-2.32}$$

which is the radiation reaction force obtained in Section 9-1. This corroborates the derivation and its consequences, and shows that they are not just artifacts of the approximations.

This derivation of the radiation reaction follows the multipole expansion of the vector potentials in Section 8-1, and suggests the following physical interpretation of the mathematical expression. A free electron has a spherically symmetrical charge distribution, centered on the center of mass and moving with it. However, as the electron enters an external field, the force is a function of position, i.e., different parts of the charge distribution experience different magnitude and direction. Such differential forces are called *tidal forces* in analogy with lunar and solar gravitational forces acting on the Earth. Additionally, differential retarded times apply across the charge distribution, especially at high velocity. The charge distribution deforms and may be resolved into multipole moments. Then, as the multipole moments change in time, they radiate as we saw in Section 8-1. This is the source of the radiation reaction force in equation (9-2.32). Then the total radiation of the electron is due to the acceleration of the electron as a whole (as in Section 8-5).

Now we can combine the symmetric and antisymmetric contributions to the force. Adding the self-forces to the external force, the equation of motion becomes:

$$m\mathbf{a} = -m_{\text{EM}}\mathbf{a} + \frac{2}{3} \frac{e^2}{c^3} \dot{\mathbf{a}} + \mathbf{F}_{\text{external}}. \tag{9-2.33}$$

In this equation there are two mass terms, the electromagnetic mass and the original mass which we may call the *bare mass m*. To an outside observer they may be combined, giving an effective mass called the *renormalized mass*:

$$m_R = m + m_{\text{EM}}. \tag{9-2.34}$$

We can rewrite the equation of motion by using the renormalized mass. The new equation is virtually identical to the Lorentz–Abraham equation:

$$m_R(\mathbf{a} - \tau\dot{\mathbf{a}}) = \mathbf{F}_{\text{ext}}. \tag{9-2.35}$$

It is difficult, in general, to distinguish between the bare and electromagnetic contributions to the mass, unless the charge distribution is known and thereby accounted for.

There are several problems associated with this classical mass renormalization. In the first place, the electromagnetic mass of the electron is larger than its measured mass. To see this, take a model for the charge distribution, i.e., assume the electron's charge is distributed uniformly on the surface of a sphere of radius R. From this shell model the electromagnetic self-energy is:

$$E_{\text{EM}} = \frac{1}{2} \int \int \frac{\rho(x_1)\rho(x_2)}{|r_2 - r_1|} \, d^3r_1 \, d^3r_2 = \frac{1}{2} \int \rho(r_2) \, \phi(r_2) \, d^3r_2$$

$$= \frac{e^2}{2R}. \tag{9-2.36}$$

But we know from scattering experiments that R must be less than 10^{-15} cm. Then the electromagnetic mass must be larger than 1.28×10^{-25} g, which is more than two orders of magnitude larger than the mass of the electron. But in the shell distribution, the charge elements are as far from each other as possible and still be inside the radius. Therefore, any other distribution would give larger values for the electromagnetic mass. This means that the bare mass must be negative, and nearly as big in absolute value as the electromagnetic mass. The limit of this process, where the electron is a point particle, amounts to subtracting two infinite masses to give a finite value. This is dubious mathematically, but is useful when treated carefully. In fact, it is so important in quantum field theories, where even worse infinities occur, that renormalization has become the criterion for an acceptable theory.

Assuming that renormalization is sensible, we are left with the question, where does the bare mass come from? Since electrons also interact with the weak and gravitational forces, these also presumably contribute to the electron mass. But there is no reason to expect any weak or gravitational contribution to the electron mass to be negative. At this point, the question of bare mass, and therefore of renormalization, is incompletely understood.

Another problem concerns the spin and magnetic moment of the electron. If we take the preceding shell model of the electronic charge, and assume that the shell rotates with an angular velocity ω, then the magnetic moment is:

$$m_e = \frac{1}{3}\frac{eR^2\omega}{2c} = \frac{1}{3}\frac{eRv}{2c}. \tag{9-2.37}$$

But the measured value is very close to the Bohr magneton:

$$m_B = \frac{e\hbar}{2mc}. \tag{9-2.38}$$

Now set the magnetic moment of the rotating charged shell equal to the Bohr magneton and substitute the classical electron radius for the radius of the shell R_C. This gives for the tangential velocity the ratio:

$$\frac{v}{c} = 3\frac{c\hbar}{e^2} \approx 411. \tag{9-2.39}$$

That is, the tangential velocity of the shell at its equator is larger than the velocity of light in vacuum. The ratio:

$$\alpha = \frac{e^2}{c\hbar} \approx \frac{1}{137}, \tag{9-2.40}$$

called the *fine structure constant*, appears frequently in quantum electrodynamics.

We are therefore left with a puzzle: scattering experiments on electrons show that the electronic charge is confined to a region smaller than a radius of 10^{-15} cm. Then, if the electromagnetic field energy is calculated, *even for the region outside this radius*, the value gives an electromagnetic mass greater than the measured mass of the electron. And if the electron is considered to rotate, the magnetic moment calculated from this model is inconsistent with the theory of relativity.

Table 9-1 lists some characteristic lengths associated with electrons. All of them, except the classical electron radius, contain Planck's constant, showing their essentially quantum foundation. Since R_C is much smaller than the Compton length, which is entirely due to quantization, it is not always easy to disentangle classical from quantum effects at these distances.

9-3   QUANTIZATION

Quantization of physical quantities was introduced by Planck in 1900 to account for the energy distribution of thermal radiation in cavities. Specifically, he found that in order to fit the known frequency distribution of energy in the cavity, the energy of the system must be an integer multiple of the frequency. That is:

$$E = nh\nu = n\hbar\omega, \qquad n = 1, 2, 3, \ldots, \tag{9-3.1}$$

TABLE 9-1.   CHARACTERISTIC LENGTHS, TIMES, AND ENERGIES

		Length (cm)	Time (s)	Energy (Mev)
Bohr radius	$R_B = \dfrac{\hbar^2}{me^2}$	$= 5.29 \times 10^{-9}$	1.76×10^{-19}	3.73×10^{-3}
Compton length	$\bar{\lambda}_C = \dfrac{\hbar}{mc}$	$= 3.86 \times 10^{-11}$	1.29×10^{-21}	5.11×10^{-1}
Electron radius	$R_C = \dfrac{e^2}{mc^2}$	$= 2.82 \times 10^{-13}$	9.40×10^{-24}	$7.00 \times 10^{+1}$
Weak Compton	$\bar{\lambda}_W = \dfrac{\hbar}{m_w c}$	$= 2.19 \times 10^{-16}$	7.31×10^{-27}	$90.0 \times 10^{+3}$

where $\nu = f$ is the frequency, and the constant h is called *Planck's constant*. In Gaussian units it has the value $h = 6.6256 \times 10^{-27}$ erg-s. As we have done, it is frequently convenient to express the frequency as angular frequency ω and the corresponding Planck constant as $\hbar = h/2\pi$.

As we have seen, the classical theory makes unambiguous statements about the wave nature of the electromagnetic field and the expression of its energy. The electromagnetic field is governed by the wave equation whose solutions can be expressed by Green function integrals; thus Huygens' principle predicts diffraction patterns and interference. By Poynting's theorem the energy density is proportional to the square of the field strength:

$$U = \frac{1}{8\pi}(E^2 + B^2). \tag{9-3.2}$$

This expression is well verified under a range of experimental conditions. Thus a major revision of physical theory was required when the electromagnetic field was found to exhibit particle-like properties, and its energy to depend on frequency.

The photoelectric effect illustrates quantization of the electromagnetic field in a particularly straightforward way. It was discovered by Hertz when testing the Maxwell theory. In his experiments, Hertz generated radiation by using a spark gap oscillator tuned to a certain frequency. The detector was another spark gap tuned to the same frequency. In the course of his experiments, he found that the sensitivity of the spark gap detectors was enhanced by light falling on the spark gap. The effect was further investigated by Elster and Geitel, Lenard, Millikan, and others. They found that light falling on a metallic surface caused electrons to be emitted from it; this is the cause of the enhanced sensitivity of the spark gap. It was found that the electrons are emitted with a distribution of energies, whose maximum E_{max} depends on the frequency of the light ω but not on its intensity. Plotting E_{max} against ω gives a straight line with the same slope for all emitters:

$$E_{\text{max}} = \hbar\omega - \phi, \tag{9-3.3}$$

where ϕ is a constant called the *work function*, the energy necessary to free the electron from the material. It was also found that the electron flux (the photocurrent) is proportional to the intensity of the light. The photoelectrons are emitted immediately, with no observable time delay. When the light intensity is low, the rate at which the electrons are emitted is small, but they have the same energy distribution.

The results of the photoelectric experiments, notably the energy dependence, are completely at odds with the classical theory of electromagnetism and light, i.e., the wave theory. Einstein provided the theoretical explanation of the experimental facts. He proposed that a beam of light is composed of a number of particles, each carrying the same energy $\hbar\omega$. This quantum of energy is absorbed by the electron and transformed into kinetic energy. The particle of light has come to be known as the *photon*. The quantum of energy can be any amount, since it depends on the frequency, but only a whole quantum is transferred to the photon. However, inventing the photon does not resolve the contradiction. Light still has wavelike properties, but the photoelectric effect implies that it has particlelike properties as well.

The photoelectric effect is observable at optical wavelengths, e.g., at energies around ten electron volts. At higher energies, e.g., 1000 electron volts, the Compton effect is observed, demonstrating some further aspects of the quantization of light. See Figure 9-4. Recall that in Thomson scattering the incident and scattered wavelengths are the same:

$$\lambda' = \lambda. \tag{9-3.4}$$

In discussing the photoelectric effect, a quantum of energy was introduced, but we know that energy is the time component of a four-dimensional momentum vector. To be consistent with relativity, the spatial components of momentum must be quantized as well. Since we know the time component, we can calculate the spatial components by using Lorentz invariance. The energy of the photon is given by the expression.

$$E = \hbar\omega \tag{9-3.5}$$

and energy and momentum are related by

$$E^2 = c^2 p^2 + m^2 c^4. \tag{9-3.6}$$

Since the photon is massless, the photon momentum must be:

$$\mathbf{p} = \frac{h\nu}{c}\mathbf{n} = \frac{\hbar\omega}{c}\mathbf{n}, \tag{9-3.7}$$

where $\mathbf{n}$ is a unit vector.

Now consider a collision between a photon and an electron at rest. In an elastic collision, the two incident particles are unaltered by the interaction except for an exchange of energy and momentum. Thus, by conservation of energy and momentum:

$$mc^2 + \hbar\omega = E + \hbar\omega', \tag{9-3.8}$$

$$\frac{\hbar\omega}{c}\mathbf{n} = \mathbf{p} + \hbar\omega'\mathbf{n}'. \tag{9-3.9}$$

Substituting these two equations into equation (9-3.6) gives:

$$(\hbar\omega - \hbar\omega' + mc^2)^2 - c^2\left(\frac{\hbar\omega}{c}\mathbf{n} - \frac{\hbar\omega'}{c}\mathbf{n}'\right)^2 = m^2c^4. \tag{9-3.10}$$

Now, solve for the final frequency:

$$\omega' = \frac{\omega}{1 + (\hbar\omega/mc^2)(1 - \cos\theta)}, \tag{9-3.11}$$

where $\cos\theta = \mathbf{n}\cdot\mathbf{n}'$. Expressed as wavelength, the *Compton wavelength shift* is:

$$\lambda' - \lambda = \frac{\hbar}{mc}(1 - \cos\theta). \tag{9-3.12}$$

We define the *electron Compton length*:

$$\bar{\lambda}_C = \frac{\hbar}{mc} = 3.86 \times 10^{-11}, \tag{9-3.13}$$

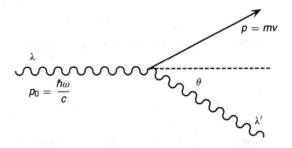

FIGURE 9-4. MOMENTUM BALANCE IN THE COMPTON EFFECT. The wavelength shift of light scattered by an electron at rest depends on the angle of scattering. The amount of shift is consistent with the light carrying a quantum of momentum.

which represents the minimum wavelength shift at a given angle. Equation (9-3.12) clearly disagrees with its classical counterpart in equation (9-3.4). This effect was discovered by A.H. Compton in 1923 during investigations of the interactions of x-rays with matter. Its importance to electromagnetism is that it gives another instance where the Planck constant appears in a physical quantity.

Although we may call the photon a particle because it has discrete values of momentum and energy, we have not justified the most fundamental property of a particle, i.e., a definite location in space and time. Consider how to determine the position of a classical particle. By passing a beam of particles through a small hole, we set the transverse uncertainty in the position of the particles to the diameter of the hole. In principle, we can confine it in a small three-dimensional volume (a classical particle in a box). An alternative scheme is to use a chopper, a high-speed rotating blade behind the hole which effectively opens and closes the hole periodically. Then those particles that the chopper passes are confined in a cylindrical volume whose length depends on the velocity of the particles and the chopper rate. This scheme is used to produce pulsed beams of low-energy neutrons for research purposes. For classical particles, the volume of confinement can be made as small as desired, unrestricted by the particle momentum.

What about photons? As an alternative to passing particles through a hole, imagine reducing the diameter of the beam using lenses. However, light has a resolution limit approximately equal to its wavelength:

$$\Delta r \approx \lambda. \tag{9-3.14}$$

As we have seen, the momentum depends on the wavelength:

$$p = \frac{\hbar \omega}{c} = \frac{\hbar}{\lambda} \approx \frac{\hbar}{\Delta r}, \tag{9-3.15}$$

which gives an *uncertainty relation* between the momentum and the range in the location of the particle. Further consideration along these lines gives a minimum joint uncertainty in **p** and **r**:

$$\Delta p_k \, \Delta r_k \geq \hbar, \qquad k = 1, 2, 3, \tag{9-3.16}$$

for each component. An uncertainty relation for energy and time holds as well:

$$\Delta E \, \Delta t \geq \hbar. \tag{9-3.17}$$

Clearly the photon is not a classical particle, for which the uncertainties in position and momentum can be made arbitrarily small, independent of each other. It is significant that in order to calculate the uncertainty relations for photons, we had to use certain wave properties of these particles." That is, the resolution limit derives from interference of waves.

How does quantization affect the energy and momentum of particles? The initial concept was provided by Louis de Broglie. We have seen a fundamental relationship between mass and energy. They can be converted, one into the other, in a kind of symmetry. But the consequences of the photoelectric effect suggests that, in some sense, light has some properties normally associated with particles. The mass–energy symmetry and particle–wave duality suggests possible wave properties for particles. That is, suppose that the energy of a particle can be represented either in terms of its mass, or in terms of a frequency:

$$E = mc^2\gamma \rightarrow \hbar\omega. \tag{9-3.18}$$

Then, from corresponding Lorentz transformations, the momentum must be:

$$p = mc\beta\gamma \rightarrow \frac{\hbar\omega}{c} = \frac{\hbar}{\lambda}. \tag{9-3.19}$$

This defines the *de Broglie wavelength*. Now assign a wave function to the particle having the form:

$$\psi = \psi_0 \exp\left[\frac{i}{\hbar}(\mathbf{p} \cdot \mathbf{r} - Et)\right]. \tag{9-3.20}$$

This may exhibit diffraction and interference like any wave, but at normal electron energies the de Broglie wavelength is extremely small; consequently, diffraction effects are not detected by casual observation.

In 1925 Davisson and Germer observed electron diffraction by a crystal, producing patterns very similar to those from x-ray diffraction. Later electron diffraction experiments with small slits have shown some truly remarkable results. Diffraction occurs even when the electron flux is so small that only a single electron passes through the slits at a given time. Individual electrons arrive at the detector in apparent random distribution, but after waiting a sufficiently long time for a large number of electrons to pass through the slits, the distribution of the electrons forms a diffraction pattern. Every electron wave apparently interferes with itself in order to cause diffraction. Thus, although individual electrons may strike the detector anywhere, the electron distribution at the detectors is determined by probability. Note that photons, whose wave properties are well established, also behave this way. These considerations establish the relationship between the wave function and probability in particle dynamics.

The de Broglie wave function provided the starting point for the quantum equations of motion. Let us assume the de Broglie function and find an equation for it that is suitable for a free particle, and then generalize to include interactions. We also require the Newtonian equations of motion to be the limit of the quantized equations of motion.

The first derivatives of ψ are:

$$\frac{\partial}{\partial x^k} \psi = \psi_0 \frac{\partial}{\partial x^k} e^{\frac{i}{\hbar} (\mathbf{p} \cdot \mathbf{r} - Et)} = \frac{i}{\hbar} p_k \psi, \qquad (9\text{-}3.21)$$

where p_k is the (constant) momentum in the exponent, and is the eigenvalue of the gradient operator. Now write the momentum in the form of an operator, that is:

$$P_k = \frac{\hbar}{i} \partial_k \rightarrow \frac{\hbar}{i} \nabla. \qquad (9\text{-}3.22)$$

For reasons of relativistic symmetry, the energy operator must be:

$$E_{op} = -\frac{\hbar}{i} \frac{\partial}{\partial t}. \qquad (9\text{-}3.23)$$

A physical argument for letting the quantum variables be operators is that they satisfy the commutation relation:

$$\left[p^j, x^k \right] \psi = \frac{\hbar}{i} \delta^{jk} \psi, \qquad (9\text{-}3.24)$$

which is a more precise statement of the uncertainty discussed above.

Now obtain an equation of motion. Replace the variables in the *Hamiltonian*:

$$H = \frac{p^2}{2m} + V \qquad (9\text{-}3.25)$$

with the corresponding quantum operators, and apply them to the wave function. This Hamiltonian is, of course, the energy, but we also have a separate energy operator. Thus it is reasonable to form an equation in which the *Hamiltonian operator* acting on the wave function equals the *energy operator* acting on it. That is:

$$-\frac{\hbar}{i} \frac{\partial \psi}{\partial t} = H \psi. \qquad (9\text{-}3.26)$$

Substituting the momentum operators into the Hamiltonian gives the *Schrödinger equation*:

$$-\frac{\hbar}{i} \frac{\partial \psi}{\partial t} = \left[\left(\frac{\hbar}{i} \right)^2 \frac{\nabla^2}{2m} + V \right] \psi. \qquad (9\text{-}3.27)$$

This equation is nonrelativistic; Schrödinger originally obtained a relativistic equation, but did not publish it, because of problems with its interpretation for electrons. However, the relativistic equation does apply to particles with integer spin, such as pions and photons. The relativistic equation is called the *Klein–Gordon equation*:

$$\left(\frac{\hbar}{i} \right)^2 \partial_k \partial^k B = m_B^2 c^2 B, \qquad (9\text{-}3.28)$$

which reduces to the D'Alembert equation for massless particles. It can be written in terms of the (boson) Compton length:

$$\bar{\lambda} = \frac{\hbar}{m_B c}.$$

(9-3.29)

Because of the central role of ψ this model is often called *wave mechanics* i.e., mechanics has become a field theory, and the equation of motion is a partial differential equation whose solutions have wavelike properties. The physical variables are operators of various kinds having constant representations. Then the particle dynamics is expressed in terms of the wave function which contains all the time dependence, and which represents the probabilities of specific trajectories. The physical interpretation of the Schrödinger equation, and its associated mathematical apparatus, is usually called the *Copenhagen interpretation*, since Bohr and his coworkers played such a large role in its development. This interpretation of quantum mechanics is summarized as follows.

The state of a given physical system is represented by its wave function (also called its state function for that reason). That is, creating a system with a particular energy, momentum, angular momentum, etc., causes its state function to reflect those values. The absolute value squared $|\psi|^2$ is the probability of the system being in the state characterized by the state function ψ. The probability is usually normalized to unity:

$$\int P(r)\, d^3r = \int |\psi|^2\, d^3r = \int \psi^* \psi\, d^3r = 1.$$

(9-3.30)

That is, the sum of probabilities of all possible outcomes is one. Quantum variables are represented by Hermitian operators. Since the operators are Hermitian, the eigenvalues are real, and their eigenfunctions can be made orthogonal:

$$Q\psi_k = \lambda_k \psi_k.$$

(9-3.31)

Eigenvalues of such an operator are the possible outcomes of a measurement of that variable. Then, from the probability function, one may calculate a weighted average called the *expectation value*:

$$\langle Q \rangle = \int \psi^* Q \psi\, d^3r,$$

(9-3.32)

where Q is the operator representing the physical quantity. We have written the expectation value as a kind of operator sandwich between the wave function and its adjoint, which is the standard expression. By making all the eigenfunctions mutually orthogonal in the Fourier series sense, the expectation value becomes the weighted average of the eigenvalues. For example:

$$\langle E \rangle = \sum E_k \psi_k^* \psi_k.$$

(9-3.33)

Quantum variables do not necessarily commute with each other. Certain sets of operators mutually commute, e.g., momentum and energy. If two or more energy eigenvalues are equal, the system is said to be degenerate. In that case, a complete set of orthogonal functions may be found by requiring the state function to be simultaneous eigenstates of the commuting variables. Then, any set of quantum variables that includes the Hamiltonian and mutually commute can all be measured exactly, and will be mutual constants of the motion. In the limit of large masses or large numbers of particles, quantum physics tends to agree with classical physics. In this sense, classical physics is a special, limiting case of quantum physics. This is called the *correspondence principle*. There are some very interesting exceptions to this principle, however, such as superfluids which have quantum properties in large samples.

The Schrödinger equation is not relativistically invariant, and the Klein–Gordon equation gives incorrect solutions for electrons. A new approach was needed. In 1928, Paul Dirac found a relativistic equation for electrons by introducing a new set of matrix operators characterized by the anticommutation rules:

$$\gamma^j \gamma^k + \gamma^k \gamma^j = 2\mu^{jk}, \qquad j, k = 0, 1, 2, 3, \qquad (9\text{-}3.34)$$

where μ^{jk} is the Minkowski metric tensor. This condition shows clearly that the γ^k cannot be ordinary numbers. However, they can be represented by matrices. For example, one common representation is:

$$\gamma^k = \left(\begin{bmatrix} 0 & 0 & 1 & 0 \\ 0 & 0 & 0 & 1 \\ 1 & 0 & 0 & 0 \\ 0 & 1 & 0 & 0 \end{bmatrix} \begin{bmatrix} 0 & 0 & 0 & 1 \\ 0 & 0 & 1 & 0 \\ 0 & -1 & 0 & 0 \\ -1 & 0 & 0 & 0 \end{bmatrix} \begin{bmatrix} 0 & 0 & 0 & -i \\ 0 & 0 & i & 0 \\ 0 & i & 0 & 0 \\ -i & 0 & 0 & 0 \end{bmatrix} \begin{bmatrix} 0 & 0 & 1 & 0 \\ 0 & 0 & 0 & -1 \\ -1 & 0 & 0 & 0 \\ 0 & 1 & 0 & 0 \end{bmatrix} \right).$$

$$(9\text{-}3.35)$$

One more definition is needed. We have not previously assigned a quantum operator to the velocity. We do so now; define the velocity operator to be:

$$U^k \equiv \frac{c}{2} \gamma^k, \qquad (9\text{-}3.36)$$

which satisfies the invariance:

$$U^k U_k = \frac{c^2}{4} \gamma^k \gamma_k = c^2. \qquad (9\text{-}3.37)$$

Following the procedure for the Schrödinger equation, we write a quantum equation of motion that includes the Dirac matrices. Recall the generalized momentum G^k in equation (7-4.24), which contains the velocity. We quantize it by replacing the classical velocity with equation

(9-3.36). That is:

$$G^k = \frac{m}{2} U^k + \frac{eA^k}{c} \rightarrow \frac{mc}{4} \gamma^k + \frac{eA^k}{c}. \tag{9-3.38}$$

But the momentum operator $(\hbar/i)\partial^k$ is also a translation generator. We obtain a quantum equation of motion by equating the effect of their operations on the wave function. After multiplying by c to give the same units, we write:

$$\frac{\hbar c}{i} \partial^k \psi = G^k \psi. \tag{9-3.39}$$

It is similar to the Schrödinger equation (9-3.27), except that the vector G^k replaces the Hamiltonian. We make it scalar by multiplying by γ_k and rearranging terms. The result is the *Dirac equation:*

$$\gamma_k \left(\frac{\hbar c}{i} \partial^k - eA^k \right) \psi = mc^2 \psi \tag{9-3.40}$$

for electrons and other fermions having electromagnetic interactions.

For free particles, i.e., in the absence of external fields, equation (9-3.40) becomes:

$$\gamma_k \partial^k \psi = \frac{imc}{\hbar} \psi = \frac{i}{\lambda_C} \psi, \tag{9-3.41}$$

where the Compton length is the only constant. Because the γ^k are 4×4 matrices, the wave functions are four-dimensional vector-like quantities called *spinors*. Then the free particle solution can be written:

$$\psi = u \exp\left(\frac{i}{\hbar} p^k x_k \right) = \begin{bmatrix} u_1 \\ u_2 \\ u_3 \\ u_4 \end{bmatrix} \exp\left(\frac{i}{\hbar} p^k x_k \right), \tag{9-3.42}$$

where the spinor components $u_n(p)$ are functions of momentum. The resulting linear equations for the components of $u_n(p)$ have four independent solutions, characterized by two half-integer spin states and two particle–antiparticle states. For example, if the Dirac equation is applied to the electron, it predicts its motion and spin. However, it also has another solution that can be interpreted as a particle with the same mass as the electron and the same spin states, but a positive charge. This particle, now called the *positron*, was observed by Carl Anderson in 1936. By convention, the charged leptons have negative charge, and the corresponding antiparticles have positive charge.

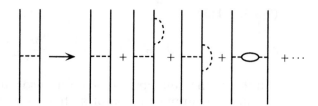

Figure 9-5. Higher-order corrections to electron–electron scattering. The simple ladder terms are augmented by higher-order terms allowed by temporary momentum nonconservation. Such particles are called virtual particles and lead to infinities in the scattering matrix.

For a particle in a given external field A^k the solutions of the Dirac equation can be found using the usual methods in partial differential equations, e.g., by separation of variables. For example, the solution for the hydrogen-like atom gives extremely good agreement with experiment.

The interaction between two particles (e.g., electron scattering) can be expressed in terms of scattering matrices and diagrams roughly similar to those discussed in Section 7-4. There are two significant modifications to the S-matrix due to quantization however: charged particle trajectories are represented by the wave function, and charged particles can be created or annihilated. The first modification puts a Green function solution of ψ into the quantum S-matrix. The second introduces many more terms into the quantum S-matrix. Like its classical counterpart, the quantum S-matrix is expressed as an infinite series of integrals, in which both electromagnetic and particle fields are represented by propagators. The method of using diagrams in detailed calculations of the S-matrix was discovered by Feynman, and the diagrams are called *Feynman diagrams*. This method reduces the calculations to finding all the diagrams corresponding to a particular process (truncating after a given order), and applying a set of rules associated with each part of the component of the diagram. Figure 9-5 shows some higher-order corrections to the ladder diagram, which lead to a set of infinities. These infinities must be resolved by renormalizing the mass and charge.

9-4 Unification with Weak Interactions

Like Maxwell's unification of electricity, magnetism, and optics, the recent unification of electromagnetism and weak interactions is a major event in the history of physics. Although many scientists contributed to electroweak unification, the main credit goes to Sheldon Glashow, Abdus Salam, and Steven Weinberg, who shared the 1979 Nobel Prize, and to Peter Higgs and Gerard 't Hooft. Since much of the theory is highly technical, we discuss unification qualitatively. Detailed discussion can be found in the references in the Selected Bibliography.

Consider a few essential facts about *weak interactions*. The initial discovery of weak interactions was made by studying beta decay of radioactive nuclides. The simplest example is the production and decay of a free neutron:

$$p + D \rightarrow +p + p + n,$$
$$n \rightarrow p + e^- + \bar{v}_e.$$

(9-4.1)

Neutron decay inside a nucleus accounts for beta decay. The nature of these decays was initially puzzling. First, the decay is anomalously weak: A typical nuclear reaction may take 10^{-22} s to create a given nuclide that decays in 10^8 s. This ratio is incredibly large for both production and decay to be caused by the same interaction. Second, the observed decay products do not conserve energy, momentum, or angular momentum! To rescue the conservation theorems, Enrico Fermi proposed the existence of neutral particles (*neutrinos*) that carry the missing momentum and angular momentum, and proposed that beta decay is caused by a new interaction, now called *weak interaction*. The measured masses of neutrino are consistent with zero. If the masses are exactly zero, the equation of motion of motion for neutrinos is:

$$\gamma_k \left(\frac{1}{i} \partial^k - g W^k \right) \psi = 0,$$

(9-4.2)

where $\hbar = c = 1$, and the second term represents the weak interaction.

Electrons and neutrino are members of a family of particles, called *leptons*. Other leptons have been discovered in the high-energy interactions. Although the word *leptons* implies small mass, the recently discovered tauons are more massive than the proton. Thus, leptons are better characterized by the conserved quantity, lepton number, which is positive for leptons and negative for antileptons. Table 9-2 gives a list of the known leptons. The tau neutrinos are shown in parentheses since they are presumed to exist by symmetry, but have not yet been directly observed.

Table 9-2.   The Leptons

e^-	e^+	μ^-	μ^+	τ^-	τ^+
v_e	$\bar{v}_e$	v_μ	$\bar{v}_\mu$	(v_τ)	$(\bar{v}_\tau)$

Neutral leptons have weak interactions only, but charged leptons have electromagnetic interactions as well. The muon decay:

$$\mu^- \rightarrow e^- + \bar{\nu}_e + \nu_\mu \qquad (9\text{-}4.3)$$

is a typical leptonic weak interaction. The related reaction:

$$e^- + \mu^+ \rightarrow \bar{\nu}_\mu + \nu_e \qquad (9\text{-}4.4)$$

is a production reaction for mu antineutrinos. In 1956 it was found that the quantum number, parity, is not conserved by weak interactions, and therefore, weak-interaction particle currents take the form:

$$J_k = \bar{\psi}\gamma_k(1 - \gamma_5)\psi, \qquad (9\text{-}4.5)$$

where $\gamma_5 = \gamma_0\gamma_1\gamma_2\gamma_3$, and $\bar{\psi}$ and ψ represent weak-interacting fermions.

Figure 9-6 shows Feynman diagrams for two possible forms of the weak interaction: the direct interaction, and an interaction mediated by an intermediate particle W^k. In terms of the currents, direct-interaction Lagrangians take the form:

$$L_{\text{int}} = J^k J_k. \qquad (9\text{-}4.6)$$

The second diagram is modeled on the electronmagnetic interaction, where a weak vector potential W^k replaces the electromagnetic vector potential A^k. That is, the weak interaction is mediated by the weak boson, just as the electronmagnetic interaction is mediated by the photon. The interaction Lagrangian is:

$$L_{\text{int}} = J^k W_k. \qquad (9\text{-}4.7)$$

Compare this with equations (4-7.10) and (6-5.3). Because the direct-interaction Lagrangian cannot be renormalized, only the second Lagrangian is an acceptable possibility.

In Section 4-5 we saw that the electromagnetic field tensor is invariant under gauge transformations of its vector potential. The quantum field ψ also exhibits invariance under transformations of a different

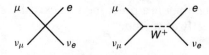

FIGURE 9-6. DIAGRAMS REPRESENTING MUON DECAY.   The first diagram is for the direct interaction between the four particles. In the second diagram, the interaction is mediated by a massive particle analogous to the photon. In the interaction shown here, the particle must be charged. However, neutral particles also mediate other weak interactions.

kind. That is, the transformations:

$$\psi \rightarrow e^{i\alpha\Gamma}\psi,$$

$$\bar{\psi} \rightarrow \bar{\psi}e^{-i\alpha\Gamma}, \tag{9-4.8}$$

where α and Γ are constants, leave the Dirac equation invariant, and consequently the probability density $|\psi|^2$ and expectation values as well. We let α, the fine structure constant, represent the coupling strength. Although these transformations should properly be called *phase transformations*, they are conventionally called *gauge transformations* also.

The gauge transformations given by (9-4.8) are global in that they are position independent. However, it is possible to consider local gauge transformations in which the constant Γ is given a spatial dependence $\Gamma(x)$. Local gauge transformations also leave the probability density invariant, but they do not leave the equation of motion invariant. For example, in the free particle equation with $\hbar = c = 1$, differentiation creates an extra term:

$$\gamma_k \partial^k (e^{i\alpha\Gamma(x)}\psi) = e^{i\alpha\Gamma(x)}(\gamma_k \partial^k + i\alpha \, \partial^k\Gamma)\psi. \tag{9-4.9}$$

Under this transformation, the Dirac free-particle equation becomes:

$$\gamma_k \left(\frac{1}{i} \partial^k + \alpha \, \partial^k\Gamma \right)\psi = \frac{1}{\lambda}\psi, \tag{9-4.10}$$

which is obviously not invariant. However, the complete Dirac equation (including A^k) can be made invariant, by having joint gauge transformations of ψ and A^k. That is, after transforming, the Dirac equation becomes:

$$\gamma_k \left(\frac{1}{i} \partial^k + \alpha \, \partial^k\Gamma - \alpha(A^k + \partial^k\Gamma) \right)\psi = \gamma_k \left(\frac{1}{i} \partial^k - \alpha A^k \right)\psi = \frac{1}{\lambda}\psi, \tag{9-4.11}$$

which is invariant. From this we conclude that local gauge invariance of the wave function demands a potential A^k, which we interpret as the electromagnetic vector potential. Consequently, invariance under this local gauge transformation implies the existence of a massless vector particle that we have called the photon.

This gauge transformation is associated with a quantum number, electric charge. In 1954 Yang and Mills discussed local gauge transformations on other quantum numbers, hoping to fit other interactions with their gauge particles into this scheme. Such gauge particles, generally associated with noncommuting groups, are now called *Yang–Mills fields*. The group associated with weak interactions, producing the weak vector bosons, is such a noncommuting group and is called *weak isospin*. This group produces three vector potentials (W^{k+}, W^{k0}, W^{k-}) corresponding to three electric charge states. Local phase invariance also requires a second neutral vector potential B^k, with quantum numbers identical to those of W^{k0}, and can form a linear combination with it. The remaining potentials, W^{k+} and W^{k-}, represent charged particles that are the experimentally observed states.

Since the two uncharged particles B^k and W^{k0} have identical quantum numbers, they are degenerate states and can be combined. That is, their linear combinations:

$$A^k = B^k \cos \theta_W - W^{k0} \sin \theta_W, \qquad (9\text{-}4.12)$$

$$Z^k = B^k \sin \theta_W + W^{k0} \cos \theta_W, \qquad (9\text{-}4.13)$$

also satisfy the gauge invariance. Expressing the coefficients as sine and cosine retains the normalization of the potentials, and represents a kind of rotation in the B–W plane, giving potentials A^k and Z^k. If we assume A^k is the electromagnetic potential, then Z^k represents a neutral particle analogous to the photon for weak interactions. The angle θ_W is a constant called the *electroweak mixing angle*. The coupling constants g_1 and g_2 are also transformed by this rotation. Expressed in these terms, the coupling constants are:

$$e = g_A = g_1 \cos \theta_W, \qquad (9\text{-}4.14)$$

$$g_z = \frac{e}{\sin \theta_W \cos \theta_W}(I_3 - q \sin^2 \theta_W), \qquad (9\text{-}4.15)$$

$$g_w = g_2 = \frac{e}{\sin \theta_W}. \qquad (9\text{-}4.16)$$

In 1971 Gerard 't Hooft showed that the theory is renormalizable. Therefore, the infinities in the calculations are manageable, and it is possible to calculate cross sections and other measurable parameters reliably. Thus, the theory passes a crucial theoretical test.

The remarkable feature of the weak coupling constants g_z and g_w is that they are so large. Since they are larger than the electric charge of the electron, why were these interactions not observed long ago? The discrepancy can be explained by assuming the weak gauge particles have very large masses. Consider the time-independent (static) case. The electromagnetic Green function is:

$$G_{\text{EM}} = \frac{-1}{4\pi} \frac{1}{r}, \qquad (9\text{-}4.17)$$

from which we got the Coulomb potential. The resulting coulomb forces can be seen at large distances. But consider the analogous Green function for the static Klein–Gordon equation is:

$$G_{\text{KG}} = \frac{-1}{4\pi} \frac{e^{-r/\bar{\lambda}}}{r}, \qquad (9\text{-}4.18)$$

which corresponds to the Yukawa potential. Its range λ depends inversely on the mass of the particle. Provided the $W^{\pm}$ and Z^0 masses are sufficiently large, the effective weak coupling constant is small at sufficiently large distances, i.e., low energies. For example, beta decay rates will be small compared to competing electromagnetic ones if $\bar{\lambda}$ is small compared to the nucleon radius. This suggests that the $W^{\pm}$ and Z^0 masses are several orders of magnitude greater than the nucleon masses.

Definitive experimental verification came in 1983 when a large team of investigators at CERN observed the $W^{\pm}$ and Z^0. The experimental measurements of the masses are:

$$m_W = 80.6 \text{ Gev}/c^2, \tag{9-4.19}$$

$$m_Z = 91.2 \text{ Gev}/c^2, \tag{9-4.20}$$

giving $\cos \theta_W = 0.88$, in good agreement with other measured parameters. The corresponding range of the weak interaction is therefore three orders of magnitude smaller than the nucleon radius. This short range accounts for the small effective coupling constants. In 1984 Carlo Rubbia and Simon van der Meer, the team leaders in this experiment, were awarded the Nobel Prize in physics.

Electroweak and electromagnetic unifications share some interesting features. In Chapters 2 and 4, we saw how the Maxwell equations unify electricity and magnetism into a single theory, by combining the electric and magnetic fields into a single tensor field, whose components depend on the velocity of the observer. The electromagnetic unification involves a gauge field, which is an angle related to the velocity, $\theta^k = \arctan \beta^k$. Then the electromagnetic field tensor is a result of making θ^k local, $\theta^k(txyz)$. In this sense electromagnetism is a classical gauge theory.

Unification of weak and electromagnetic interactions has occurred almost exactly a century after Maxwell's unification of electricity and magnetism, and marks the end of an era in the study of electrodynamics. Although features discussed here would likely be incorporated into a grand unified theory that includes strong and gravitational interactions, there is not yet a definitive prescription for such unifications.

9-5 SELECTED BIBLIOGRAPHY

I. Aitchison and A. Hey, *Gauge Theories in Particle Physics*. Adam Hilger, 1982.
P. Dirac, *The Principles of Quantum Mechanics*, Oxford University Press, 4th ed., 1938.

P. Dirac, Proc. Roy. Soc., **167A**, 1938, 148.

R. Feynman, Phys. Rev., **74**, 1948, "A Relativistic Cut–Off for Classical Electrodynamics," Part A, 939, Part B, 1430.

C. Eliezer, Rev. Mod. Phys., **19**, 1947, 1447.

H. Goldstein, *Classical Mechanics*, Addison-Wesley, 1950.

D. Jackson, *Classical Electrodynamics*, Wiley, 1967, Chapter 17.

G. Kane, *Modern Elementary Particle Physics*, Addison-Wesley, 1987.

E. Konopinski, *Electromagnetic Fields and Relativistic Particles*, McGraw-Hill, 1981.

W. Moore, *Schrödinger*, Cambridge University Press, 1989.

D. Perkins, *Introduction to High Energy Physics*, Addison-Wesley, 2nd ed., 1982.

F. Richtmeyer, E. Kennard, and J. Cooper, *Introduction to Modern Physics*, 6th ed., McGraw-Hill, 1955.

F. Rohrlich, *Classical Charged Particles*, Addison-Wesley, 1965, Chapter 6.

J. Wheeler and R. Feynman, "Interaction with the Absorber as the Mechanism of Radiation." Rev. Mod. Phys., **17**, 1945, 157.

Index

Action, 178, 179, 331
Ampère's law, 78, 93, 94
Analytic continuation, 318
Angular distribution of
    radiation, 365, 366
Antisymmetric tensor, 145
Associated vector, 144
Attenuation length, 266, 267
Axial gauge, 174
Axial vector, 147

Bare mass, 412
Basis vectors, 27, 30, 32, 59,
    203
Bessel equation, 255, 294
Bessel functions, 294
Biot–Savart law, 80
Bohr radius, 415
Boundary conditions, 111,
    126, 130, 138, 196
    mathematical aspects, 130
Boundary surface, 45, 73
Boundary values, 73
Bremsstrahlung, 386, 390
Brewster angle, 272

Cable termination, 273
Canonical momentum, 180
Cathode rays, 313
Cauchy conditions, 131, 134
Cauchy–Riemann equations,
    209, 210
Causality, 159, 404
Cerenkov radiation, 373

Characteristics, 135, 136,
    137
Characteristic time, 385, 403
Charge, 99
Charge conservation, 7, 81,
    87
Charge density, 19, 74, 81
Charge of the electron, 314
Charge to mass ratio, 314
Circular polarization, 252
Classical electron radius,
    281, 333, 355, 385, 415
Classical models of the
    electron, 405
Coercive force, 124
Compton effect, 355, 416
Compton length, 416
Conductive media, 261
Conductivity, 118
Conformal mapping, 208,
    209, 210, 212, 213
Constitutive equations, 125
Contravariant coordinates,
    26
Contravariant tensor, 143
Contravariant vector, 28, 32,
    33, 143, 144
Convection current, 118
Conversion factors, 100
Copenhagen interpretation,
    421
Correspondence principle,
    422
Coulomb gauge, 174
Coulomb's law, 73, 74, 79
Coupling constant, 333

Covariant coordinate, 30
Covariant tensor, 143
Covariant vector, 29, 32, 33,
    142
Critical angle, 273
Cross product, 17, 23
Curl, 23, 24, 25, 36, 37, 45,
    50
Current, 5, 9, 10, 78, 79, 99,
    117, 118
Current density, 80, 81, 87,
    88, 128, 168
Curvilinear coordinates, 26,
    34, 38, 40, 50, 59
Cyclotron frequency, 312,
    332

D' Alembertian, 149
D' Alembert wave equation,
    135, 136, 188
de Broglie wave function,
    419
de Broglie wavelength, 419
Delta function, 46, 47, 368
Diamagnetic, 122
Dielectric constant, 6, 116,
    117
Dielectric media, 2, 111,
    112, 116, 247
Difference equation method,
    315
Differential cross section,
    338
Diffraction, 282, 283

Dipole moment, *see* electric and magnetic dipole moments

Dirac delta function, 46

Dirac equation, 13, 423

Dirac matrices, 422

Directional derivative, 20, 22, 43, 50

Direction factor, 373, 381, 382

Dirichlet conditions, 131, 134

Dispersion relation, 262, 265

Dispersive media, 261, 262, 263

Displacement, 116, 127

Divergence, 23, 24, 25, 35, 50, 75

Divergence theorem, 45, 46, 152

Dot product, 15, 16, 20

Double refraction, 3, 274

Dual tensor, 147, 149, 171

Dust models, 324

Eigenvalue equations, 197, 254

Eigenvalues, 197, 421

Einstein summation convention, 28

Elastic moduli, 279

Electrets, 115

Electric dipole moment, 91, 112, 191, 353

Electric field, 19, 74, 75, 99

Electric field model, 92

Electric generator, 328, 331

Electric potential, 75, 76, 78

Electric susceptibility, 115, 262, 263

Electromagnetic energy, 274, 275

Electromagnetic equations, 88

Electromagnetic field, 90, 164, 167, 187

Electromagnetic field equations, 46, 126

Electromagnetic field tensor, 374, 376

Electromagnetic force density, 278

Electromagnetic mass, 410

Electromagnetic momentum, 274

Electromagnetic potentials, 173

Electromagnetic self-energy, 409

Electromagnetic stress tensor, 277, 306, 337

Electromotive force (emf), 76

Electron, 9, 10, 313, 405

Electron Compton length, 417

Electron diffraction, 419

Electron radius, 415

Elliptical polarization, 252, 253

emf, 76, 77, 78, 85, 86

Energy, electromagnetic, 274

Energy balance, 276

Energy operator, 420

Equation of continuity, 81, 195

Equations of motion, 13, 164, 166, 180, 311, 312

Euler's method, 214

Expectation value, 421

Extended electron, 406

Family of surfaces, 30

Faraday's law, 84, 86, 87

Ferroelectric, 115, 116, 125

Ferromagnetic, 122

Feynman diagrams, 335, 424

Field charge, 73

Fields, 71, 72, 73

Field tensor, 167, 168

Fine structure constant, 414

Finite-element analysis, 214, 215, 342

First law of thermo-dynamics, 76, 276

FitzGerald contraction, 9, 12, 98, 154, 158

Fluid models, 324

Flux, 72, 73

Flux tubes, 73

Force density, 278

Fourier coefficients, 200

Fourier integral, 350

Fourier series, 198, 200, 201, 231

Fourier space, 203

Fourier transform, 48, 201, 361

Fraunhofer diffraction, 290

Free particle stress tensor, 337

Frequency analysis of radiation, 360

Fresnel's equations, 271

Galilean relativity, 153

Gauge condition, 83, 84, 89, 90, 173

Gauge function, 89, 173, 174, 329

Gauge theories, 174, 426

Gauge transformation, 89, 173, 426

Gaussian pillbox, 127

Gaussian units, 98, 99

Gedankenexperimente, 153

Global transformations, 153, 157

Gradient, 20, 21, 22, 23, 25, 34, 35, 50, 148

Gram–Schmidt procedure, 202, 224

Granular model of space, 90

Green function, 189, 282, 283

    advanced, 286

    Helmholtz, 284

    relativistic, 371

    retarded, 285, 371

Green's theorem, 46

Group velocity, 262, 263

G-vector, 328, 330, 422

Hamiltonian, 420
Hankel functions, 256, 294
Heaviside units, 98, 99
Helmholtz equation, 249
Helmholtz–Kirchhoff
    integral, 287
Huygens' principle, 282, 283
Hysteresis, 124, 125

Impact parameter, 338
Impedance matching, 273
Inclination factor, 288
Index of refraction, 262,
    269, 270
Induction, 84
Induction field, 376
Inhomogeneous wave
    equation, 188, 349
Inner product, 32
Integral equations of
    motion, 333, 335
Integral theorems, 149
Integral vector identities, 46
Integration of vectors, 37
Interference pattern, 96
Interferometer, 9, 96, 97, 98
Intrinsic impedance, 266,
    267
Inverse Fourier transform,
    201

Jacobian, 28, 41, 151
Jacobian matrix, 28, 29, 33

Kinetic energy, relativistic,
    163
Kinetic field, 311
Kinetic field equations, 324,
    328, 332
Kirchhoff diffraction
    integral, 282
Kirchhoff's theorem, 287
Klein–Gordon equation, 420
Kronecker delta, 34

Ladder diagrams, 335

Lagrange equations, 179,
    180
Lagrangian, 179, 331
Lagrangian method, 178
Laplace equation, 133, 134,
    135, 188, 196, 197, 198,
    208, 215, 240
Laplacian, 25, 35, 36, 48, 75,
    207
Larmor power formula, 383
    relativistic, 383
Law of elasticity, 279
Law of reflection, 269
Legendre expansion, 205
Legendre functions
    (polynomials), 192, 193,
    205, 294
Lenz's law, 86
Levi–Civita tensor, 146, 147
Liénard–Wiechert potential,
    372
Light cone, 160
Linear current density, 128
Lines of flux, 6, 72, 73, 82, 83
Lines of force, 72
Local gauge
    transformations, 427
Local properties, 22
Local transformation, 141,
    164, 166
Lorentz–Abraham equation,
    404
Lorentz force, 11, 82, 167,
    168, 311
Lorentz gauge, 89, 173, 329
Lorentz transformation, 12,
    141, 155, 157
Lorentz transformed fields,
    175
Lowering operator, 33

Magnetic dipole moment,
    119, 195, 353
Magnetic domains, 125
Magnetic field, 19, 78, 93,
    94, 98, 99, 217
Magnetic field intensity, 121
Magnetic field model, 92
Magnetic flux, 82, 83, 84, 85

Magnetic induction, 7
Magnetic induction field, 80
Magnetic materials, 111, 121
Magnetic moment, 13, 119,
    120, 413
Magnetic monopoles, 172
Magnetic saturation, 125
Magnetic scalar potential,
    218
Magnetic susceptibility, 122,
    124
Magnetization, 119, 120
Magnetization current, 119
Magnetization curve, 124
Magnitude, 14, 16
Mathematica, 22, 49, 50,
    205, 212, 320
Maxwell's equations, 8, 11,
    87, 88, 125, 130, 170
Maxwell's mechanical
    model, 90
Method of images, 216
Metric, 144
    Minkowski, 145
Metric tensor, 33, 34, 144
Michelson–Morley
    experiment, 8, 9, 11, 95,
    152, 154
Minkowski space, 145
Minor hysteresis curve, 125
Mixed tensor, 143
Momentum
    electromagnetic, 274
    relativistic, 162
Monochromatic, 250
Monopole moment, 191
Multipole expansion, 189,
    193, 350
Multipole moments, 192
Multipole radiation, 300

Neumann conditions, 131
Neumann functions, 255
Neutrino, 14, 425
Numerical solutions, 214,
    314, 315

Ohm's law, 117

Orthogonal coordinate
   systems, 32
Orthogonal functions, 199
Orthogonality, 32, 199, 203
Orthogonality of radiation
   fields, 251
Orthonormality, 16, 17, 31
Outer product, 142, 147

Paramagnetic, 122
Particle optics, 320
Path integral, 37, 42, 43, 44,
   45
Permeability, 122, 123
Permittivity, 116, 264
Phase transformations, 427
Phase velocity, 261, 263
Photoelectric effect, 11, 415
Photon, 14, 416, 417
Pillbox, 127
Planck's constant, 414
Plane polarization, 252, 253
Plane wave, 250
Poincaré transformations,
   157
Point model of the electron,
   406
Poisson equation, 188
Polarization, 6, 8, 112
   elliptical, 252, 253
   plane, 252, 253
Polarization charge, 114,
   115, 118
Polarization current, 118,
   119
Polarization of radiation,
   251
Positron, 423
Potential, 75, 76, 77
Potential energy, 76, 77, 78
Poynting's theorem, 276
Poynting vector, 275, 276
Preacceleration, 405
Predictor–corrector, 320
Principle of superposition,
   189
Propagation vector, 249
Pseudoscalar, 147
Pseudovector, 147

Quadrupole moment, 191,
   192
Quantization, 11, 13, 414
Quantum, 416
Quantum variables, 421

Radiation, 247
Radiation by charged
   particles, 371
Radiation by fast charged
   particle, 390
Radiation by moving
   charges, 349
Radiation field, 376
Radiation from accelerating
   charge, 359
Radiation from electric
   dipole moment, 353
Radiation from electric
   quadrupole moment,
   353
Radiation from free
   electron, 354
Radiation from magnetic
   dipole moment, 353
Radiation gauge, 174
Radiation patterns, 386
Radiation reaction, 402, 403
Radiation zone, 260
Rayleigh scattering, 358
Recursion relation, 215, 315
Reference frame, 97, 153
Reflection at boundary, 267
Reflection ratio, 271
Refraction at boundary, 267,
   268
Relative tensor, 151
Relativistic kinetic energy,
   163
Relativistic Larmor power
   formula, 383
Relativistic Maxwell's
   equations, 170
Relativistic momentum, 162
Relativity, 152
Relaxation solution of
   Laplace equation, 222
Relaxation time, 188, 265
Renormalization, 413

Renormalized mass, 412
Resistivity, 117
Resonance cross sections,
   368
Runaway solution, 404
Runge–Kutta methods, 316

Scalar, 16, 32, 45
Scalar field, 19, 22, 23, 44,
   72, 75
Scalar potential, 82
Scalar triple product, 17
Scalar wave solution, 259
Scale factors, 27, 31, 32, 36,
   59
Scattering matrix, 334, 335
Schrödinger equation, 420
Schwartz transformation,
   213, 226
Separation of variables, 196,
   201, 253
Simultaneity, 160
Single slit diffraction, 305
SI units, 98, 99
Skin depth, 266
Snell's law, 3, 269
Solenoidal, 82, 83
Solid angle, 47, 48
Source charge, 73, 74
Spacelike, 159
Spectral analysis, 361
Spectral distribution, 366
Spherical Bessel functions,
   256
Spherical harmonics, 253,
   254
Spherical waves, 252
Spin, 13, 413
Spinors, 423
Static fields, 2, 4, 75, 187
Steady state, 187
S-tensor, 336, 337
Stokes's theorem, 42, 43, 45,
   46, 87, 151
Strain, 279
Strain tensor, 279
Stress, 279
Stress tensor, 277, 279

Summation convention, Einstein, 28
Surface charge density, 113
Surface integral, 17, 39, 42
Symmetric tensor, 145
Synchrotron radiation, 387, 390
Systems of units, 98, 99

Taylor series solution, 317, 345
Tensor, 143, 144
Tensor notation, 142
Tensor weight, 151
Test charge, 72, 75
Thomson experiment, 313
Thomson scattering, 355
't Hooft gauge, 174
Thought experiments, 153
Time dialation, 158
Time independent, 187
Timelike, 159
Total internal reflection, 273
Total radiated energy, 366
Trace, 143

Transformations
  gauge, 89, 173, 426
  global, 157, 158, 164
  local, 157, 158, 164, 166
  Lorentz, 155, 157
  Poincare, 157, 164
Transmission lines, 273
Transmission ratio, 271
Transverse electric mode (TE), 259
Transverse magnetic mode (TM), 259
True charge, 116, 127

Uncertainty relation, 418
Unification, 5, 8, 10, 12, 424
Unified electromagnetic field, 172
Uniform fields, 311
Units, 98, 99

Vacuum polarization, 7, 88, 91
Vector calculus, 18

Vector fields, 72
Vector identities, 25, 46
Vector integration, 37
Vector operators, 25, 50
Vector potential, 82
Vector product, 147
Vectors, 14, 15
Vector triple product, 18
Velocity field, 325
Velocity field model, 324
Velocity operator, 422
Volume charge density, 113
Volume integral, 41

Wave equation, 88, 135, 137, 247
Wave front, 250
Wave guides, 273
Wave mechanics, 13, 418
Weak Compton length, 415
Weak interactions, 13, 14, 424
Weak isospin, 427

Yang–Mills fields, 427